Management for Professionals

More information about this series at
http://www.springer.com/series/10101

Jürg Kuster • Eugen Huber •
Robert Lippmann • Alphons Schmid •
Emil Schneider • Urs Witschi • Roger Wüst

Project Management Handbook

Springer

Jürg Kuster
Winterthur, Switzerland

Eugen Huber
Sargans, Switzerland

Robert Lippmann
Männedorf, Switzerland

Alphons Schmid
Winkel, Switzerland

Emil Schneider
Warth, Switzerland

Urs Witschi
Ennetbaden, Switzerland

Roger Wüst
Dänikon, Switzerland

This is the first English language edition following three German editions 2006, 2008, 2011

Translated by Nick Rosenthal and Judith Townsley, Salford Translations Ltd, UK

ISSN 2192-8096 ISSN 2192-810X (electronic)
Management for Professionals
ISBN 978-3-662-50027-9 ISBN 978-3-662-45373-5 (eBook)
DOI 10.1007/978-3-662-45373-5

Springer Heidelberg New York Dordrecht London

Softcover reprint of the hardcover 1st edition 2015

Printed on acid-free paper

Springer-Verlag GmbH Berlin Heidelberg is part of Springer Science+Business Media (www.springer.com)

Preface

The German edition of this book has found great approval in the project manager community in Switzerland, Germany, and Austria. We are very pleased that now an English-language edition is published for the international market in order to bring maximum practical benefit to many project managers in their daily work.
In this first edition in English, following three German editions, we provide updated and extended coverage of many topics, especially in the area of project leadership. We also share the experience and challenges from our own project management work, both as practitioners and as teachers. During more than 2,000 seminars and workshops in the past 10 years, we have constantly refined and optimised the discussed theories, models, and checklists.
We are particularly pleased that the result of our work does not come across as simply a collection of articles by different people. We worked on this book – this project – as a team, working together to structure the book and to ensure a consistent approach. That is why this book flows smoothly, as a cohesive entity, proof in itself of good project management.

Why We Wrote This Book

The complexity of projects has increased significantly in recent years: Time and resources are becoming more limited, product cycles are getting shorter, and project contents are becoming more varied, more interdisciplinary, and more interconnected. A wide range of stakeholders are expressing their wishes more assertively, and regional and cultural differences have become the norm.
In our experience as practitioners and trainers, we see again and again that countless projects only partially fulfil their objectives, or even fail completely. This has three key reasons:

1. In many organisations, there is not enough understanding of project management and not enough support for it. Many project managers simply don't have the right project management skills.
2. Even though this is how tasks of strategic importance for the organisation are handled, the complexity of projects is not fully recognised.

3. Projects involve change around the organisation's existing hierarchical structures. The need to clarify relevant functions and roles is often overlooked.

This book addresses these key issues. Building on the success of our previous publications, we extensively revised the material and included information about the latest trends. The central theme revolves around the main challenges that project managers face, as process architects and moderators who have to define the work processes involved in a successful project, in implementing the project, in checking, and in effectively managing a project.
Our understanding of project management is based on Systems Engineering concepts, on management psychology, on interpersonal skills in a team context, and on process dynamics. This systemic approach and "networked thinking" matches the complex reality of projects and goes beyond the mechanistic and causal approach. In that respect, this book can be considered to be at the leading edge of project management developments.

The Structure of This Book

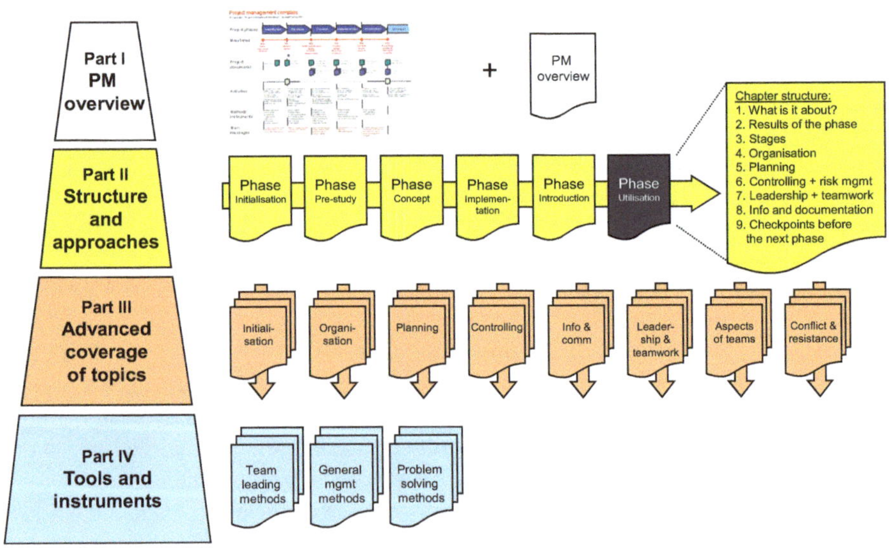

We have favoured a very broad approach which should be suitable for most projects, and it is not tailored for any particular specialist area. One feature that differentiates this book from other literature is how we have structured the material. We have introduced the concept of a project management compass to help guide you, and we have split the book into four distinct sections. This enables the reader to use the book in a way that is targeted to their own role and function and to build on their existing knowledge.

How to Get the Best from This Book

For **project managers** and **managers of sub-projects** who are about to start a project, or who are in the middle of a project: Part II contains checklists that show what needs to be done in the current phase, broken down by different aspects:

- Purpose of this phase
- Results of this phase
- Steps in this phase
- Things to watch
- Checklist of important points

For **members of project teams, students** or **beginners** who want to learn more about the subject: Part I gives you a basic introduction and explains why this approach is useful and necessary. Part II shows you a project and gives you an overview of the relevant steps. In Parts III and IV, you can choose individual topics and methods that will be useful in your specific situation.

For **experienced project managers** who want to learn more about specific topics or refresh their knowledge of particular procedures:

Choose the relevant topics that are of interest to you from Parts III and IV. When you prepare each phase, use Part II as a reminder and as a checklist. You can also use it as a starting point for creating and developing project-specific checklists.

For **decision-makers** with management functions such as HR managers, department heads, quality managers and product managers, Parts I and II will give you an overview of the important factors and approaches for project managers and provide you with insights that enable you to give optimum support to your project and to the project team.

We have deliberately kept the material very general so that it is suitable for application in industry, the service sector, public administration, non-profit organisations, and scientific bodies. The book is also suitable for use as a standard textbook in colleges and universities.

March 2015

Jürg Kuster
Eugen Huber
Robert Lippmann
Alphons Schmid
Emil Schneider
Urs Witschi
Roger Wüst

Contents

Part I

An Overview of Project Management

This section offers an introduction to the broad topic of project management. It provides a summary of all key considerations and approaches to project work. Part I is also essential reading for managers and decision makers.

1 Why Project Management?

In recent years, the pace of change and the complexity of the business environment have increased rapidly. And this has often left operational structures outdated, no longer suitable for the new challenges. Existing organisations are often fragmented, their structures too hierarchical. This gets in the way of quick decision making, and makes interdisciplinary teamwork difficult. In situations like these, using established procedures to manage projects is rarely a recipe for success. We need new organisational forms and structures in order to facilitate more efficient management and communication, new forms and structures that are suitable for the challenges of modern business life.

Project management was born in the 1950s, in the space and construction sectors. Special planning methods were developed for these projects, such as the network planning method. Nowadays, these are used not just to manage complex tasks in technical environments, but also for problem solving and crisis situations in all management fields, such as marketing, personnel, finance and corporate organisation, in both businesses and public sector organisations. In the context of internal projects involving structural, organisational or personnel issues, project management is also often called change management.

Characteristics of Project Management

- A simple, flexible and quickly adaptable temporary organisation, suited exactly to the needs of the project.
- Simplifies and encourages direct, interdisciplinary cooperation.
- The project organisation clarifies the management team's areas of responsibility.
- Direct communication channels, both internal and external, are readily accessible.
- Teamwork and a stimulating environment unlock existing performance potential.

J. Kuster et al., *Project Management Handbook*, Management for Professionals,
DOI 10.1007/978-3-662-45373-5_1

- A clear sense of belonging to the project team makes conflicts of loyalty more visible, and thus easier to deal with.
- Resource management is concentrated in one place, and is therefore more manageable.
- By involving the relevant people, the organisation can learn new skills; it becomes a learning organisation.

2 What is a Project?

There is no generally accepted definition of what we mean when we use the terms "project". Different organisations define projects in different ways. But if we look carefully for points that projects have in common, we can identify the following features:

- Projects involve change: This can trigger many different reactions (from euphoria to resistance, from scepticism and anxiety to joy and motivation), and requires project managers to deal with major psycho-organisational challenges.
- Projects have time constraints: They have a time limit (deadline), and there is pressure to complete them within a fixed timeframe.
- Projects are innovative: They will either push the boundaries of what is achievable in technical or organisational terms (e.g. new information and communication technologies), or they will involve something completely new for the organisation, something which means the organisation must first develop new knowledge (e.g. performance-related pay).
- Projects are complex: They cross the lines of normal organisational structures, and bring together different disciplines and different areas of responsibility.
- The project's character (vision, concept, implementation) changes from phase to phase, and therefore calls for a range of management skills.
- Projects are hard to plan and to guide, they need special organisational measures and clear, unambiguous decisions.
- Projects need exceptional resources in terms of knowledge, personnel, and finance.
- Depending on their size and complexity, projects involve a range of financial, personnel, subject and deadline risks.
- In order to work, projects need their own project organisation: "Projects are organisations"

J. Kuster et al., *Project Management Handbook*, Management for Professionals,
DOI 10.1007/978-3-662-45373-5_2

Many one-off initiatives that fall outside of routine work cannot really be called "projects". But the basic principles and methods of project management can still be applied.
So we could use this as a general definition:

> If a one-off initiative extends across departments, has a limited time frame, is focused on a specific objective, is interdisciplinary and is so important, critical and urgent that it cannot be easily managed by the existing line organisation, but instead needs special organisational measures to be taken, then we can call it a project.

Non-projects where individual project management elements might be applicable could include:

- **One-off special orders** which can essentially be fulfilled by one person (i.e. without a project organisation).
- **Processes** such as learning processes, change processes, etc., are ongoing, they have no defined end, they flow like a river. They may contain other projects, for example, devising and introducing a quality management system is generally handled as a project, in order to implement feedback and learning cycles for future use at the same time.

2.1 Types of Project

There are various different ways to characterise projects. This is often a useful way of helping the project manager to determine how to structure the project, how to define the project organisation and which resources will be needed.
We can define projects on the one hand by the type of goals (closed/open), and by the social complexity (low/high). Here is a description of these dimensions:

Type of Goals

- closed = known, clear goals with a limited range of solutions (e.g. a building extension for a specific purpose).
- open = numerous possibilities in terms of content and approach, not based on a specific solution (e.g. improving an organisation's flexibility and reaction speed).

Social Complexity

- low = no problems with team dynamics (e.g. no significant differences of interest, cooperation is mainly within one specialist field)
- high = interdisciplinary, politically sensitive, different user interests, high potential for conflict

By using a matrix, we can identify four basic types of project:

- **Standard projects** can draw on a wealth of experience, can therefore be standardised, and are simple to manage (examples: a technical project for a client, investment in replacing outdated equipment).
- **Acceptance projects** are tasks with clearly defined goals. Experience means that, to a certain degree, methods and tools can be formalised and standardised. They are also referred to as complex repeat projects. Since these projects often have associated acceptance problems, information and communication are vital (examples: road building, complex ICT projects).
- **Potential projects** are projects with lots of open questions, but which are not (yet) heavily interconnected to the project environment and which are therefore relatively low-risk. The project organisation here is generally fairly simple, and small in size. This category includes preliminary projects, potential assessment, feasibility studies, and often also research projects (example: An initial study into improving customer relations).
- **Pioneer projects** are interventions in the organisation that have far-reaching consequences, that extend across several different areas of the organisation, that involve a high degree of innovation and that are perceived as threatening and high-risk. It is difficult to assess the scale of the project (Example: Two companies merging)

As they progress from a preliminary study to a fully-fledged project, many projects undergo a change of project type. In most cases, they progress from being a potential project to a pioneer project, and then to an acceptance project – or even to a standard project (Fig. 2.1).

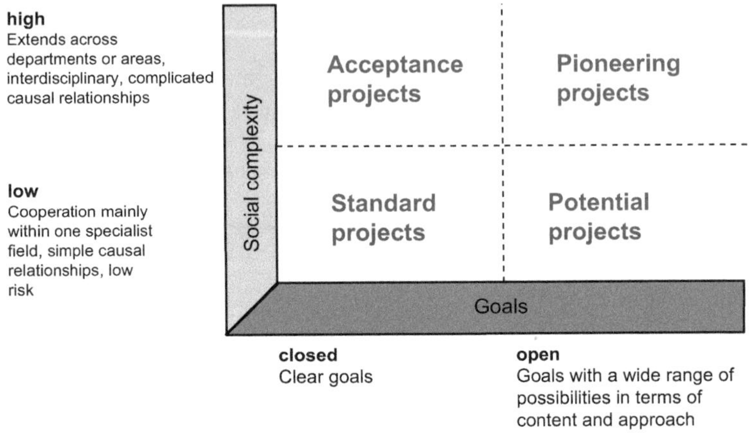

Fig. 2.1 Project types

As well as offering a starting point for deciding on the best project management approach, this typology also helps in selecting a project's organisational structure, in identifying the key communication or methodology issues, and in identifying the key skills and qualifications that the project manager will need. For example, the skills and qualifications needed by a project manager running a construction project will be different to those needed by a project manager running a change management project, or a development project or customer project.

2.2 Classifying Projects

Another way of classifying projects is to categorise them according to their purpose. The relevant standards bodies have agreed and standardised a number of project procedures for classes of this sort. Typical project classes would be:

- Investment projects
- Infrastructure projects
- Product development projects
- Organisational projects
- Organisational development projects
- Information technology projects (ICT projects)
- Construction projects

What is Project Management? 3

Every business process is ultimately based on a business case that is intended to achieve specific objectives. The way that this is implemented, in operational terms, will vary according to the company's corporate culture and leadership style. One approach that is widely used these days is management by objectives (or MbO). A fundamental principle of this approach is to delegate tasks, competencies and responsibility to the lowest possible level. In addition to simple line management, in terms of leadership style this approach calls for other special forms of operational management. This is where the following aspects of project management have proved to be particularly useful:

- Approach is based on structuring workload into phases and work packages.
- Specifying the decision-making, management and subject competencies separately for each phase.

Project management is used as a high-level term that covers all the planning, monitoring, coordinating and controls that are required when developing or redeveloping systems or processes, or in problem-solving. The approach used to achieve the solution, the required resources, and how the resources are deployed and coordinated are more important than the solution itself. In contrast to project management, line management relates more to the ongoing business and to the management of the relevant organisations.

3.1 Hierarchy in Project Management

The "project management" method extends right through the whole organisation. Different tasks are carried out at different hierarchical levels in the business (Fig. 3.1).

J. Kuster et al., *Project Management Handbook*, Management for Professionals,
DOI 10.1007/978-3-662-45373-5_3

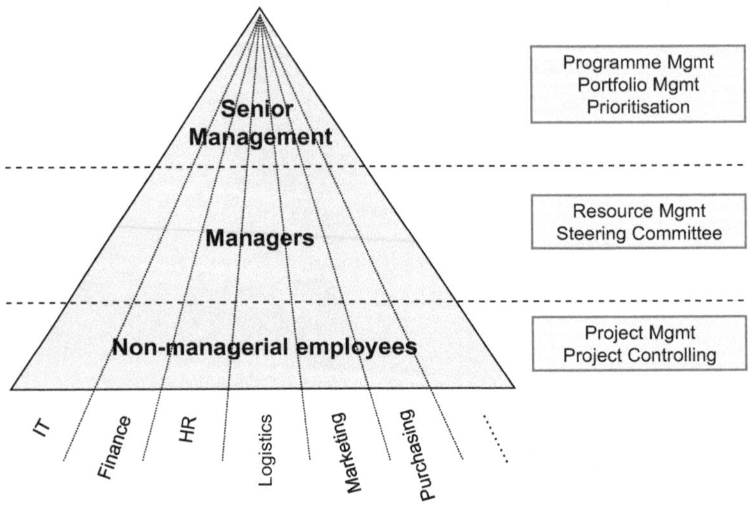

Fig. 3.1 Project management tasks in a company hierarchy

When we speak of **programme management** in a project management context, we mean the management of all the projects that are collectively intended to achieve the same strategic objective. It can be based at any of several different levels of the business, and may involve just some of the company's projects (e.g. a development project), or it may include all of the company's projects.

Programme management is about coordinating several different projects that may be relatively unrelated, about determining priorities and assigning resources such as contracted services and budgets. Examples: Research programmes, development programmes, etc.

The most important tools for this are a company-wide **project portfolio** and a model for **project evaluation. Product management** means all the strategic and operational activities carried out by a person (or a team) who is responsible for a product or service in all areas of the organisation. This person or team usually acts as a contact point for customers. Some aspects of product management lend themselves well to being managed as projects. These include new product development, product launches, or resolving product-related issues (Fig. 3.2).

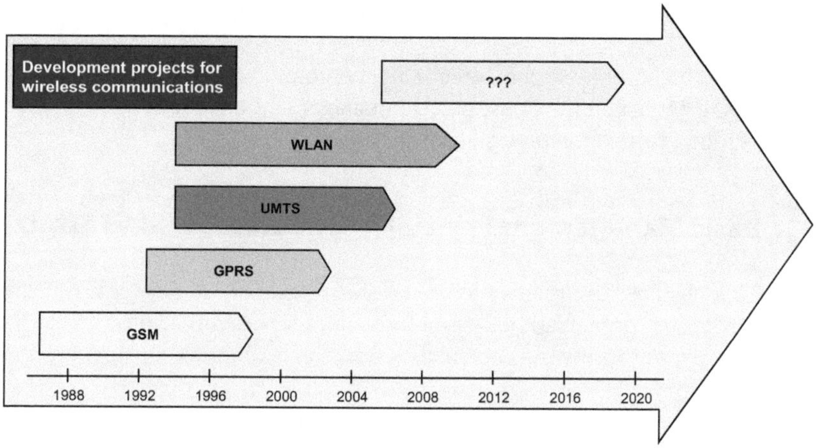

Fig. 3.2 Examples of a project programme

3.2 Dimensions in Project Management

The **functional dimension** looks at what needs to be done. It looks at the procedural steps involved in the individual life phases of a project:

- Initialisation: Getting the project going
- Project work: Keeping the project going
- Concluding the project

The **institutional dimension** focuses on the interdisciplinary project organisation and the way it interconnects with the rest of the business and with its environment:

- Identify the project bodies
- Specify roles and functions
- Set up a project team
- Identify the areas of responsibility, and agree who has authority

The **human, psychological and social dimension** primarily addresses the social skills of the project manager, and relates to the following tasks:

- Recruiting personnel, and checking their skills and qualifications
- Leading the project team
- Encouraging and supporting cooperation
- Managing conflict
- Shaping social processes

The **instrumental dimension** covers tools and techniques:

- Information technology support (planning, communication, documentation, etc.)
- Processes, established methods, project management handbook
- Support tools, forms, templates

3.3 Basic Management Principle: "Structure Follows Strategy"

In order for a business or organisation to be able to achieve its strategic objectives, it defines various organisational structures and operational processes. When running projects, the following dimensions must be taken into consideration:

- The **procedural structure** defines the amount of time, the space requirements, the volume or quantities, and the logic structures that are needed to achieve the goal. It describes the necessary activities, the sequence of these activities and assigns the activities to the responsible persons according to their skills and expertise. Typical tools would be a project schedule, a list of activities, a critical path analysis, a monitoring structure, a decision table and a material flow plan.
- The **organisational structure** generally determines the ongoing relationship structures in the line organisation. Jobs or posts are created, and links are established between them both through communication processes and through the chain of command. They are formed into functional relationships with defined roles. Important tools here include organisational charts, function diagrams and job descriptions. A process-oriented organisational structure focuses on core processes that normally start and end with the customer.

The central question is: What should be defined first? As early as 1962, Alfred Chandler stated in his book "Strategy and Structure" that structure (the organisation) must follow strategy (the processes to achieve the objective). Today, this is a widely accepted process logic. Work processes are derived from the strategy – in other words, from the objectives or goals. Then we identify the project resources who are best suited. In turn, they are incorporated into a relationship structure that reflects their roles and their duties.

One exception to this is in research projects, where qualified teams (often called potential teams) are often set up. They then largely determine the objectives and the procedural structure themselves.

In practice, this is not quite as easy as it sounds, because there will already be existing structures and relationships in place in an organisation, and these cannot always be rearranged at will. And "ownership issues", often involving senior management, frequently prevent the right people being assigned to the right project roles.

Project Management: Key Theoretical Concepts

4

A complex methodology such as project management should not be based on only one theory or school of thought. Modern project management has to be capable of handling a very wide range of project types and situations. It must be possible to apply it on a situational, differentiated basis. If we think of the project manager as a musician, he needs to be capable of playing a range of instruments. Which means that he needs to be comfortable using a range of different intellectual approaches, some of which may at times appear contradictory or mutually exclusive.

The following thoughts on structuring methodology and on planning methodology are based broadly on the theory of Systems Engineering. Economic aspects, such as cost-benefit analysis and return on investment, are based on business principles. For issues relating to cooperation, communication, conflict management, leadership and organisation, we draw on psychology and on psycho-organisational modelling approaches. This relates, for example, to behaviour, to relationships, and to team development and team dynamics.

4.1 A Systemic Approach

Over recent years, with an increasing awareness that all these aspects contribute to integrated project management, and also that individual projects are not actually isolated, but should themselves be viewed as part of an integrated system, the systemic approach has proved its worth in the field of project management. Using this approach, we view projects as social systems which, whilst having their own tasks and goals, their own internal structures and system boundaries, also have significant interactions with the wider world. Social systems construct their own reality (radical constructivism), are largely self-regulating and have an ability to learn. As social systems, projects cannot be "controlled". But they can be developed and, to a limited degree, managed (for example, by creating good framework conditions).

J. Kuster et al., *Project Management Handbook*, Management for Professionals,
DOI 10.1007/978-3-662-45373-5_4

- Social systems are interlinked. This influences their behaviour. It enables communication links between project members and stakeholder groups to be established and managed.
- Social systems are self-referential: the management of reflection and feedback has central significance.
- The path is also a goal: The process (procedural process, team development, learning processes, etc.) becomes a central topic. For example, it is important that there is close identification with the project objectives, that the framework parameters are clearly defined and that areas of authority and responsibility are agreed and understood (Fig. 4.1).

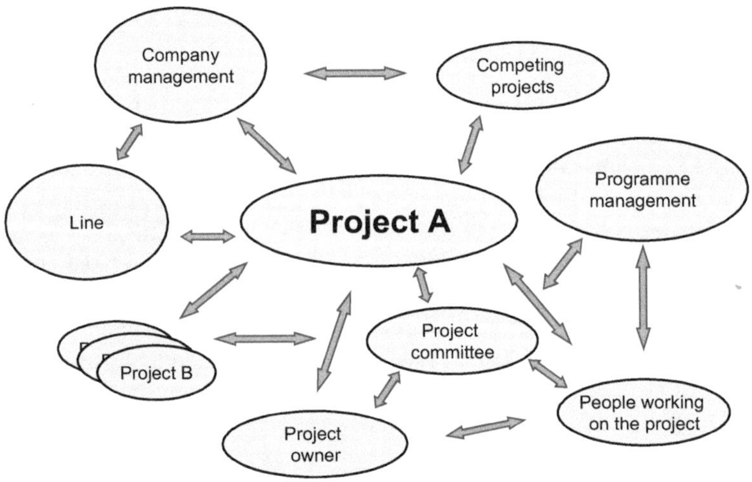

Fig. 4.1 A systemic view of a project

The main benefit for projects is that systems which are self-regulating and which are capable of independent learning are able to handle significant degrees of complexity. They can handle change processes more effectively and more sustainably than systems that are externally regulated.

4.2 Systems Engineering

Systems Engineering (or SE) offers a systemic approach, both when developing new systems and when adapting existing systems. SE facilitates structuring the project, and dividing it into phases. It combines the following principles in a single system (Fig. 4.2).

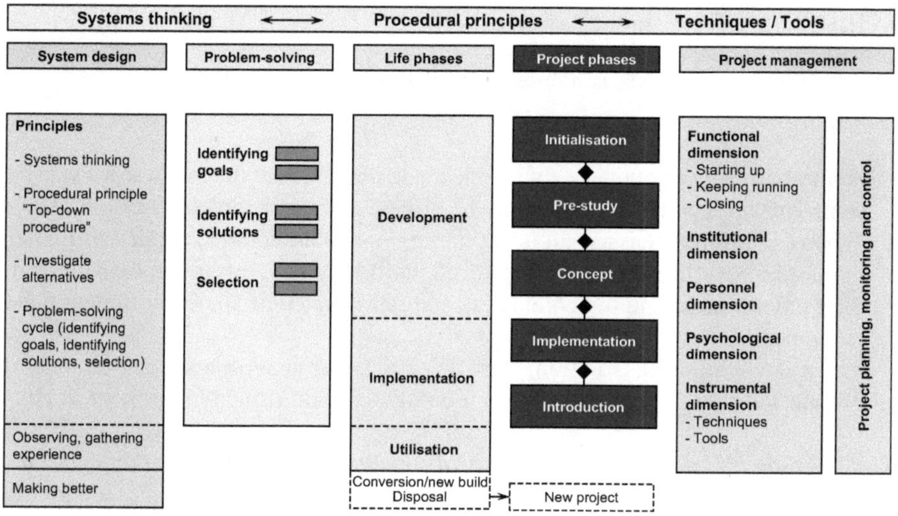

Fig. 4.2 Project management in the context of Systems Engineering (Daenzer, 2002)

Systems Thinking

A system is a group of elements that, in some manner, form a whole. A systemic approach is useful in resolving the actual problems: the elements to observe, influencing factors, interrelationships and boundaries. Systems thinking is also helpful in devising appropriate solutions. There are various different ways of forming groups:

- The context in which questions are asked
- According to their internal structures, processes, causal mechanisms, etc.
- Different perspectives of parties affected by the project
- Hierarchical view (top down or bottom up)
- Networked thinking (causal relationships)

Procedural Principles

The following principles and approaches have developed into a core element of Systems Engineering:

- Top-down procedure
- Investigate alternatives
- Arrangement of phases
- Problem solving methodology

Let's now take a look at the "top-down procedure" and "investigate alternatives" principles. The other two principles (work in phases, and problem solving) are so central to project management that they must be explored separately.

Top-Down Procedure

The "top-down" principle is a core concept in running a project, and is described below:

- At the start of a project, take a wide view and then narrow the focus down step by step. This applies when examining the issues, and when developing solutions.
- When examining the issues, it is important not to start detailed studies until after the issues have been roughly structured, until they have been contextualised in their environment and the boundaries and interfaces with the environment have been identified.
- When developing the solution, start by setting out general goals and a general outline solution. The details can then be filled in and firmed up in more depth.

The "**Top-down**" principle also has a mirror image, a "**Bottom-up**" approach. This can be very useful in certain circumstances, such as improving existing, functioning solutions (empirical approach). In the case of more extensive changes, or when developing projects from scratch (conceptual approach), it is usually better to develop an overall concept that starts from the broad outline. This has the benefit of creating an orientation framework for the individual steps that need to be carried out.
In any event, it is accepted that during implementation a circular approach using both "top-down" and "bottom-up" will lead to all stakeholders reaching the same viewpoints, which is essential for success. This iterative agreement process also significantly people's level of commitment and their feeling of ownership of and responsibility for the jointly developed structures and plans (Fig. 4.3).

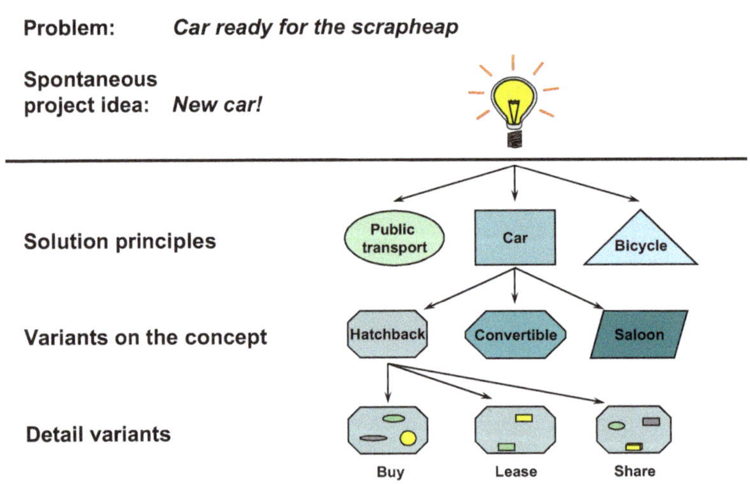

Fig. 4.3 Example of investigating alternatives at every stage

Investigate Alternatives

The principle of investigating alternatives, of considering different options, is an essential part of good planning. It is a methodical basic approach. And if you use the "top-down" principle, it need not involve much additional planning workload. If you do not follow this principle, there is a serious risk that other appropriate solutions might be introduced into the discussion at an advanced planning stage.

Problem-Solving Methodology

Alongside the use of phases, which represent the macro cycle for an entire project, there is also a micro cycle: The problem-solving cycle (identifying goals, identifying solutions, selection). This is a useful tool throughout the entire lifetime of the project, and one that project managers keep returning to. The only differences between using it at the start of a project or at a later stage in the project lie in the degree of detail. The form it takes, and the way that it is used, will differ according to the project environment.

The Phase Concept

5

The "top-down" and "investigate alternatives" principles result in the following workflow when resolving a problem: Idea, development, implementation planning and delivering a solution. Each of these must be broken down into individual work packages, or phases. Each phase must be distinct from the others in terms of time and logic. The purpose of this is to break down the development of the solution into chunks, thus giving a clear overview. This makes it possible to draw up a graduated planning, decision-making and implementation process, with predefined breakpoints (milestones) or intervention points.

The following diagram shows the phase model in its most simple (and idealistic) form (Fig. 5.1).

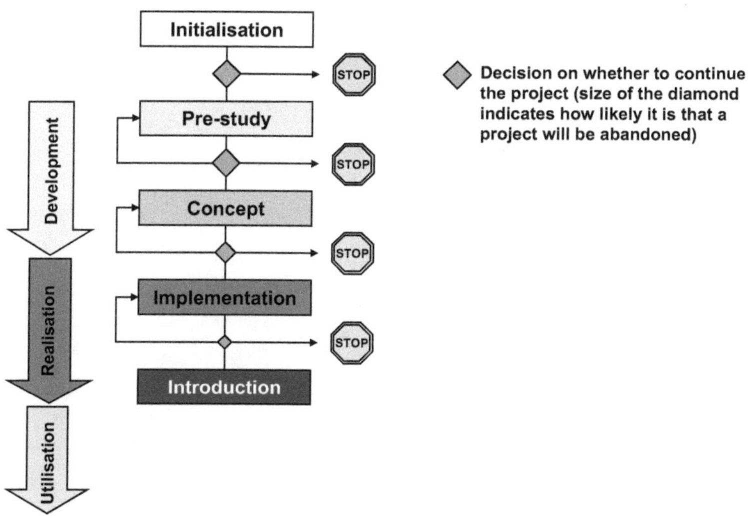

Fig. 5.1 The ideal phase concept

J. Kuster et al., *Project Management Handbook*, Management for Professionals,
DOI 10.1007/978-3-662-45373-5_5

Examples of Abandoned Projects

One of the most spectacular uncompleted large-scale projects in Switzerland is the Kaiseraugst nuclear power plant. Between 1965 and 1989, an international consortium led by Motor-Columbus planned to build a nuclear power plant in the Swiss village of Kaiseraugst, near Basel. Even a quick glance at the length of the planning period is enough to show that "Kaiseraugst" lay well outside the normal parameters for a technical project. The plans for the nuclear power plant were well received on all sides in the mid-1960s. But over the following decade, the project developed into an epic political power struggle. Instead of splitting the atom, the project began to split Swiss society. But even these changes in society were not enough to sound the death knell for the project. The Kaiseraugst project was finally laid to rest in 1988, when it lost all political support in the light of the Chernobyl nuclear disaster. Within just 1 year, in 1988/1989, the decision to build the plant was reviewed by the Swiss Federal Council and by the Federal Assembly (both houses of the Swiss parliament).

On 8th March 1989, in response to a petition by the advisory commission, the National Assembly (the Swiss House of Representatives) voted by 105 votes to 29 to pay 350 million Swiss Francs of government money in compensation for abandoning the nuclear power plant project which the Swiss Federal Council had contracted Kernkraftwerk Kaiseraugst AG to build. The project was then terminated without any significant additional costs.

Things were very different in the case of a similar project in Austria. In a referendum in 1978, Austrians decided that the Zwentendorf nuclear power plant should not be brought into service, even though construction was already completed. As late as 1984, some 50 staff were still employed to conserve the plant, costing the government around 80 million Swiss Francs a year. Later, parts were sold abroad and the plant was scrapped at a cost of around one billion Swiss Francs. Overall, Zwentendorf cost 14 billion Swiss Francs, with conservation alone costing 600 million.

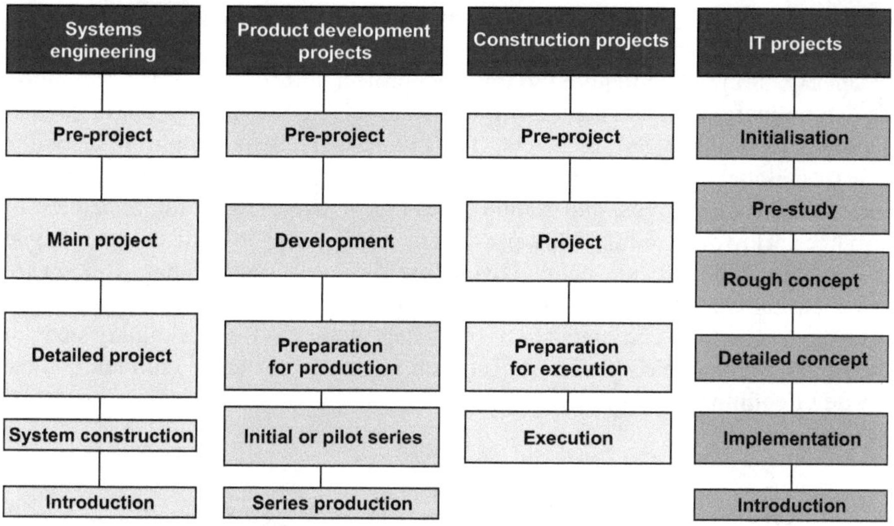

Fig. 5.2 Phase models and phase descriptions

The number of project phases, and the formalism used for their implementation will clearly depend on the nature, scope, risks and importance of the project. They will also depend on the degree of influence that the client wishes to have (Fig. 5.2).
Generally speaking, smaller projects can be completed with fewer phases and with less formality. The theoretical model can also be extended through additional phases, such as a preliminary feasibility study, a prototype phase, a testing and acceptance phase, etc.
How the phases are shown (as a block diagram or as a "waterfall model"), and how they are labelled, is of secondary importance, and will in any event depend on the sector, the goals, and the terminology used within any given organisation. The diagram above shows a few common phase models. The important thing is that the use of deliberate segmentation of the work packages into individual planning stages and delivery stages, with decision-making meetings between each stage, enables project managers to reduce the complexity of the problem, and to reduce the risk of an incorrect decision.

5.1 The Initialisation Phase

This phase is generally fairly unstructured, and covers a time frame between identifying the problem and taking a decision to do something concrete about it. The problem may already be formulated in a relatively concrete way, or it may be based on vague assumption, on gut feelings. It doesn't actually matter where the initial incentive to create something or to restructure something comes from. The

important thing is that it is understood and accepted by the key players who are responsible for and have authority over the resources (personnel, financial, organisational) that are required to initiate a project agreement.
The preliminary work and activities involved in this first "definition phase" should ideally lead to a project agreement. The project agreement should describe the project's global objective and the project priority, the goals, the procedures, responsibilities and roles, and should give a clear overview of the allocation of resources. However, during this phase it is often only possible to give vague information about specific points covered by the project agreement. This means that a feasibility study must be carried out, or that the missing items from the project agreement have to be completed in the next stage, which is the preliminary study or pre-study. For example, the full project team might not be named until later, in the course of defining the project's scope.

5.2 The Preliminary Study Phase (Pre-Study Phase)

The main purpose of this phase (often also referred to as the preliminary project or the feasibility study) is basically to determine that solving the problem is a realistic proposition. For example, the field of study should be consciously identified. More information should also be given about the environment in which the solution will at a later stage have to function (and with which it will have a two-way relationship). The following points should be clarified:

- The relationships and mechanisms that are involved in the problem
- How deeply to go into the subject (delimiting the problem, environment)
- Whether this is the right problem to tackle, and whether just one symptom has been observed so far
- How much need there is for a new or revised solution (i.e. estimate the impact of the current problem, and develop the definitive project priorities from that assessment)
- What requirements the solution needs to fulfil (description of desired results)
- What alternative solutions could be considered, and whether they appear to be achievable in technical, economic, political terms (feasibility)
- Which solutions are the most promising (choose from the alternatives and determine the extent to which the organisation has the appropriate competencies, or whether external resources will be required)
- What risks are associated with the project, how tolerable they are, and how they can be minimalised

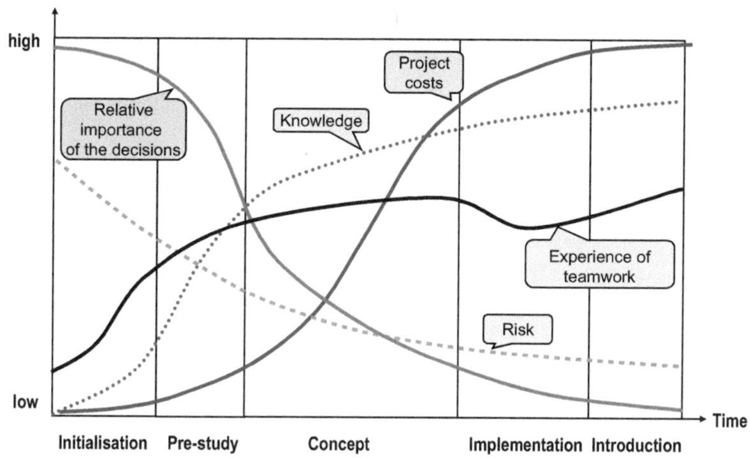

Fig. 5.3 The impact of decisions, knowledge and cooperation

By clarifying all of these points, it should be possible to produce a first comprehensive process plan for the project. This will form the basis for the detailed project plans in terms of work packages (sub-projects) and organisational units (type and size of teams).

The preliminary study creates "decision points" that determine the way the project runs from there on in. So even though it involves a relatively high investment of time and overheads, it is of central significance. The preliminary study is used to precisely identify the problem, to develop and clarify goals, to discuss fundamental approaches to solutions, to make a decision on the best way forward (whether just one option is proposed, or a choice is made from several options), and to plan the next steps: project organisation, scheduling, resources, methods, etc.

From an organisational psychology perspective, the preliminary study also places high demands on the project's management (on the project manager and on the project owner). Fundamental decisions about the project have to be made at a point where there is not yet enough information, and where the team has also not yet got much experience of working together (Fig. 5.3).

In some cases, a decision may be taken at the end of the preliminary study to abandon the project. This should not be interpreted as a mistake or failure. It is a conscious decision, taken at an appropriate time based on the knowledge that has been acquired.

5.3 The Concept Phase

The purpose of this phase (which is also known as the main project) is to develop an overall concept with a range of possible solutions, based on the solution principles and framework concept developed during the preliminary study. This involves making informed decisions about achieving the goals, about functionality, about

effectiveness and about economic considerations. During the concept phase, the focus narrows even further. The main focus is on developing a range of possible solutions, a number of options.
The concept phase results in the decision about which of the possible solutions to use. For each possible solution, a concept must be drawn up for subsequent detailed planning at a later stage, and for implementation. This perspective results in the following tasks:

- Create a framework plan (milestone plan, master plan) for the next phases
- Indicate opportunities for savings (parallelisation of activities)
- Define the sub-projects
- Formulate investment decisions

The next thing to do is to plan the selected solution in detail, and to develop it. At this stage, subsystems and individual aspects are often explored, areas that can be isolated from the overall system and handled separately. Typical activities in detail concept planning include:

- Developing individual solution concepts and making decisions about the structure of the different options
- Firming up the individual partial solutions so that they can then be "built" and implemented as easily as possible
- Planning the successor organisations and maintenance organisations that will manage the system after completion of the project
- Developing training and familiarisation plans
- Defining an approach for Change Request Management. This involves agreeing an approach (and the relevant procedures) with the client for handling any change requests while the system is being developed.

5.4 The Implementation Phase

In this phase, the solution is implemented in its widest sense. In reality, this phase does not necessarily follow the concept phase; the two phases generally tend to overlap to some extent. Typical tasks in the implementation phase include:

- Building facilities and equipment (or even prototypes)
- Developing software
- Producing user-friendly documentation and instruction manuals
- Specifying organisational rules (information, faults, etc.)
- Sorting out maintenance organisations, servicing schedules, etc.

Partial systems are also often built during this phase, ready to be integrated into the overall solution.

5.5 The Introduction Phase

Introduction

Only relatively small and simple solutions can be introduced in their entirety without risk. For large and complex systems, the wide range of unpredictable side effects and interrelated dependencies means that launching everything at once is not sensible. It is often better to use a staged approach: Feedback gained from the experience of rolling out the first elements can be fed into how the later stages are launched, always with the focus on the overall objective.
At first glance, this phase appears to be very technical. But that is just a superficial perception, and it often proves to be very delicate and time-consuming. By this stage, the project team has spent a long time becoming accustomed to the change and innovation that the project involves, and they often don't realise how striking the changes that the launch involves are for other people. This discrepancy in the timelines of the two systems (project and line) once again calls for excellent cooperative skills on the part of the relevant managers.

Hand-Over

Another decisive factor in introducing a new system is how well the knowledge and expertise is transferred, whether it is possible to train and inform the system's new owners, operators and users quickly and comprehensively. The aim here is for the development and the implementation team to make themselves redundant as quickly as possible.

Concluding the Project

All projects come to an end. Even projects that have been abandoned involve close-off work. In many cases project close-off work is not consciously completed, which means that nobody really knows whether the project is actually finished or not.
Concluding a project involves the following tasks:

- Closing off the project work. In other words, completing any remaining work or postponing it to a future release
- Finalising all cost accounting issues
- Completing the project documentation and archiving (especially important in the information technology and plant engineering sectors)
- Handing over tasks, competencies and responsibility to the operator
- Giving the project documentation to the maintenance organisation
- Project completion with the project owner (in the sense of project hand-over) and with the project team (disbanding the team). In both systems, it is often useful to

have a critical debriefing or project review session. On the one hand, this helps in letting go off the project; on the other hand, and perhaps more importantly, because things that have gone wrong represent big learning opportunities, in the sense of a learning organisation. Possible questions to explore might be: What went well? Where did we experience problems? Were the planned budgets and deadlines met (personnel, costs, time)? What could we do differently in future?

5.6 The Utilisation Phase

After completion of the project, the utilisation phase begins. After an agreed time, the results of the project should be assessed. Depending on the nature of the project, work may be required under a guarantee, or there may be a need to produce an improved release. This stage also generally involves carrying out a project evaluation: how well does the business or commercial prognosis meet expectations?

5.7 Milestones

In each phase, there are certain predefined results which have important information that is used for making decisions about the project's direction from that point on. These decisions are generally referred to as milestones.
Milestones are significant events in a project, and they are of central importance for the ongoing project planning and project management. It is very important that the steering committee, in collaboration with the project manager, discusses and resolves the central conflicts in terms of time, resources and quality (result) in a solutions-oriented manner. If they don't, it becomes very hard for the project manager to continue managing the project and to deliver it on schedule.
Information from the project team and the project manager's status reports enable the project client or the steering committee to take decisions such as:

- Accept the project status
- Close off a project phase
- Continue or abandon the project
- Start the next project phase or work package
- Identify any key changes in the goals
- Make key changes to the way the project is run
- Introduce any additional measures (e.g. information sessions) to suit the project dynamics
- Sort out major staffing issues such as additional resources, redeployments, changes to roles, setting up new teams
- Authorise additional investment or project loans

The steering committee is a decision-making body for the project. It consists of people from different disciplines, and brings together the different interest groups.

Working and making decisions on an interdisciplinary basis at this level has a positive impact on the culture of the whole project.
If people on this committee just represented their own interests, if people relied on power and authority using hierarchical organisational structures, then this committee would have a negative impact on the project (Fig. 5.4).

Phase	Activities	Results
1. Initialisation	- Clarify roughly what it is about (content) - Which organisational units are involved	- Rough description of the problem - Rough definition of the goals - Application for project to Projects staffing unit
♦	**1. Milestone - Decision**	**- Decision on project or mandate** **- Initiate project start**
2. Pre-study	- Draw up project plan (structure + workflow) - Devise + clarify project mandate - Communication concept	- Procedural concept (workflow planning + project org.) - Project order - Application for a project committee
♦	**2. Milestone - Decision**	**- Approval of the project order** **- Agreement on the schedule**
3. Rough concept	- Devise overall concept for the alternative solutions with variants - Check cost effectiveness	- Present overall concept - Demonstrate & assess possible approaches (variants) - Application for a project committee
♦	**3. Milestone - Decision**	**- Approval of the project status** **- Select variants**
4. Detailed concept	- Flesh out solutions that are ready for implementation (from variants) - Draw up detailed cost effectiveness review - Plan implementation, introduction (including training), and subsequent support / service	- Plan and draw up detailed project solution - Detailed plans - Application for a steering committee
♦	**4. Milestone - Decision**	**- Approval for implementation**
5. Implementation	- Construct, test and introduce system - Training - Establish service organisation	- System introduced - Training arranged and provided
♦	**5. Milestone - Decision**	**- Acceptance of the system** **- Approve project close**
6. Introduction	- Hand over project to the line	- Project report - Project organisation broken up
♦	**6. Milestone - Decision**	**- Project manager discharged** **- Project ended**

Fig. 5.4 Example of a phase model with activities and results

As well as the milestones that are provided for in the phase model, serious setbacks or crises might result in additional meetings, with the project owner having to take additional decisions. Either the project manager or the project owner may call such meetings.

Other Process Models 6

As well as the classic, sequential phasing (or waterfall) method, there are also specific process models that are better suited to certain specific types of project, such as the V model for software development. The following process models have become established as alternatives, and also as enhancements to the classic phase concept.

6.1 Simultaneous Engineering

The origins of Simultaneous Engineering lie in product development. The main impetus for this idea is the desire for shorter development times. The use of parallel processes speeds up the project. The different groups that are involved in the product's development should be involved at the earliest possible stage. By using phases which are designed to overlap, some of the work can be done simultaneously (Fig. 6.1).

J. Kuster et al., *Project Management Handbook*, Management for Professionals,
DOI 10.1007/978-3-662-45373-5_6

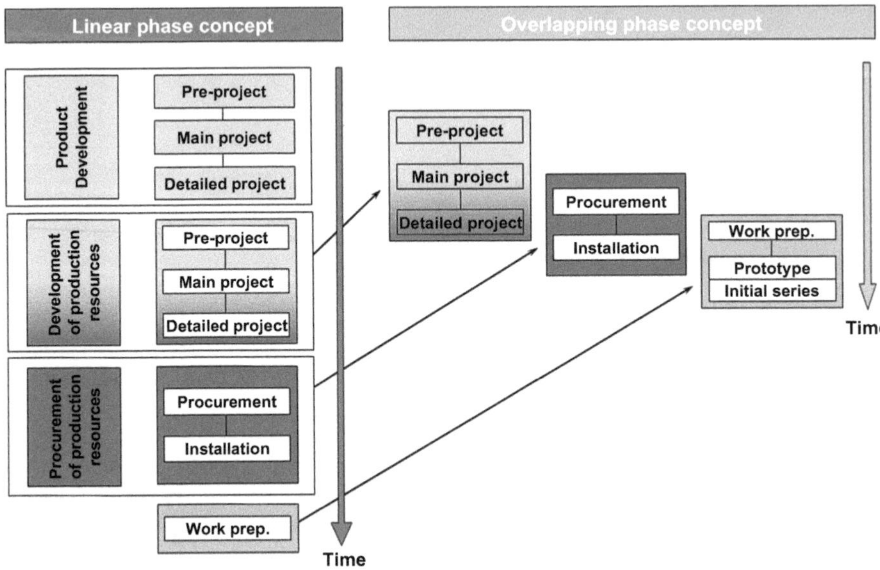

Fig. 6.1 Simultaneous Engineering as an overlapping phase concept

Today, virtually all projects are managed on this basis. The fact that very diverse activities are taking place simultaneously means that the project manager has to constantly check that the goals and the plan are being met. This is made all the more difficult when the project manager is involved in the project as a specialist, or if they are also participating in other projects.

If the client specifies or wants a parallel project approach (Simultaneous Engineering), delivered to a tight time frame, the project manager should ideally be free of other work commitments, and able to devote themselves fully to the project's process control.

6.2 Prototyping

The concept of "prototyping" as an approach was first used in data processing in the mid 1970s. It involves starting with an abstract concept, and quickly making it into a firm, concrete plan, in order to bring about more efficient communication between the developers and the users, for example. In this sense, prototyping is a development tool, since it helps to identify the user's requirements more quickly.

In concrete terms, the idea is to develop some form of "prototype" with relatively little effort. This can then be used to better evaluate the concept used up to that point, and it may also be used for testing under working conditions.

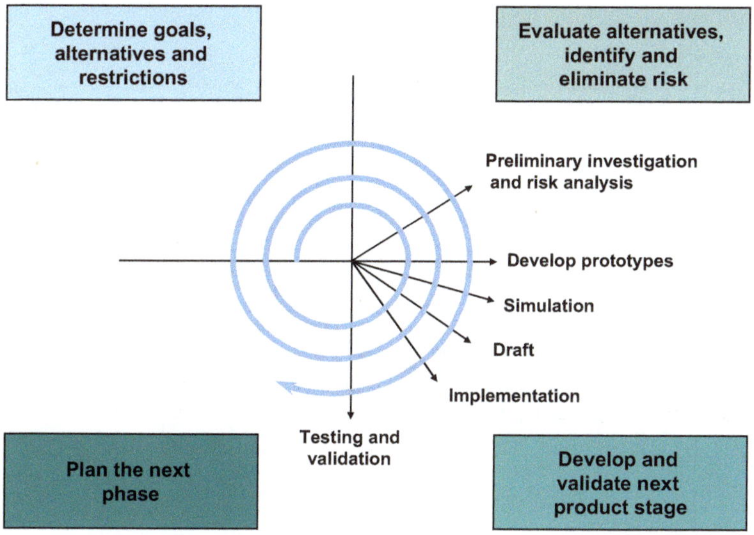

Fig. 6.2 The spiral model

As a development aid, prototyping is especially helpful during the final phases as implementation draws closer. However, it does not actually replace any of the individual phases – none of them is rendered redundant. In particular, the preliminary study and the concept phase remain essential (Fig. 6.2).
However, one major change is that it is possible to introduce an iterative process between the concept and implementation phases. There is a risk though that this may lead to "quick and dirty" solutions, which then end up finding their way into the project in the absence of a real solution, and which end up being developed further as a piggyback solution. Or the client may think: "The solution is already there. Why are you spending so much more time and money on this?"

6.3 Versioning

Versioning has some similarities with prototyping, and it can be used for a wide range of development work (machines, plant, information technology).
This approach does not try to find a perfect solution in one go. Rather, a first version is developed and produced, and made available to the user. Using this as a starting point, operational experience is used to make improvements from one version to the next ("slowly growing systems"). This cyclical approach is also known as the spiral model.

This approach offers both benefits and drawbacks:

Benefits	Drawbacks
Often an "emergency solution" Solution is available quickly	Less careful planning, because it is tempting to put problems and improvements back until the next version
Fast development pace Progress is quickly visible	Places high demands on the documentation and project administration, because there is a need to show at any time which version is valid, and how the solution's individual components have been developed and how they interrelate
Offers a way of examining problems without needing to know much about them	Extremely difficult to plan and to predict project costs
If budget is limited, the range of goals can grow with the budget (perfecting the solution)	Constant new demands from the user can exceed the system's boundaries, the project can grow on an uncontrolled basis and never reach an end

6.4 The Agile Approach

If we take the prototyping and versioning approaches to their extreme, what we end up with is agile project management. This approach has also become popular as a response to the need to have functioning, customer-specific or user-specific software available in the shortest possible time, without needing the precise requirements to be specified in detail at the start of the process. Agile project management means having a dynamic, responsive, quick, process-oriented, reflective, learning approach. Its basic principles are set out in the Manifesto for Agile Software Development (2001), which was drawn up by the 17 founders of the Agile Alliance.

- Individuals and interactions are valued more than processes and tools
- Working software is valued more than comprehensive documentation
- Customer collaboration is valued more than contract negotiation
- Responding to change is valued more than following a plan

There are some very different approaches in agile project management, such as "extreme programming" or "SCRUM" (Fig. 6.3).

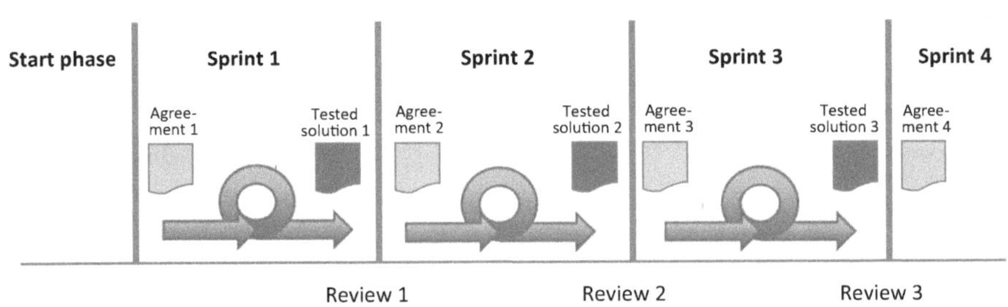

Fig. 6.3 The SCRUM approach to agile project management

After the start phase, there are iterations (or sprints) that are short, scheduled time frames (timeboxes). The teams, which are self-organizing, work independently on the agreed tasks.
The basic principles of this approach are the relatively short iteration cycles (or timeboxes), during which one or more highly motivated teams take responsibility for developing and testing solutions. Their learning points from the experience, and new user experiences, are fed into the next cycle. In the complex software development sector, this approach has proved to be more flexible, faster, and more economical than planning-based project management.
Many individual principles used in agile project management are eminently transferable to other projects, such as having flexible goals and a process-oriented approach in research and development projects, or self-organizing teamwork in change projects.

6.5 Change Projects

All projects involve change. When we talk about "change management projects", we specifically mean projects which aim to bring about radical, far-reaching and cross-sector changes to the organisation. This could be introducing new processes, mergers, implementing new strategies, etc.
This sort of project often involves trying to bring about new ways of working and new cultures, such as new communication cultures or ways of handling faults. A distinction is made between change projects and "continuous improvement" (CIP, Kaizen). Since the actual changes in change projects have to come about through the management system itself, they are rendered particularly difficult due to the involvement of the organisation's members on the one hand, and on the other hand due to the tendency to underestimate the challenges involved. There is a need to overcome barriers, deal with fear and anxiety, with resistance to change, but also recognise when expectations are unrealistic, and manage them. This means that the best approach is draw on organisational development, which focuses on social processes.
Any change project requires an awareness of change if it is to succeed. Otherwise, there will simply be too much resistance. The following formula offers a guideline:

$$D \times V \times F > R$$

D = Dissatisfaction with how things are now
V = Vision of what is possible
F = First concrete steps that can be taken towards the vision, achievable first steps
R = Resistance to change (the energy to keep things as they are)

According to Doppler and Lauterburg, the key factors to note for change projects are:

- Tap into people's enthusiasm and build trust
- Think in terms of processes, rather than structures
- Shape the organisation to fit its surroundings

So for change projects, the important things are to have a vision, clear objectives, to share information transparently, good communication and a process-oriented approach that enables people to learn together and to have new experiences together. John P. Kotter presents an 8-step model that is based on the formula quoted above:

1. Establish a sense of urgency
2. Form a powerful coalition of people willing to drive change
3. Create a vision for change, and a strategy to execute the vision
4. Communicate the vision
5. Empower people to act on the vision and the goals
6. Create short-term wins
7. Consolidate and build on the gains
8. Anchor the changes in corporate culture

Previous models have still tended to be focused on direct management of change. Even when applied with enthusiasm and determination, this approach often falls short in practice. Organisations prove to be resistant to change; it is hard to unlearn "the way things are done round here", and existing barriers remain stumbling blocks. Newer approaches also draw on systems experience. They focus on the organisation's reflexive dynamics. If these are used properly, they can produce significantly more leverage in bringing about fundamental change than if we are simply people-oriented. This puts the focus on the macro level, on working on the system rather than working in the system, on indirect management, on shaping the project's context and its interrelationship with the everyday work environment. In other words, the focus is on the requirements and on creating the right climate to encourage the organisation to change itself in a particular direction. One idea that arises out of this is "WAVE" (Petersen et al., 2011), which turns the systems approach into a repeatable procedure, but does so without losing sight of the non-determinability of social systems.

But any company or organisation that wants to bring about change is well advised to make a clear decision about which approach they want to use, and to work with a neutral, external person (or external system) as a consultant or change agent. It is very hard for any organisation to drive and manage this change on its own. Using the complementary consulting approach (Königswieser et al. 2006), this could also be a consultancy team that draws together people from process consulting and from subject consulting, but who get on well together, have good understanding of each others' skillsets, and who work well together.

Project Evaluation and Project Portfolios 7

Many organisations are currently experiencing a boom in projects, which stretches their resources to the full. For reasons that are not always clear, there is a tendency to make all problem solving into a project, or at least, to call the solving of the problem a project. Often, this is to do with the fact that many senior managers feel that they face too many big or difficult management decisions. So they delegate them as projects, or they are deliberately vague about the criteria that should be used when deciding to "outsource" an issue from the core organisation to a project. Management of the various projects is often underestimated. There is frequently no overview of all the company's ongoing projects. This makes it impossible to prioritise individual projects, and stops the organisation being able to effectively allocate the available resources. This is where a project portfolio is useful, to help provide an overview and for use as a decision-making tool.

7.1 Project Worthiness

Many companies and organisations have developed decision-making tools in the form of evaluation charts that help them to determine when a planned project is likely to be worthwhile or viable, and when it is not (Fig. 7.1).

J. Kuster et al., *Project Management Handbook*, Management for Professionals,
DOI 10.1007/978-3-662-45373-5_7

Criteria	Description	Estimate = 1	Estimate = 2	Estimate = 3
People involved	Job roles and organisational units affected by the solution	A few, from same department	A few, work together	From virtually the whole company
Interdisciplinarity	Number of organisational units working on the development process	Just one	A few	Lots
Complexity	Extent to which the various problems are interconnected	Low, OK	Medium,clear overview	High, complex interconnections, no clear overview yet
Importance	The project's strategic importance for the department or company	Unimportant, low	Medium, strategically important for parts of the company	High, strategically important for the company, key function
Urgency	Time pressure (how quickly are solutions needed?)	Not an issue	Firm deadline Enough time available	Time-critical, tight firm deadline
Financial cost	Investments (Over what time frames? ROI?)	Small, approx. < 0.5 million CHF, readily available	Medium, approx. < 0.5 million CHF, needs special provision making	High, > 0.5 million CHF, exceeds budget authority, needs Board decision
Financial parameters	Investments: How long is amortisation period (ROI) and benefit period?	Fast, within current year, insignificant	Bearable, within 1-2 years	Significant cost to company, will take several years
Effort	Duration of implementation period internal staff overhead	Low, will not impact on day-to-day business	Medium, can be done with existing resources and with day-to-day business	High, needs additional resources beyond day-to-day business
Knowledge	Does the company have the necessary skills and experience?	Available, routine project, standards available	Partially available for some key people, few standards	Not available, will need to be developed, no standards
Risk	To implementation, size of damage in the event of failure, viability	Low	Medium	High, could endanger the organisation
Predictability	How well can the process, the individual steps be planned?	Good, easily	Average, with difficulty	Poor, almost impossible to plan and predict
Motivation	Commitment to project by project owner and staff	Good, little conflict is expected	Varies, needs management to be alert	Critical, crises and resistance are to be expected

Fig. 7.1 Assessing the project worthiness

Or they may divide projects into various categories, depending on their complexity and on their strategic importance for the organisation.
The assessment of how a project fulfils these criteria gives grounds for arguing in favour of or against the project's worthiness, and a suitable basis must be selected. If it reaches a score of 40 %, for example (14 out of a possible 36 points), then a project and the relevant procedure should be examined more closely. Depending on the weightings assigned to the different criteria, the temporary project organisation may take different forms (Fig. 7.2).

Type of process	Repeated process	One-off process		
	Low complexity	Low to medium complexity	Medium to high complexity	High to very high complexity
	Short / medium-term	Short-term	Short-term	Short / medium-term
Type of organisation	**Ongoing business:** Permanent process organisation	**Special initiative:** Temporary person or working group	**Project:** Temporary project organisation	**Programme:** Temporary programme organisation

Fig. 7.2 Project characteristics and organisational forms

The diagram shows a simple method for deciding on the organisation's form, based on various project characteristics. Each of the different project categories will place different demands on the project manager.

7.2 Project Portfolios

With increasing frequency, we are seeing "strategic" projects started, in order to implement the organisation's strategy and to directly safeguard the organisation's long-term interests.

The organisation's senior management need to ensure that the available resources are used in the best possible way. In order to provide a link between the organisation's strategy and the management (or prioritisation) of the organisation's projects, we need to draw up a comprehensive overview of "planned" and "ongoing" projects.

A project portfolio is an overview of all the projects sorted by different criteria, presented either as a structured list or in the form of a diagram (Fig. 7.3).

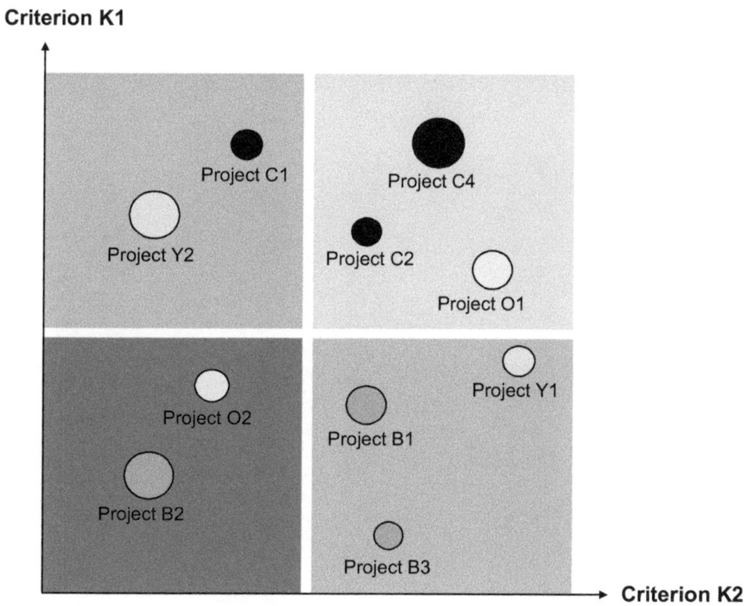

Fig. 7.3 Example of a project portfolio

Depending on the type of project this might simply be a list of paired criteria, such as opportunity versus risk, cost versus benefit, etc. But a project portfolio can also be multi-dimensional, using:

- Strategic criteria (e.g. a project score card)
- Economic criteria (forecasts, market)
- Ecological criteria

- Ethical criteria
- Opportunity/risk criteria
- Mandatory criteria (new legislation, technologies)

It is also normal for a number of employees at several different hierarchical levels in the organisation to be involved in drawing up the project portfolio. Senior management typically only draw up a portfolio of the 10 most important projects, with the other projects being evaluated and prioritised at departmental level.

Project Management Trends 8

As a field, project management is still developing. It has not yet reached the pinnacle of its "career" – that will be when we no longer speak about it. It will continue to evolve, and to develop its surrounding environment.

8.1 Increased Differentiation

Nowadays, project management includes a very wide range of areas. A universal "one size fits all" approach is no longer useful for all projects. This section will suggest the possibility of using different project types, so priorities can be established in project management and the approach can be tailored to the different requirements. Some typical special areas are:

- Basic research projects. Special requirements for this sort of project include establishing objectives, a process-oriented approach, and creativity.
- Change projects: Combining aspects of organisational development and project management.
- Projects in the political field: Dealing with a wide range of different interest groups, establishing shared visions, dealing with stakeholder groups over which you have no direct influence, mediation.

The differentiation system will need to be developed further. This also means that project management training and the profession of project management will also become more specialised.

J. Kuster et al., *Project Management Handbook*, Management for Professionals,
DOI 10.1007/978-3-662-45373-5_8

8.2 Agile Project Management

Projects (and especially ICT projects) are often challenged by turbulent environments, and constantly changing goals. This means that project processes cannot be planned in the conventional sense. It needs a flexible (or agile), adaptable, responsive form of project management. The classic phase models and problem-solving methods are too rigid for use here. The project process must be made more dynamic. The important thing is to make the process faster, more evolutionary, more recursive (with feedback loops). Some suitable approaches are prototyping and versioning. But the methods can be developed further, and give new emphasis to several established project management principles:

- More self-organisation, networks of interdisciplinary teams, less emphasis on hierarchical project organisations.
- Individuals and interactions are more important than embedded processes and tools.
- The dynamic interaction between the project and the stakeholder groups (especially the customers) is more important than guarantees, agreements and contracts.
- Responding to change is more important than sticking to a plan.

In dynamic projects, the form of project management is based more on the concepts of framework agreements, established rules and cultures.

8.3 International Project Management

As globalisation becomes more common, more and more projects involve working across different countries or regions. This can give rise to some additional complexity:

- Face-to-face communication is either very expensive, or simply not possible.
- The participants' different cultural backgrounds can make communication and cooperation more difficult.
- There may be a language barrier, and different time zones.

These need not all be negative points. Studies of intercultural teams show that their variety and different approaches can actually give a better result than monocultural groups, provided appropriate expertise in dealing with different cultures is developed and available.

8.4 Staff Development and Qualification

In a project-oriented organisation, the HR approach to staff development increasingly recognises "project manager" as a career description, alongside conventional management career tracks or subject specialist career tracks. In that sort of environment, project managers are systematically encouraged to gain qualifications, and promoted. Project managers fit in to the organisation's career model, based on their experience. A project manager's role offers enough opportunities to develop themselves in challenging management situations, in dealing with complex systems and in terms of their social skills. A career as a "project manager" offers opportunities to switch to some more conventional form of career path at a later stage, according to the manager's interests and inclinations (Fig. 8.1).

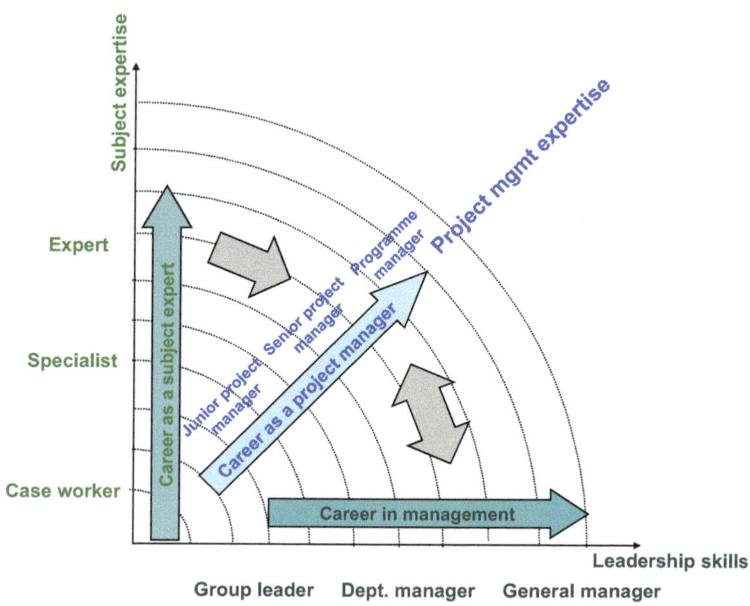

Fig. 8.1 Project management career development using the three-tier personnel development model

On a more general scale, certification models for project managers are also increasingly common. The IPMA model is described in this book. The American PMI model and the British Prince2 models have also both helped raise awareness. Universities are offering courses right up to Masters level, although project management is barely even mentioned at high school level. However, some first steps are being taken towards integrating project management into the curriculum.

Part II

Structure and Approaches

This part of the book describes the project process in detail, from the initial idea right through to implementation of the project's results. For each project phase, this part also offers tried and tested approaches and useful checklists, which project managers can make immediate use of in their projects.

The "Initialisation" Phase 9

9.1 What Is This Phase About?

A company always seeks to identify and start the few projects that are actually important. There are normally more requests for new projects than the company can deliver. There must always be a link between projects and the organisation's aims and objectives. The main challenge faced by the project idea during the initialisation phase is to make the cut, to be included in the list of projects.

The project owner and the project manager must jointly decide whether a new initiative is worth developing as a project. In order to properly start a project, the project manager draws up a project request. This sets out the project's goals and framework conditions. The necessary resources are planned and made available. A successful preparation phase ends with the project agreement being signed by the project owner. At that point, the request becomes an order. This formal approach protects the organisation against being overwhelmed by an unmanageable flood of projects, and provides a clear focus for the project team.

9.2 Results of the Initialisation Phase

The initialisation phase ends with an agreed project order. The initial ideas, either concrete or still being developed, lead to a more solid commitment: Who is going to achieve what goals, by when, and at what cost?

The deadline, and the rough estimate of the resources that will be needed, will answer two questions:

- Can this project be carried out with the available resources, in addition to the projects that are already included in the project portfolio?
- Given the costs of this project, and the anticipated benefits, is it worth carrying out this project?

J. Kuster et al., *Project Management Handbook*, Management for Professionals,
DOI 10.1007/978-3-662-45373-5_9

By the end of this phase, it will be possible to answer the following questions:

- When will the results or intermediate results be available?
- In broad terms, how many person days will the project involve and how much financial investment is required throughout the project?
- What critical resources will be needed, and when?
- Who needs to give input during the preliminary study, and when? How much of their time will be needed? What resources are needed from line management?

9.3 Stages of the Initialisation Phase

Check Project Worthiness

The following criteria help decide whether an initiative should be managed within the line management environment, or if it should be run as a broader project:

- Does it affect other departments?
- What resources will other departments need to contribute?
- What significance and consequences does the initiative have for the future of the organisation?
- What are the risks? What happens if the initiative is stopped?

Put the Project Agreement in Writing

The following points must be clarified and agreed:

- Starting point, the problem
- Project goals
- Area of action and boundaries: Identify system boundaries, delimit and analyse subsystems and partial systems, and identify any commonality
- Influences and dependencies: Identify influencing variables
- Framework conditions
- General parameters and concepts
- Results
- Project costs and expected benefits
- Risks, and what happens if the project is not carried out
- Procedure and deadlines, milestone plan
- The project's priority
- Project organisation, available resources
- Information and communication
- Signatures by project owner and project manager, possibly by project controller

9.4 Organisation

When the project agreement is being drawn up, there isn't normally a project organisation in place. One person (possibly the future project manager) is generally tasked with drawing it up. For large or very complex initiatives, it may be useful to view the project definition as a "mini preliminary study", and to set up an interdisciplinary team. That organisation need not be identical to the future project organisation. However, it is recommended that at least the key people should be involved in the subsequent project, in order to provide continuity.
If the project is initiated "from the bottom up" (that is, there is not yet a project owner), it is vitally important that the initiative has "connections" at the decision-making level. If not, there will be nobody with decision-making authority to support the project. This decision-making role should be defined by the time the project agreement is signed, at the very latest.

The Project Agreement Contains Important "Decision-Making Points" for the Project Organisation

The key roles (at the very least, the project owner and the project manager) are identified and contractually agreed.
The project order should also specify the broad outline of the project organisation:

- The type of organisation: project coordination, pure project organisation and matrix organisation?
- As well as the project manager and project team, what advisory boards will be needed: steering committee, advisory board, etc.?
- Should key stakeholder groups be represented in the project organisation? For example: organisational units, associations, interest groups, etc.

Skills Needed in This Phase

Even as early as the project definition, decisions that are made can have far-reaching consequences. These decisions should therefore be reached by people who bring substantial experience:

- With similar projects (similar approach, possible methods used)
- In assessing the possible risks
- In estimating approximate overheads (staff, possibly financial)
- In developing process concepts (procedures, social processes)

9.5 Planning: Preliminary Estimate

At a very early stage, as soon as an idea begins to take shape, the project's initiator will start to think about what will be involved in implementing the idea, about what will be needed. During this stage it is not really possible to speak of planning but more of rough estimates. The uncertainty of a rough estimate can vary significantly, for example +100 %/−50 %. For projects that involve a high level of innovation, the variation could be even greater.
If this develops into a worthwhile project idea, and if that in turn leads to a project request, the project owner will want a rough estimate of the necessary resources: staff resources, other critical resources, financial costs, follow-on costs, operational costs, etc. The project owner will also want to know when the project will be completed, and when intermediate results should be expected. The preliminary estimate is often drawn up by staff with extensive experience of projects, or by a core team, possibly led by the designated project manager. They often draw heavily on experience gained in similar projects. This enables the project owner to establish whether the critical resources are available, which in turn determines whether implementing the project is realistic.

Planning the Next Phase (Preliminary Study)

- What results must be achieved at the end of the preliminary study?
- What intermediate goals must be achieved in order to ensure that the preliminary study is completed on time?
- Are there any special risks or problems that need addressing? Should a risk analysis or feasibility study be carried out?
- How should the key activities in this phase be arranged in order to make best use of the resources?
- What special expertise will be needed? Where will critical resources need to be used?
- Who is available? What experience do they have? Has line management given a firm commitment yet?
- Are there any activities that need starting earlier because they will take longer?
- What internal and external costs will be incurred in the next phase?

9.6 Controlling and Risk Management

During the initialisation phase, project controlling decides which projects contribute to achieving the organisation's goals, and to what extent. The purpose of that is to broaden the view, to avoid simply looking at the merits of one individual project. A project that looks to offer a good return on investment when viewed in isolation may not actually offer real benefit to the company as a whole.

This might be the case, for example, where critical resources used by a project could actually achieve higher margins if used elsewhere. During the initialisation phase, strategic controlling asks project owners and project managers the following questions:

1. Company Strategy
- Does this initiative fit with the company's strategy?
- Is it part of the core business?
- Is this initiative essential? Does it have to be done?

2. Cost Effectiveness
- What will this initiative cost? (Full cost calculation)
- What are the short-term, medium-term and long-term benefits?

The following outcomes are essential in order to take professional decisions:

- Cost-benefit analysis with a meaningful cost structure
- A market study for development projects: market potential, benchmarking
- A rough risk assessment: SWOT analysis and context analysis
- Decision about carrying out the project
- Establish the project's priority level
- Planned and clearly specified controlling intervals
- Next controlling measures specified: milestone meeting, review

9.7 Leadership and Teamwork

All projects involve a variety of people from different specialist fields working on a solution together. The project initiator needs to be aware that this wide range of perspectives and expectations in the project will need to be reconciled. From a content perspective, the project has interrelationships and dependencies relating to a wide range of internal areas on the one hand, but also with a wide range of environments such as customers, project owners, colleagues, suppliers, the authorities, competitors, etc. It is interconnected with all of these elements.

Even the smallest changes will be felt in all areas of this network. Intensity and commitment will impact on the stability of the network. So it is important to analyse this network right at the very start, and to be fully aware of it (Fig. 9.1).

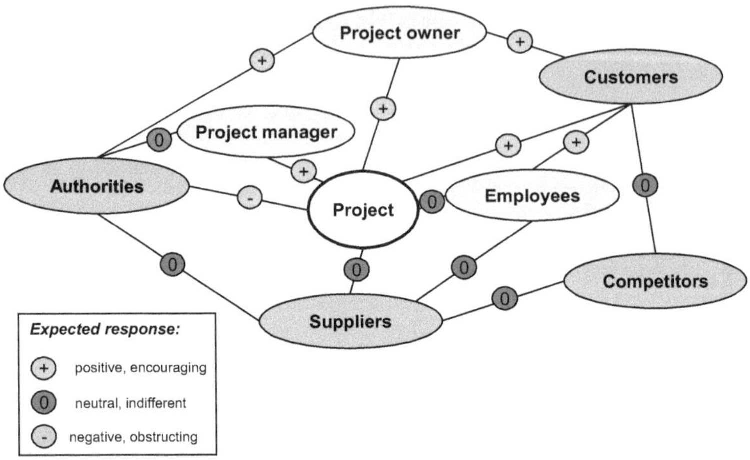

Fig. 9.1 Network of project influences

Project management is a form of relationship management. These relationships involve interpersonal, factual and organisational aspects. The project owner initiates the project by starting out with the interpersonal side of things. He needs to be very conscious that his choice of "the right project manager" is a crucial decision.

9.8 Information and Documentation

In the initialisation phase, the project is not yet public. In other words, communication is restricted to a very small number of involved people or bodies. Especially amongst these groups, care will be needed with the communication:

- For strategically important projects, the decision-makers will need to decide whether they wish to clearly demonstrate their commitment to this project to the outside world, or whether strict secrecy is more appropriate (for example: development projects).
- The project agreement is developed through a communication process or a negotiation process, which ensures that both parties share the same understanding of the project and the same high degree of commitment.
- For new developments, organisational projects, etc., thought must be given to whether the communication culture should be set up on a new basis, for example for team-oriented cooperation, for communication between the project team and the line organisation, between customers and developers.
- During resource considerations for the project agreement, the overhead that information and communication entails is generally underestimated. This applies to project marketing or for communications officers, and also to keeping full and up to date records and to recording management information.

Process-Focused Documents

The following documents are important decision-making tools for the project owner and the future project manager:

- A project request or project order as a document that confirms what is agreed between two parties.
- A procedure plan: Methods, steps, milestones, perhaps even a rough schedule for the preliminary study.
- Staff resources, project costs and project organisation.
- Communication plans, along with other strategies and guidelines.

Content-Focused Documents

Although most documents during the early phase tend to be process-focused documents, factual results and documents are also used:

- Business case: Deliberations about whether the project is commercially viable, meaningful and implementable from the market perspective, in terms of company strategy, etc.
- Depending on the project, this phase might result in a report containing the starting point, rough analysis results, risk considerations, goals or specifications, possible solution strategies.
- Results of earlier projects, investigations, and issues that might be relevant for the project.
- Market studies, surveys, staff interviews, etc. (if available).

9.9 Checklist "Conclusion of Initialisation Phase"

- How worthwhile is the project, and is its priority clear?
- Where does the project rank in the project portfolio?
- Are the roles of the project owner (decision maker) and project manager clear, and are they understood?
- Is the project's global objective clearly formulated? Do the project owner and the project manager agree on the goals and framework conditions?
- Have the milestones and the expected results at each milestone been agreed?

(continued)

- Has someone with experience of project management drawn up realistic estimates regarding expense, scheduling and availability of resources and expertise?
- Has a cost assessment been carried out, and is it based on realistic assumptions?
- Have anticipated follow-on costs been identified and included in the request?
- Have all currently known risks been identified, and do the decision-makers accept them?
- Has support been obtained from management and the project owner?
- Are there any views that have not been expressed at this stage?
- Have alternative scenarios for carrying out the project been explored?
- Has the project organisation for the preliminary study been drawn up?
- Have lines of authority and responsibility been agreed?
- Does the request identify who derives the main benefit from the project?
- Are the future users involved in the project?
- Is it clear who will be responsible for managing the solution that is developed? Do the relevant managers know when they will assume responsibility for the project's results?
- Has the written project request been signed by all important parties involved in the project, and thereby become an order? Have necessary resources been planned and approved?
- Does the plan state who must be kept informed about the project, when, and by whom?
- Is there an idea of how the project should be documented?

The "Preliminary Study" Phase 10

10.1 What Is This Phase About?

During this preliminary study, binding statements relating to feasibility, risks and benefits must be drawn up. These are all based on an analysis of the current situation, and on clearly agreed goals.
At the start of the project, there is limited knowledge about the project contents and the solutions. But more knowledge is gained as the project progresses. The risks are greatest at the start, and they must be reduced as quickly as possible and as far as possible. If the requirements (goals) push the limits of what is possible, or if the possibilities are not precisely known (technological limits, politically sensitive goals), then it is useful to carry out a preliminary study (also known as a feasibility study, pre-study or pre-project) before starting the main project. If it becomes clear that there are not enough resources available to realistically achieve the goals, a decision to abandon the project can be taken when this milestone is reached. That avoids devoting valuable resources to a project that has no realistic prospect of success (Fig. 10.1).

J. Kuster et al., *Project Management Handbook*, Management for Professionals,
DOI 10.1007/978-3-662-45373-5_10

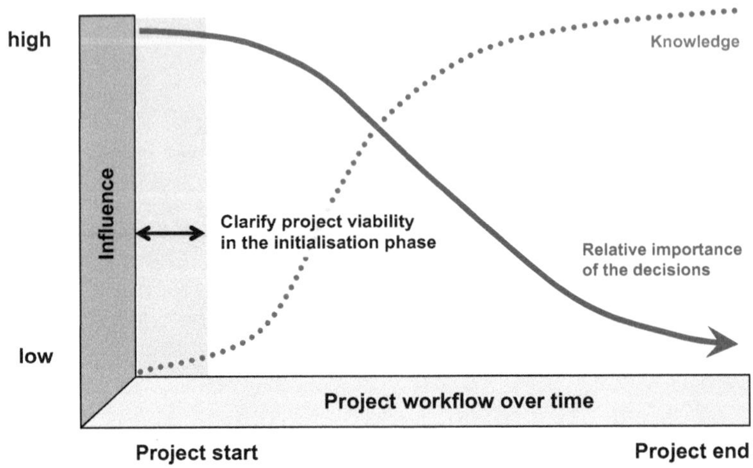

Fig. 10.1 Possible influences and risks in a project

10.2 Results of the Preliminary Study Phase

The preliminary study establishes whether the project is feasible or not, and whether it will generate sufficient benefits without causing any harm (feasibility study). If it becomes clear that the project should be carried out, the next phase is planned. That goes into greater detail, and the cost and scheduling estimates become more precise. In turn, that enables the necessary resources to be planned more accurately, and the project's benefits to be identified more clearly.

10.3 Stages of the Preliminary Study Phase

Preliminary studies must be carried out thoroughly and carefully, because the decisions taken at this stage have the biggest impact. As a rule, the preliminary study is often too superficial and there is always a price to pay for that later in the project.

- Start the project: hold a kick-off meeting.
- Identify and analyse the current position. Draw up a list of problems. Analysis results are based on working hypotheses.
- Set goals: Identify the results and knowledge that should be available at the end of the project.
- Check that the project fits in with the corporate strategy.
- Cost effectiveness estimate.
- Look for, evaluate and select broad solution approaches.

If the preliminary study ends with the conclusion that the project is worth pursuing, it is essential that the green light to continue the project is obtained from the decision maker. If not, the initiative must be ended as swiftly and smoothly as possible. It is good to thank everyone who has been involved for their work, and taking a bold decision to protect resources is a commendable thing.

10.4 Organisation

Preliminary studies are of great importance to a project, because the decisions taken at this stage are decisive in setting the direction of the project. Characteristically, preliminary studies often take place in "protected environments", and also involve a high degree of teamwork. Integral solutions and plans need to be developed, and this is best done when a broad range of viewpoints are taken into consideration (Fig. 10.2).

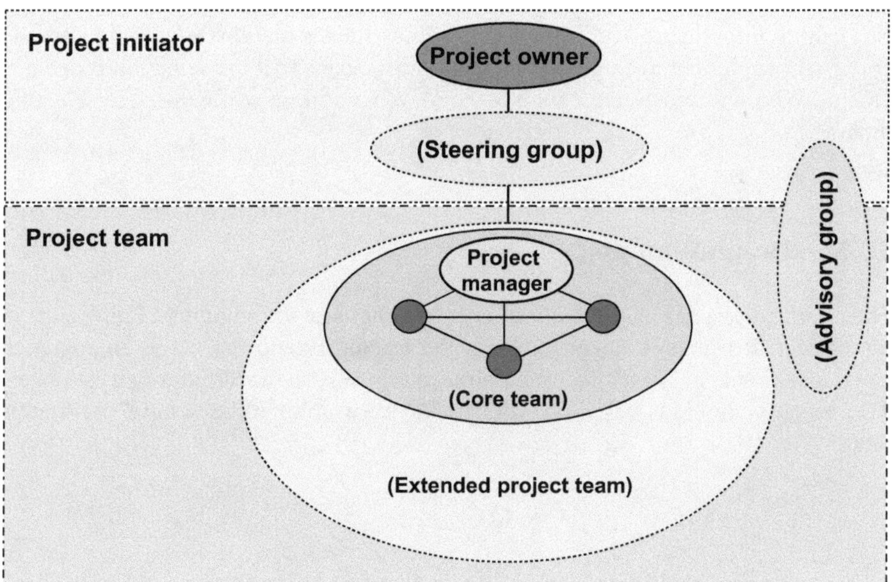

Fig. 10.2 An ideal project organisation for the preliminary study phase

The important points here are that the roles of the project owner, team leader and team members must be clearly set out, and that they must be mutually understood and accepted. If "creative teams" are used with no identified leader, the team must assign their own roles (or role flexibility) within the team.
Who will moderate? Who will present? Who will update the plans? It is worth giving high priority to issues relating to cooperation and communication.

The project manager should bring the following skills:

- Joined-up thinking: Encouraging networking between the different specialist areas.
- Moderating: Unleashing potential, sensitivity to group processes.
- Recognising financial and commercial connections.
- Thinking that is in line with the organisation's policies.
- Good time management skills, since there is often not much external pressure at this early stage.

The project team should bring the following skills:

- Creativity, an ability to think outside the box.
- Thinking in terms of alternatives, and in terms of the big picture.
- Making assumptions and taking decisions.
- Working "independently" of individual interests.

The central importance of the preliminary study means that this is a time to use the very best people, not average ones. That in turn means that the selection process is crucial. Who will select the team members? What criteria will be used in selecting them?

10.5 Rough Planning

The clarifications during the preliminary study increase the amount of knowledge in the project. It becomes easier to assess the feasibility and risks, and an approach starts to take shape. That makes it possible to improve on the initial rough estimates. This rough planning is based on work packages, which will be defined during this phase (Fig. 10.3).

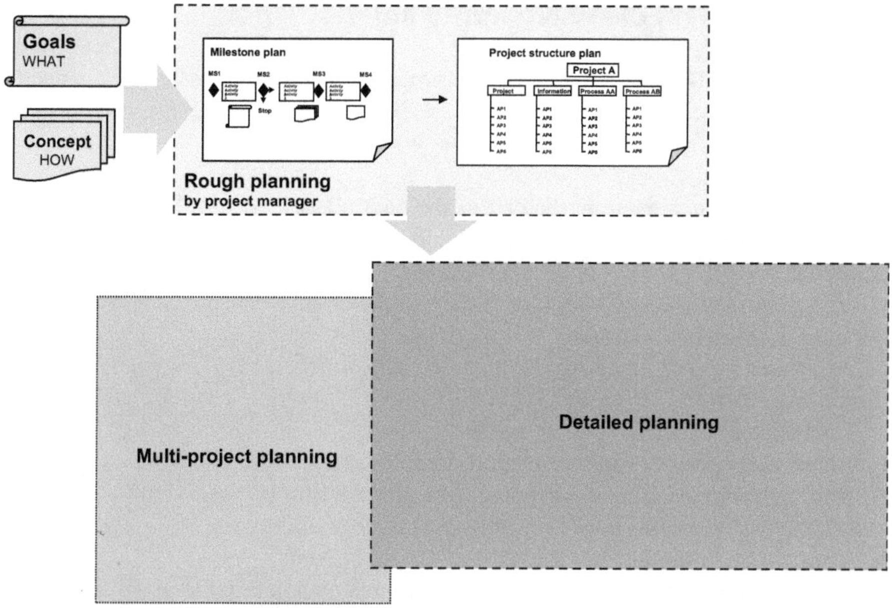

Fig. 10.3 Rough planning as part of the project planning process

At the end of the preliminary study, there will be a meaningful rough plan. This consists of:

- A milestone plan (specifies the phases and milestones)
- A project structure plan (defines the work packages, and any sub-projects)

Procedure for Rough Planning: The Steps

- Identify the project phases and milestones.
- Identify the decisions that need to be made at each milestone.
- If necessary, break large projects down into sub-projects. Identify the boundaries of each sub-project, and appoint managers of sub-projects, then hold initial discussions.
- Define the work packages, and their boundaries. Identify the structure of the work packages, and agree who is responsible for each one.

The results must be documented in the milestone plan/project structure plan.

Planning the Next Phase (Concept Phase)

In order to continue with the initiative, a project structure plan, schedule, capacity plan and cost forecast will be needed for the next phase. The same information will be needed for a high-level overview of the whole project:

- What must be achieved by the end of the next phase?
- What information must be available by the next milestone so that decisions can be made?
- What intermediate goals must be achieved in order to ensure that the concept phase is completed on time?
- Does the project need dividing into sub-projects? If so, what will the sub-projects be?
- What special expertise will be needed?
- Where will critical resources need to be used?
- Who is available? Has line management given a firm commitment?
- Are there any special risks or problems that need addressing?
- What are the internal and external cost implications?
- How must the key activities in this phase be arranged in order to make best use of the resources?

10.6 Controlling and Risk Management

Carrying out the preliminary study thoroughly helps to minimise the essential project risks, and helps in establishing how to run the project effectively. During the preliminary study phase, the main purpose of project controlling is to clarify and make decisions: Is the project feasible? Subsequent issues relating to how to proceed cannot be addressed until this question has been answered.

In the preliminary study phase, the following questions are crucial:

1. Feasibility
By the end of the preliminary study it must be clear whether the initiative can be successfully carried out, whether it is feasible. If the initiative is still in the balance, a simulation can be used to check various parameters in the form of different scenarios. If the simulation shows that the relationship between costs and benefits is not good, then the initiative should be stopped.

- Is the project feasible, both technically and politically? Will the project help reduce the various different uncertainties?
- Are sufficient resources and expertise available for this initiative?
- Is the project feasible in terms of finance and deadlines? Are the resources being used correctly in this project?
- What are the key scenarios to achieve the overall objective?

2. Cost effectiveness

The key objective of any organisation is to ensure the organisation's long-term survival. In the commercial world, survival is directly related to liquidity, to having money in the bank. Every deal, every planned project, should therefore be assessed to see if carrying it out will increase the organisation's survival prospects.

The cost-effectiveness of the project covers project costs and investment costs, and compares these with the anticipated result at the point when the solution is in place (planning horizon). At the start of a project, all data must be estimated or based on assumptions. The cost-effectiveness assessment can therefore only be as good as these assumptions. That is a simple fact, which cannot be avoided. For the planner or project manager, this means that they should assign a reliability grade to each aspect of their cost-effectiveness assessment. At the start of the project, for example, the rough estimates will probably have an accuracy of ± 35 %. Even so, in many cases these estimates offer a better assessment of the risks and a more informed decision about financial investment in the project. The degree of uncertainty reduces as the project progresses. Before implementing the project, it should be possible to work with an accuracy level of ± 10 %.

The project manager must learn to cope with the uncertainty and inaccuracy of these forward-looking assumptions. In assessing the cost-effectiveness of the project, two key questions should be answered:

- What profits, cost reductions or savings will the project deliver?
- What are the risks or returns on capital that is invested in the project?

The project cannot be evaluated on the basis of financial criteria alone. It is a good idea to also use "benefit criteria". Evaluating the project's cost-effectiveness involves an assessment of the economic efficiency on the one hand, and an assessment of the economic returns on the other:

- How can the benefits be quantified?
- What form do the individual cost factors take?
- How sensitive are the project costs in relation to the various alternatives?

The following outcomes are essential in order to take clear decisions:

- Carefully checked overall cost-effectiveness of the project
- Comprehensive feasibility study
- Verified project risks
- Approval for the next phase (budget, resources) or a decision to abandon the project
- Confirmation of the project's priority
- Next controlling measures specified: Milestone meeting, review

10.7 Leadership and Teamwork

The greater the social complexity, the more interlinked the project, and the more open the project's goals, the more challenging the project will be for the project manager. He must set up the processes in such a way as to encourage the team to deliver its full potential. The project manager usually has little formal authority in leading his team. His leadership will come from his own personal authority. He must develop his new team in a manner that encourages cooperation, that develops a feeling of "we" rather than "me", that progresses to a jointly developed and mutually shared goal, and to shared values and standards.
The project manager must cultivate this feeling of "we" before, during and after the kick-off meeting (Figs. 10.4 and 10.5). His objectives are to:

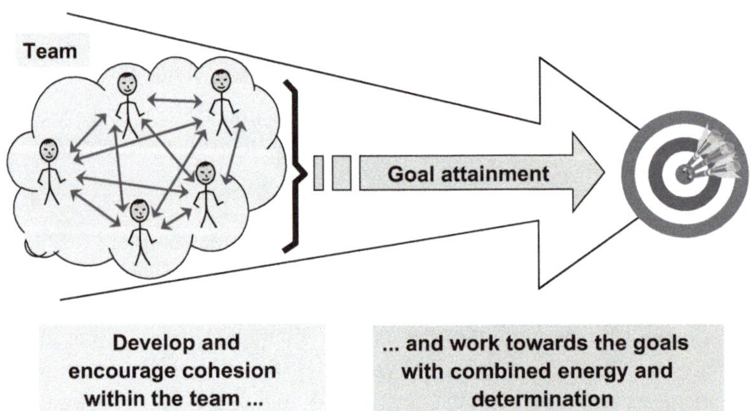

Fig. 10.4 Team building and combined energy

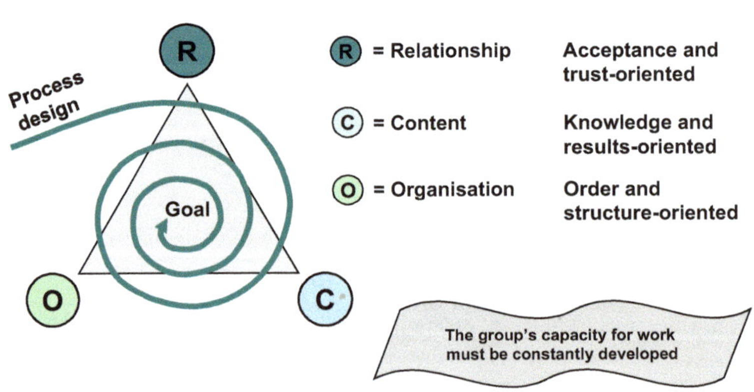

Fig. 10.5 Aspects of process design

- Promote a climate of acceptance and trust at the cooperation level.
- Develop and encourage a knowledge-based and results-based approach at the content level.
- Provide order and structure at the organisation level.

Right from the start, the cooperation phase is a very sensitive time that is socially complex and characterised by uncertainty for the participants. The project manager can develop and encourage a positive culture of cooperation through the following steps:

- By being aware that the newly formed project team may wish to focus at the start on binding framework conditions, on rules and on the participants' working methods.
- By agreeing appropriate rules for working together, and by enabling tasks and areas of competence to be distributed across the team.
- By being aware of his own social and personal skill set, and by using them sensitively.

10.8 Information and Documentation

General

Even though the preliminary study will only involve a small project organisation, the mutual exchange of information and internal communication should be set out in a plan:

- How will meetings and workshops be organised?
- What notes will be produced?
- What will be documented, and how?
- How will the project owner be kept informed?
- How will the project team maintain a dialog with the project owner?
- What will be outsourced?

For socially complex projects, it is helpful to use a project environment analysis to identify expectations, fears, interests, support, rejection points, etc. Making the project environment's interrelationships visible can be helpful when designing the information and communication plan (Fig. 10.6).

Receiver	WHO Person responsible for reporting	WHAT Subject	WHERE Location Medium	HOW Format	WHEN Deadline, frequency
Project owner	Project manager	Decision about further course of action	Board meeting	Presentation	End of pre-project
Steering committee					
Project manager					
Project team					
User					

Fig. 10.6 Example of an information and communication plan

Process-Focused Documents

The project handbook or project folder should contain the following documents:

- Project agreement for this phase, as an absolute minimum.
- Updated procedure plan: A plan showing deadlines, resources and costs.
- Updated project organisation.
- A rough plan for resource requirements (staff resources, and material resources).
- Information and communication plan (internal and external).
- Status reports, minutes of meetings.
- A procedure plan for the next phase (the preliminary study covers concepts for the future actions and for process design; these strategic visions are often more important than the actual concrete results, especially in very open projects).
- A critical review of the process is particularly valuable during this phase: What lessons can the team learn for the next phase?
- A phase report, and an application for the next phase.

Content-Focused Documents

Documents that relate to content, to outcomes, to results:

- Analysis results.
- Assumptions and conditions.
- Weaknesses and shortcomings in the current situation.

- Goals and challenges.
- Revised, detailed goals (specifications)
- Feasibility study: Assessment of feasibility, costs, financial return, risks and detailed economic efficiency calculation.
- Rough approach to a solution.
- Concept: During the preliminary study phase, solution concepts are not the top priority. The analysis results and the goals are more important. For now, solutions should be limited to "rough concept", to ideas for solutions, "alternatives", etc.
- The phase report, status report, review should bring all the items listed above together as a close-off report.

10.9 Checklist "Conclusion of Preliminary Phase"

- How have the goals or system boundaries changed when compared with the project order?
- Are the changes justified? What are the effects of the changes?
- Are the goals expressed in a positive and solution-neutral manner? Are they complete, measurable, consistent and achievable? How will you recognise whether your goals are achieved?
- Have any contradictions in the goals been clarified through discussion? Are the goals classified and weighted?
- Is there credible proof that the goals can be achieved?
- What results came out of the market study or competitor analysis?
- Is the project still shown to be cost-effective at this stage?
- Will the project have relevant and positive impacts on the balance sheet and the income statement?
- Have the risks been realistically assessed, and any discrepancies in the assessment resolved?
- Have all of the possible approaches to solutions been identified?
- Have the benefits and drawbacks of the solution approaches been discussed in sufficient depth, and properly evaluated?
- Has the feasibility been demonstrated in the market, and also in the context of the political and technical environments?
- Has a project structure plan been drawn up? Have sub-projects been properly set up, or provision made for them? Are there clear dates for critical decisions?
- Have costs, deadlines, the required skills and availability of key people with the new expertise been checked, and are they realistic?
- Are all estimates (quantities, frequency, milestones, costs, time) realistic? Is the accuracy of any estimates indicated?

(continued)

- Will subject experts and the relevant resources be available at the required times?
- Is the project owner supporting the project team with all the resources at his disposal?
- Does the project follow quality guidelines?
- Are there concrete guidelines for the periodic exchange of information and for data security?
- Are there regular discussions about teamwork and cooperation?

11 The "Concept" Phase

11.1 What Is This Phase About?

During the concept phase, different solutions are developed and evaluated. Implementation plans are drawn up for the selected alternatives. To the extent that it is possible, the needs of all interest groups should be covered. In doing so, it is important to be aware of your own social conventions. For large projects, the "concept" phase can be split into a main project and a detail project. Splitting the phase up creates an additional milestone. This prevents the project team from straying away from the path. At each milestone, the project owner decides which of the alternatives should be pursued. He also authorises the resources for the next phase.

11.2 Results of the Concept Phase

At the end of the concept phase, the following results will be available:

- A range of possible solutions will have been developed.
- One of the alternatives will have been selected, with implementation plans.
- The selected alternative will have been worked out in detail.

J. Kuster et al., *Project Management Handbook*, Management for Professionals,
DOI 10.1007/978-3-662-45373-5_11

11.3 Stages of the Concept Phase

- Specify detailed goals or goals for sub-projects and sub-systems, and revise the goals if necessary.
- Develop a range of alternative solutions (this phase needs the most creativity), and check that they match the goals.
- Evaluate the different alternative solutions, and have the project owner select one of them.
- Draw up a more detailed solution for the selected alternative.
- Check and adapt the required resources.

11.4 Organisation

The relevant stakeholder groups are usually represented during the concept phase, either in the project team or sub-project teams, or in special working groups or advisory groups. There is a tendency to involve all the representatives of a stakeholder group, which can make the project organisation difficult and time-consuming. By using a suitable structure, it is possible to have specific bodies meet only when they are really needed.
From the range of alternative project organisation options, the best one must be identified. The following criteria are useful in that respect:

- What significance does the project have in the organisation?
- What does the project change?
- What subject expertise does the project need, and what support does it need?
- What should the culture of the project be (and thus possibly the future culture)?
- What level of commitment is expected from the affected parties?
- What resources are available? How much can be spent or committed?

Points to Pay Particular Attention to in the Concept Phase

- Use a project environment analysis to identify the key indicators for the organisation.
- Discuss, agree and write down the roles (tasks, authorities) for each individual advisory board, and make them transparent.
- Draw up clear plans for communication between the advisory boards, and manage them: two-way exchanges of information and networking, joint discussion meetings, reviews, etc.

Skills Needed in This Phase

- For projects where the emphasis is on a technical solution: Subject expertise.
- For change projects: Process expertise (procedural expertise, social and management skills).

- A balanced mix of creative people and realists (pragmatists).
- Project manager: Facilitation skills, teamwork, team development.
- Include roles for lateral thinkers: For example, people from different specialist fields often bring a valuable external perspective to the project.

11.5 Detailed Planning

In terms of content, the concept phase is when a solution concept is drawn up. This indicates how to proceed in order to meet the goals. Detailed planning is then carried out, drawing on the rough planning done in the preliminary study and incorporating the selected solution. Planning for implementation normally has an accuracy of ± 10 %. The degree of accuracy that is possible (or necessary) will depend very much on the content of the project and on its environment. For customer projects with a fixed-price quote, a high degree of accuracy is even more important than for internal projects.
At the end of the concept phase, a comprehensive and helpful detailed plan will be available:

- A procedural plan and a schedule in the form of a Gantt chart or network plan.
- A detailed resource usage plan, with approval from line management.
- A cost plan, showing which financial resources are needed, and when (Fig. 11.1).

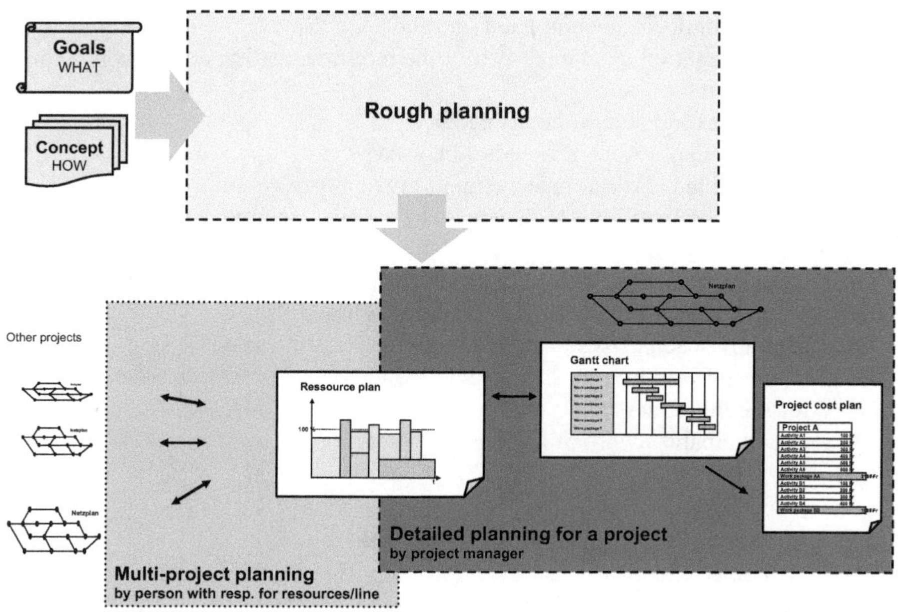

Fig. 11.1 Detailed planning as part of the project planning process

Procedure for Detailed Planning

- Update the project structure plan.
- Identify the activities that must be carried out within the work package.
- Identify the interdependencies between activities.
- Identify the skills and experience that are needed to efficiently complete the activities, and check the availability of the relevant specialists.
- Estimate the time required and the throughput time for resource allocation.
- Draw up a schedule, and optimise the throughput time if necessary.
- Get firm agreement from line management for the availability of resources.
- Identify resource conflicts, and work with line management to agree solutions.
- Plan the timings of financial resources: Liquidity, cash management.
- Resolve any financial or scheduling conflicts.
- Communicate the planning results to all participants and affected parties.

This first round of planning is called initial planning. If necessary, this plan can be adapted as the project progresses.

Planning for the Next Phase (Implementation)

- What must be achieved by the end of the next phase?
- What information will be available by the next milestone so that decisions can be made?
- What intermediate goals must be achieved in order to ensure that the implementation phase is completed on time?
- Is all the information available? Is it of the required quality, and does it contain sufficient detail?
- What special expertise will be needed?
- Where will critical resources need to be used?
- Who is available? Has line management given a firm commitment?
- How must the key activities in this phase be arranged in order to make best use of the resources?
- Are there any activities with particularly long throughput times that should be started earlier in order to meet the overall deadline?
- Are there any special risks or problems that need addressing?
- What financial resources will be needed, and when? Is that the latest possible time for the investment?
- Are all participants aware of the critical points?

11.6 Controlling and Risk Management

Project controlling is particularly concerned with decision-making activities during the concept phase: The available range of alternative solutions must be examined to determine the extent to which they contribute to the organisation's strategic and commercial objectives.

During the concept phase, the following areas are the most important:

1. Investigate Alternatives
- Which scenarios or alternative solutions are commercially interesting?
- Do the different alternatives involve different risks for the project?

2. Cost Effectiveness
- How can the benefits be quantified?
- How can the value created by each alternative be quantified?
- How sensitive are the project costs in relation to the different alternatives?

The following outcomes are essential in order to take professional decisions:

- Cost-benefit analysis
- Detailed risk analysis (e.g. FMEA)
- Carefully checked overall cost-effectiveness of the project
- Approval for the next phase (budget, resources)
- Decision to abandon the project
- Confirmation of the project's priority
- Progress report
- Next controlling measures specified: Milestone meeting, review

11.7 Leadership and Teamwork

Just as each person has developed over the course of their own journey, groups also develop their own "life story". Topics, situations, the environment, the context and, above all else, the personalities of the individual team members and of the project manager mean that every group process will be different. The "we" feeling in a group must be developed one step at a time. Disquiet, resistance or conflict in a team can cause problems (Fig. 11.2).

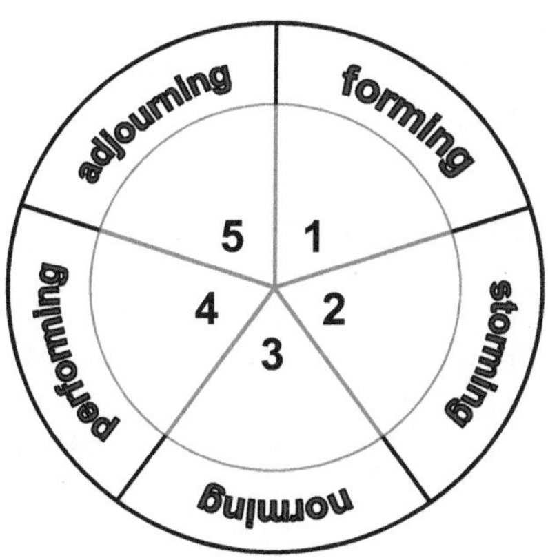

Fig. 11.2 A group's development phases

These disruptions make the work more difficult, or even impossible. Resolving the disruptions must then be a priority. If they are not resolved, the group's productivity is affected, and there is a risk that the goal may not be achieved (Fig. 11.3).

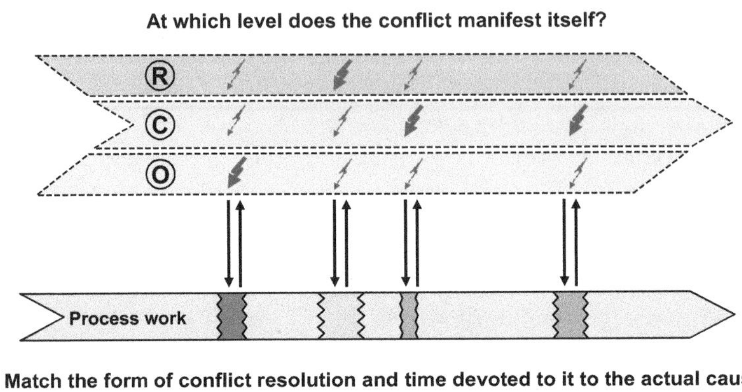

Fig. 11.3 Taking time for process work in the project

In such situations, it is better for the project manager to assume the role of a neutral moderator. If he helps the group to resolve their issues, the solutions tend to last. But if he takes the approach of "laying the law down" as the group's leader, the team members will withdraw and reject his authority. The team must recognise relationship issues as a top priority.

The project manager must be aware that each individual team member will have their own viewpoints, values, knowledge and experience, based on their own backgrounds. If there is disharmony at the Relationships (R), Content (C) or Organisation (O) levels, the project manager must work on the team's own development.

The team's full effectiveness often only comes into play after the group has met a few times. This time is necessary for the team to develop a "we" feeling, a team spirit. It takes time for everyone to settle in and get to know each other, to get used to one another and to learn to work well together.

11.8 Information and Documentation

During the concept phase, the main focus should be on "how". Communication with the stakeholder groups needs particular attention in order to develop trust, identification and support. Important tools to that end include a project environment analysis, an information and communication plan and project marketing. By the concept phase (at the latest) the documentation system should also be in place.

Process-Focused Documents

These are the same documents as in the preliminary study. Special emphasis should be placed on controlling aspects, i.e. on status reports and change reports.
Here again, an evaluation of the project work or a phase review can be very valuable.

Results-Focused Documents

The central document is the concept document, which describes what the solution looks like. It could be a product, a hardware or software product, a procedure plan, an organisation specification, a marketing concept, a training course, etc. Detailed plans must also be available, as they will be required for the implementation phase.

Phase Report (Status Report, Review)

This document contains an evaluation or a summary of the concept phase, and forms the starting point for the next phase:

- Requirements, assumptions, any problems that occurred, consequences.
- Goals, any new framework conditions.
- Solution, alternative solutions, deliverables.
- Effects on stakeholders (e.g. customers, shareholders, staff) and the environment.
- Updated cost-effectiveness analysis, particularly in terms of any changes since the preliminary study.
- Evaluation of the project: Advantages, drawbacks, threats, the degree to which the goal is met, scalability, summary evaluation, possibly position statements from stakeholders.
- Updated project organisation.
- Overall project planning, planning the next phase, financial management.
- Communication plan: internal and external.
- A plan for resource requirements (staff resources and material resources).
- Submission of application (solution, and procedure from here on).

11.9 Checklist "Conclusion of Concept Phase"

- How have the conditions that were set been complied with?
- Have the alternative solutions that were selected in the preliminary study been followed through properly?
- Since the preliminary study, have there been any changes in terms of goals, system boundaries and system design? Are the changes justified? What are the effects of the changes?
- How have the proposed changes been reflected in the project plan?
- Are the estimates in terms of quantities, frequency and scheduling firm enough to be implemented?
- Are the benefits and drawbacks of the different solutions, and their associated risks, clearly indicated?
- Has the risk assessment been checked against the outcomes of the concept phase? Have the necessary measures been identified, and the actions prepared?
- Is the project still shown to be cost-effective at this stage?
- Have the costs, resources and deadlines been fully and realistically planned in detail, and has a plausibility test been carried out?
- Are the financial resources and liquidity guaranteed for the entire duration of the project?
- Is the availability of the necessary financial and human resources assured?
- Is the necessary expertise available? Are the relevant resources guaranteed?
- Has the approach for the implementation phase been agreed with all involved parties?
- Is a transition to new project team members planned, prepared for, or has it already taken place?
- Have lines of authority and responsibility been set up appropriately for the requirements?
- How have the guidelines for project controlling and reporting been followed?
- How have the quality management guidelines for the running of the project and for project content been followed?
- Are the planning documents complete? Or at the very least, are there concrete plans for completing them?

The "Implementation" Phase

12

12.1 What Is This Phase About?

In the implementation phase, the plans from concept phase are made into reality.

12.2 Results of the Implementation Phase

The solution, or the system, should be built and tested. In order for the project to be concluded, the following goals must be achieved by the end of the implementation phase:

- The system, product or service should be created.
- The solution should be tested.
- The product or service should be deliverable as soon as the users are ready for it.

12.3 Stages of the Implementation Phase

- Draw up a budget.
- Put material resources, personnel and financial resources in place.
- Adapt the project organisation.
- Produce the solution, and test it.
- Plan the training for future users.
- Manage the running of the project.
- Compare the plans against how things are actually going (financial management, controlling).
- Communicate clearly about any differences or discrepancies.
- Keep the involved parties informed.
- Keep external contacts (e.g. customers) informed.

J. Kuster et al., *Project Management Handbook*, Management for Professionals,
DOI 10.1007/978-3-662-45373-5_12

12.4 Organisation

For infrastructure projects, the implementation phase is about production, installation and assembly. For organisational projects, it is about implementation, trial runs, pilot tests, etc. The project is then much closer to physical production, or to everyday operation. This has implications for the project organisation, which is generally more hierarchical than in the concept phase, for example, which was largely based on a team culture. So it is characterised by a more marked leadership structure, and tighter management. This is often necessary because there will be more people involved than in the earlier phases.

Clearly identifying and resolving the roles, duties and lines of authority is especially important at this stage. For example, cost overruns can often be traced back to the project controlling being set up in an "untidy" way, in organisational terms.

- Doers are more useful here than visionaries
- Organisational talent
- Clear, consistent, approachable management style
- Ability to solve problems
- Able to resolve conflict
- Conduct and behaviour, commitment

12.5 Adapting the Plans

Often, planning only becomes real when it is carried out: Who is doing what? Which company is responsible for which component? Which team is implementing which area?

During the concept phase, the initial plan is drawn up and approved. At each milestone, critical questions are asked and re-evaluated each time: Are the assumptions that were made during the initial planning still fully valid? Are there any new framework conditions? If necessary, the existing plan must be updated in terms of resources, scheduling and costs. That may also involve reaching new agreements. Any changes to the project plan must be documented, and must be communicated to the affected parties.

If completely new framework conditions or issues are significant enough to mean that the existing plan is no longer suitable for a "planned/actual" comparison, then the plan should be cut off at the project's current status. A new plan should then be produced from that time forward (time-to-complete and cost-to-complete).

Any resource conflicts arising from the new situation must be resolved.

Planning the Next Phase (Introduction Phase)

Now (at the latest) plans should be drawn up for the transition to the utilisation phase and for closing off the project, or the existing plan should be reviewed.

- What must be achieved by the end of the project?
- How will the product or service be introduced for the user?
- Are any special measures needed to address risks or problems?
- What unresolved issues still need resolving before the end of the project?
- What milestone decisions, reviews or acceptances must be prepared for?
- What activities are required in order to hand over and close off the project? When, and at what cost?
- Are the authorised resources sufficient to allow all work to be completed?

12.6 Controlling and Risk Management

Project controlling is particularly concerned with monitoring and management activities during the implementation phase: Is the project running as predicted in terms of scheduling, costs and resources? If any change requests come in during this phase, the implications for quality, costs and scheduling should be established (see the "magic triangle"). The project owner will use this information to decide whether the change request should be accepted or not.

During the implementation phase, the following areas are the most important:

1. Management of the Project
- Where have things not gone in accordance with the project plan, and what happened?
- What activities or work packages are on the critical path?
- Where is support needed from management or from the steering committee?

2. Changes to the Project
- Which changes need implementing? At what cost?
- Which changes can be put back until later?
- Where should additional demands be anticipated?

The following outcomes are essential in order to take professional decisions:

- Carefully checked overall cost-effectiveness of the project
- Updated risk analysis: New risks, changes to the framework conditions
- Approval for the next phase, including budget and resources
- Confirmation of the project's priority
- Progress report
- Next controlling measures specified: Milestone meeting, review

12.7 Leadership and Teamwork

The implementation phase is a time when problems often occur in the project team. These can affect the team's performance, and thus place the team's ability to achieve its goals at risk. The project manager then needs to intervene at the appropriate level and address the cause. He must consciously interrupt the work process in order to clarify who needs to do what in order to achieve desired results. At the relationship level, coaching measures that clarify, support, structure and develop are useful (Fig. 12.1).

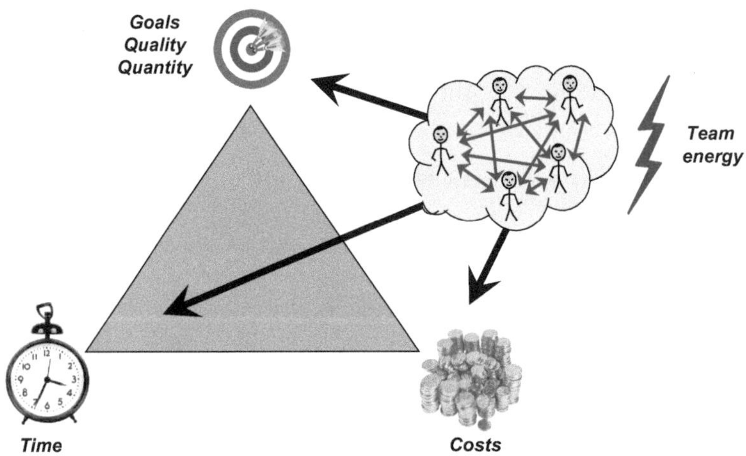

Fig. 12.1 Influence on the magic triangle

As well as acting appropriately, the project manager must also ensure that his interventions result in an improvement in the problems being experienced. He can do this by approaching the issues from an objective level and focusing on the goals, scheduling and costs.

12.8 Information and Documentation

In the implementation phase, the emphasis is on checks and controls: Communication about the project status (compared with plans) and any planned changes. This phase normally involves a lot of people, so systematic communication and exchanges of information are essential.

During the implementation phase, the following documents are the most important:

Process-Focused Documents

- Project progress reports (progress reports and expense reports)
- Management summary
- Change documents
- Tenders, contracts, accounting
- Quality assessment
- Introduction plan
- Review report

Results-Focused Documents

- Test reports
- Documents relating to the launch and training: Introduction plan
- Goals for the project, assessment of how the goals have been met
- Cost effectiveness
- Implementation plans

12.9 Checklist "Conclusion of Implementation Phase"

- How well are the goals that were agreed in the project order being met?
- What successes will be demonstrated by the planned trials and tests on the new product, system and organisation?
- How well do the results that have been achieved match the specifications?
- Has the introduction of the solution for the user been planned in a way that makes success a realistic outcome?
- How are accompanying measures such as training, adaptation to the organisation, and documentation for the launch being provided?
- Is a pilot run needed before the full-scale rollout?
- How broad is acceptance for successful introduction? Is that sufficient?
- Which results were not achieved? What are the consequences of that?
- Are there any defects for which corrective measures are not available?
- What should special attention be paid to during final checks? Which open issues remain unresolved?

The "Introduction" Phase

13

13.1 What Is This Phase About?

When the new service or product has been rolled out, the project organisation must be dissolved. This includes recognising what has been achieved, and learning any lessons from the project. The best way to do this is to review the project jointly with the project owner and the user once the solution has been in use for an agreed time. After completion of the project, the utilisation phase begins. Experience is gained at this stage which can be used to improve the existing solution, and which can be fed into the design of similar systems in the future.

13.2 Results of the Introduction Phase

Managing the introduction of the new solution well, and holding a comprehensive close-off discussion with a full evaluation of the project, enables the project to be brought to a clean end and leaves behind a positive impression. The project team should be careful not to slip into "almost finished" mode, where they mentally start to move on to other things, until all of the unfinished work has been completed. It is important to learn lessons from anything that didn't go right. Other projects will be able to benefit from the knowledge and experience if it is discussed and documented.

When the project is finished, the following will apply:

- The system, product or service will have been carefully handed over to the line organisation.
- The users will be able to make productive use of it.

J. Kuster et al., *Project Management Handbook*, Management for Professionals,
DOI 10.1007/978-3-662-45373-5_13

- The acceptance report will have been signed.
- Final costing will have been carried out.
- Documentation will have been produced for subject knowledge and for learning experiences.
- A final assessment will have been carried out.
- The project team will have been wound down.
- The date of the follow-up review will have been agreed.

13.3 Stages of the Introduction Phase

Concluding a project involves the following tasks:

- Enabling the user to use the new solution productively.
- Introducing the solution, and bringing it into full use.
- Checking that the goals have been achieved.
- Making preparations for maintenance and repairs, designing the successor organisation.
- Finalising all cost accounting issues and final costing.
- Drawing up the report and application for final assessment.
- Handing over the project documents to the maintenance organisation.
- Handing over the system to the user or the customer (acceptance).
- Completing the archiving of the project documentation and storing it safely (especially important in information technology, and plant engineering for product liability issues).
- Dissolving the project team in a managed way.

To assess how well the goals have been achieved over the longer term, a meeting should be held with the project owner after the system or product has been in use for an agreed time, in order to review the effectiveness and the benefits delivered by the project goals. In some situations, it may be more useful to review this at the handover stage.

13.4 Organisation

There should be an official end to every project, and to every project organisation. This discharges the project bodies from their duties, and leaves them free to focus on new jobs. Of course, the project never actually comes to a complete finish. There could be some follow-up work or warranty work; documentation may need completing, etc. Organising and carrying out this work can be delegated to specific individuals.
In the same way as a kick-off meeting was held at the start of the project, there should also be a close-off meeting. The close-off meeting should involve the same three levels (relationship, content, organisation) as the kick-off meeting: The

process should be brought to an end in appropriate ways for the different levels, and successes should be celebrated. The following organisational topics will need addressing:

- Introducing the solution, and planning training for future users.
- Structure and evaluation of the successor organisation: Who will operate the solution?
- Critical review: How effective was the project organisation?
- What can the organisation learn from this for the future?
- Praising the team's performance, letting the team members go.
- Reporting back to management on the performance of the team members.
- Explicitly dissolving the project organisation.
- If necessary, offering support to help reintegrate the team members into the main organisation.
- Organising any follow-up work and subsequent evaluation of success.

13.5 Planning: Using the Experience of How the Project Went

There is often a degree of stress at the end of the project due to deadlines or unexpected problems occurring. People might even be distracted by an interesting new project, which moves their focus away from concentrating on properly resolving any outstanding issues, on completing the documentation and on a smooth handover. But these important activities must be planned, and those responsible for them must be encouraged to stay focused.

To help improve the accuracy of future cost and time estimates, the project leader should at this point compare the initial planning estimates against the actual overheads that occurred during the project. The results of this comparison should be discussed with the relevant affected parties. This will enable lessons to be learned about implications for the future. These may relate to the actual overhead estimates, or to the working methods and conduct of the relevant people in relation to these overheads.

The project manager must document the realistic cost, how he would estimate it for future projects given the experience from this project, and also the framework conditions that influenced the costs and overheads for example, the scope and complexity of the work packages, the experience levels of the staff involved, or particularly sensitive political environments. If project managers carry out this review systematically after each and every project, the company can build up a valuable reference system that will help the company when it comes to estimating the overheads involved in future projects. It will enable them to put together cost and overhead estimates more quickly, more accurately, and at less cost. This fits well with the concept of a "learning organisation".

13.6 Controlling and Risk Management

During the introduction phase, the main purpose of project controlling is to check and verify. It supports a smooth introduction of the product or system for the customer or user, and the complete close-off of the project.

The following areas are the most important during the introduction phase:

1. Achieving the Goal

- To what extent have the goals that were defined at the start been achieved? Have the project goals been used as control objectives for the final assessment?
- Have the actual data from the final assessment been compared against the plan data, and any differences been analysed?
- What deliverables were not produced or not completed (list of shortcomings)? What are the implications?
- Does anything learned from the project's final assessment have consequences for other ongoing or future projects?
- Have the guidelines for project controlling and reporting been followed?
- Have the quality management guidelines for the running of the project and for project content been followed?
- Has proof been provided to show that the specified quality was achieved?

2. Cost Effectiveness

- How big are the differences from the project costs defined at the start of the project?
- To what extent has the expected benefit from the project changed?

The following outcomes are essential in order to take professional decisions:

- Carefully checked overall cost-effectiveness of the project.
- Internal acceptance in order to minimise the risks around customer acceptance.
- Project acceptance by the customer: Guarantee period and payment terms begin.
- Binding list of shortcomings.
- Final report, final billing.
- Document learning outcomes and aspects offering potential for improvement.
- Finalise all accounting issues after commissioning. Define responsibility.

13.7 Leadership and Teamwork

Generally speaking, not enough attention is focussed on the closing down phase. Instead of setting a clear end date, many projects just gradually wind down and fizzle out. That may simply be because in our culture, we find the whole concept of

parting, of ending, of saying goodbye difficult. Terminating the work on a planned basis, bringing the cooperative exercise to a controlled end (kick-out) is just as important as having a planned start.
As well as reviewing the content of the project, organisational and administrative wrapping-up should also include a specific review of and managed ending of teamwork and cooperation:

- How did the team work? Were personal goals achieved?
- What has been learned from the experience?
- Are there any bad feelings, or unresolved issues that need discussing openly and bringing to a conclusion?
- By this point at the very latest, the project manager must decide what the future holds for the project's team members: What roles are they going back to? Is their place in the "original organisation" secure? Do they need references or active support?

Project teamwork is often recommended as an important element for personal development. And yet, ironically, staff development is an area that is often lacking in project teams. Working together as a team began with a special start meeting. So it should be brought to an end in an equally appropriate manner. This is especially important if the project's results are not completely positive. Clearing the air at the end of a project creates space for new, improved cooperation (Fig. 13.1).

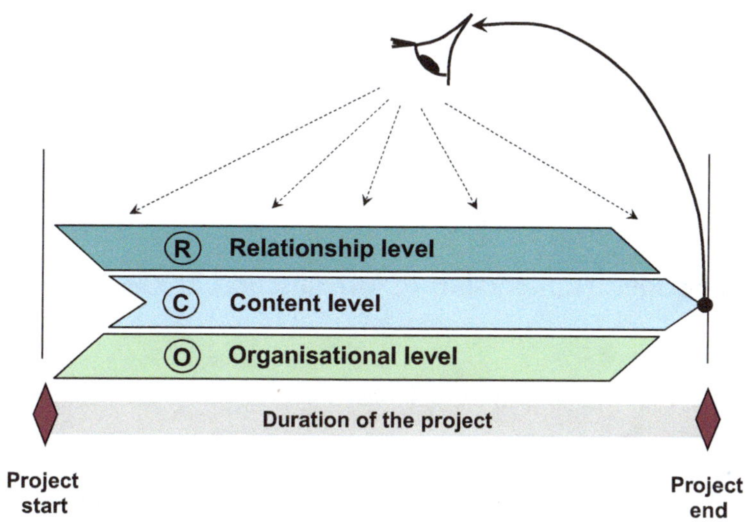

Fig. 13.1 Project review

13.8 Information and Documentation

When introducing a new system or product, the information should be compiled with the operator and users in mind. Updating and completing these documents is often seen by those involved in the project as unattractive, because the project objective has been achieved and they feel there are no more brownie points to be earned by producing documents for "life after the project". The introduction phase allows the usability of the documents to be tested. By the time revisions are made, at the latest, any shortcomings will be apparent. So history always catches up with the project participants!
The things learned during this project should be set out in an accessible manner for future projects. It is especially important to focus on what went well, on the success factors: What characteristics and skills did team members contribute, and what did they do in order to successfully carry out the project?

The project documents must also be reduced to the essentials, and archived:

Process-Focused Documents

- Internal final report: Notes on the project process, administrative close-off work
- Final cost accounting issues
- Acceptance report

Results-Focused Documents

- Training documents, updated instructions for users, manuals
- Final report: Results, administrative close-off work
- Updated outcome documents, e.g. architectural plans, programming documents, product descriptions
- Archiving

13.9 Checklist "Conclusion of Introduction Phase"

- Have the achieved results been checked against the project owner's specifications?
- Does the product or service offer the right functionality?
- Does the product's or service's effectiveness match the objectives?
- Is the organisation that is responsible for correct use of the product or service working correctly?

(continued)

- Have all the project documents been produced, and completed? Has all the necessary information been provided to the future users?
- Has the completeness of the documentation been checked, and has it been archived?
- Has an acceptance report been produced, and signed by the project owner/ client? Has the final report been approved?
- Has the project manager reviewed the performance of the project team members?
- Have the project team members had an opportunity to analyse their teamwork, and to give each other feedback?
- Has there been a joint review to look at both positive and negative experiences in respect of cost, methodology of approach, and teamwork?
- Have measures been put in place to ensure that the knowledge gained is transferred to other projects, and to ensure that there is systematic process improvement?
- Have special performances been identified, and recognised appropriately?
- Has a list of all unresolved issues been drawn up, and plans made to resolve them? Are all participants clear about who needs to do which close-off work, and by when?
- Have team members found suitable employments after the project ends?
- Have contact persons for future questions or problems been defined, and have all beneficiaries of the project been informed who they are?
- Is it necessary to check that the benefits set out in the project order have been achieved? Have plans been made to do so?
- Who will check sustainability and effectiveness of the project, and when?
- Who will work out whether the results that were planned in the cost effectiveness calculations have been achieved, and what financial return the project brought? When will they do that? How well does that match the initial predictions?

Part III

Advanced Coverage of Topics

This section goes into more depth about the key topics in professional project management. This section covers the methodical aspects, but also focuses in particular on the topics that are relevant in leadership terms. They contribute significantly to developing the social skills of all participants. The topics are comprehensively described, and in depth. This means they can be used even in complex projects in difficult environments.

Project Initialisation 14

14.1 Managing an Organisation's Projects Properly from the Start

Companies and organisations have to constantly adapt to changing economic and market conditions. Incentives to innovate can come from all manner of places. Professional project management channels these ideas and ensures that the best ones are carried through.

From an Idea to a Project

Ideas Go Beyond the Organisation's Normal Boundaries

Some events go beyond an organisation's everyday capacity. They are best managed as projects:

- Changes to the regulatory environment and market changes necessitate increasing adaptation by companies and by the public sector.
- Change is often initiated by an organisation's customers. It may take the form of suggested improvements, complaints or concrete orders.
- These ideas often come from within the organisation: New products or improvements to existing systems are initiated to support the organisation in achieving its long-term strategic aims or its short-term operational objectives. Who actually identifies the need for change, or draws up the plans and implements them doesn't actually matter at this stage.

Proactive Project Start

Projects can either be started on a strategic, forward-looking basis (proactively), or reactively, like the fire brigade being called out. For any organisation, the objective

J. Kuster et al., *Project Management Handbook*, Management for Professionals,
DOI 10.1007/978-3-662-45373-5_14

must be to undertake all projects on a forward-looking basis. Examples of starting a project proactively would be:

- Establishing whether more can be done with existing skills.
- Identifying flows and assessing their speed (trend scouting).
- Harnessing new technology.
- Making the most of competitive advantages.
- Thinking about successor products during the mature phase of a product's life cycle.
- Systematically releasing better solutions than your competitors do.
- Learning lessons from benchmarking, and wanting to outperform the best in the sector.

Helpful questions to ask around this process would include:

- What are the latest developments?
- How attractive is this for our customers?
- How can we benefit from that?
- How should we introduce our product or service to the market?

Reactive Project Start

Ideas for projects often come about from everyday activities, or from a desire to improve current processes. Examples of starting a project reactively would be:

- Solving problems that have cropped up.
- Satisfying individual needs.
- Highlighting and following market trends.
- Implementing ideas for improvement that come from the corporate suggestion process.

Helpful questions to ask around this process would include:

- How is the problem manifesting itself now?
- Who is affected by the problem, and which processes does it affect?
- How could it be different, and better?
- What expectations will be met by solving the problem?
- How can the company get from where it is now to the objective?

Accidental Project Start

Creativity or pure chance can also give people ideas, which subsequently turn into projects. A lot of successful projects started like that. "Post-It" notes are a good example: A glue that wouldn't stick, and a scientist who wanted to stop the strips of paper that he used as bookmarks from falling out of his hymn book. But the "unusable" adhesive kept the strip of paper in place for as long as necessary, without damaging the book.

Ideas may be vaguely formulated to begin with, but it is important that they become clearly defined as they progress towards implementation. Whether the changes should be handled in the line organisation as a project is something that should be clarified by the ideas manager. First, the framework conditions should be identified. The need for change should be identified in the proposals. Proposals are the best means of ensuring that no ideas or external orders are lost. They also offer a complete overview of everything that is in the pipeline. Somebody with good access to senior management should be tasked with processing the proposals and customer requests. This role can be called the "ideas manager". Where ideas come from practitioners within the organisation who may not be experienced at writing proposals, the ideas manager can help them draw up the proposal.

The following questions may be helpful:

- What needs changing? What should stay as it is?
- What expertise is required? Does the company have that expertise?
- What risks are associated with the project idea?
- What will the consequences be if the project is not carried out?

For investment expenditure, the following questions should also be resolved:

- What sort of investment does this relate to (new acquisition, replacement, expansion, infrastructure, etc.)?
- What is the amount of the anticipated investment?
- What additional income or savings will this generate?

There are several different ways of implementing ideas. Each organisation will have its own rules on which criteria need to be fulfilled in order for an initiative to be assigned to a particular category. The categories should be clearly defined, and must be clearly communicated:

- Large projects and key projects that have to be authorised by the board.
- Medium-sized projects that can be authorised by individual members of the board.
- Small projects that affect different organisational units.
- Orders that can be handled directly by the line organisation, alongside its everyday work.

Suitability and Readiness for Becoming a Project

For each project initiative, it is important to establish whether the organisation is in a position to make the necessary resources available, and whether the key people are able and willing to take on responsibility for it. The possibility of a worthwhile project has been put forward. Alongside the criteria for checking how worthwhile the project would be, the following questions may also be helpful:

- Is the organisation capable of running this project?
- Is there sufficient desire to carry out this project?
- Are the stakeholders clearly identified?
- Is the time frame short, medium or long?
- What sort of project is it? (e.g. capital, ICT or organisational project)
- What stage of its life cycle is the initiative at? (e.g. research, development, implementation, disposal)

These are key points in establishing how the project order can be best developed, and what agreements it must contain.
Planning the resources that will be required to complete the project and meet the objective is part of the business planning process. In order to ensure its long-term success, every organisation must develop and make available qualified expertise and powerful technology solutions. For each project initiative, the organisation must check which of these resources will be required, and the extent to which they will be required. The organisation must then decide whether using the limited resources for this project is worthwhile. When assessing whether the organisation is in a position to carry out a project, the following criteria should be considered:

- Does the organisation have the necessary financial resources available? What proportion of the overall investment budget will this project require? What financial bottlenecks will the organisation experience because of this project?
- Can the right people be recruited for this project? Does the project owner have suitable people, and do they have enough time available?
- How accessible is the relevant expertise and experience? To what extent will qualifications need to be developed during the project?
- Are there sufficient standards and methods for carrying out the project? How familiar are project participants with these methods and standards?
- Has experience with similar projects been documented and is the information available? What leadership experience does the intended project manager have?
- To what extent will the project involve new and unknown territory for the participants? What experience of dealing with new situations can be used for this initiative?
- How easily can technical resources and infrastructures be provided in good time for the project?

Time and again, research shows that one of the main reasons that projects fail is because senior management don't relate closely enough to the project. It is therefore essential to understand how much commitment there is to a project before approval is given.

- How strong is the customer's desire for the outcome of this project? For internal projects: How strong is the support from those affected by the project idea? Did the initiative come from the future users?

- How many senior managers are personally interested in the project idea? To what extent are the senior managers in agreement? Who will be the project's sponsor or patron?
- How impressed are middle management by the project idea? Are these middle managers fully aware of the changes that will directly affect them? Do they have any constructive ideas about dealing with the changes? How crucial do middle managers consider the project to be? How prepared are they to make staff available to work on the project?
- How eager are the future users to see concrete improvements? Do they understand the future benefits for themselves and for the organisation? What are they prepared to contribute in order to help with the innovation? How will they be involved with the project? Where is active or passive resistance likely to come from?
- What personal development opportunities does the project offer for project staff? What learning opportunities will there be? How do members of the line organisation feel about working on the project? How will the line organisation cope with the additional pressure?
- How will the organisation ensure that the whole of the affected system cooperates with the project? Can all the different cultures be actively involved in the project? Is there enough time and professional support available to integrate the different system elements and for team development?

14.2 Legal Aspects of Project Management

Even if they're not actually using a "product push" strategy, companies develop products or services, which have to meet criteria that need to be agreed with their clients. The two or more parties involved in the project have different objectives and are probably active in different market sectors. A good procurement process enables organisations to enter into contracts with good suppliers, resulting in good quality and in good profits. For example, if something costs significantly less than other similar products, the customer may well ask himself whether the cheap price means quality is compromised in some way. Customers protect themselves against this risk through the use of penalty clauses. The following risks in particular should be borne in mind: Licences, patent rights, product liability, complaints, legal action, services to be performed by sub-contractors, contractual penalties, contractual defects and even fraud. The project manager should also consider what level of patent protection is appropriate for any new products or services.

Most European legal systems assign responsibility for the project risks to the party that has the most influence over the project. Negotiating a contract is a complex matter that calls for professional legal expertise. The nature of the contract is decisive for the costs and benefits of a project. The differences between legal systems in different countries represent an additional layer of complexity; so even lawyers find it challenging to draw up international contracts. It is not really possible to delegate client negotiations in the preliminary phase of a project to an internal or external

legal task force, because bid managers generally prefer to deal directly with the client. And it can't simply be left to the project manager to resolve such complex issues. This is why organisations have a procurement process, to cover "procurement and contracting". In centralised "procurement/legal services" teams, lawyers and other procurement experts draw up the relevant contracts, and refine them.
And in the same way as the external legal situation has to be clarified, there are also internal legal aspects. If all the staff is directly employed by the company, the results of the project will belong to the company. But if external staff or even an external team are used, ownership of the results will normally be defined in the work contract. The external contractor is obliged to deliver the work to the commissioning party. The commissioning party is obliged to pay the contractor the agreed fee for the work. Most such work involves a measurable result, such as the agreed project result.
How does this work for software? Who owns it? As well as the actual program itself, software also involves various by-products from the software development process. Program flow charts, development documentation, structural programs and all of the accompanying materials can be transferred in the same way as goods. But the computer program itself cannot. The use of computer programs is covered by assigning them under licence agreements, in return for a financial consideration. If the company wishes to further develop the software that they are using, this must be explicitly agreed in the licence conditions. Patent rights and trademark rights must also be taken into consideration, along with copyright issues. Special copyright law applies to computer programs, under which the creator of the software always holds the copyright.

Conclusion for project managers.
Make contact with legal specialists very early on in the project!

14.3 Internal Projects (Inside Out View)

Internal and external projects are generally very different in character. Internal projects are changes to the "processing system" itself, and can generate a lot of resistance because they often involve a wide range of different interests or involve changes in the power structure. External projects are less problematic in terms of conflicts of interest, they are normally customer projects.
The challenge here relates to precision in terms of quality, scheduling and finances, and to the standardisation of processes. Internal projects are challenging in that they involve engaging with social stakeholders.

Ideas and Improvement Suggestions from Staff

The ideas manager helps to identify and prioritise good ideas from staff, and helps them come to fruition. It is helpful to combine the ideas management and project portfolio functions with the strategic project controlling role. This enables an organisation to commit its resources to implementing the best ideas. Incentive schemes can increase the staff's willingness to develop and implement ideas or patents. Bonuses can be offered for good ideas, although tax will have to be paid on them. But for creative staff, it is far more satisfying to see what becomes of their ideas, and to receive recognition for their initiative.

- The entire management team encourages individuals and teams to involve themselves in improvement activities.
- Management creates opportunities that encourage innovative and creative behaviour.
- Managers authorise their staff to act, and encourage teamwork.
- The ideas manager systematically gathers ideas and supports individuals and teams as they work on improvement activities.
- The ideas manager prepares the ideas that have come forward, and regularly submits them to the relevant project owner for a decision about whether to develop the idea further.
- The project owners take steps to implement the selected ideas.

The Business Case

Many organisations want to see a "business case" before a project starts, usually containing the following information:

- An outline of the problem and reasons for the initiative
- SWOT analysis
- Customer needs and market potential
- Analysis of the environment, framework conditions
- Competitive position
- Strategic options
- Marketing mix
- Quality criteria
- Different implementation options
- Non-monetary benefits for the organisation
- Financial analysis

14.4 External Projects (Outside in View)

Unlike internal projects, which tend to be one-off and very varied, external projects are normally customer projects. In other words, they are related to a service or a product that the organisation is selling. This might take the form of construction projects, technical installations, ICT projects, customer-specific product developments, etc. If these projects are repeated in a similar fashion, they are

called standard or repeat projects. In an ideal world, the relevant project management will have been professionalised, with standardised processes, methods and tools that are covered by written guidelines and updated on an ongoing basis. The project managers should also be professionals as opposed to internal projects, where these tasks are generally undertaken by staffs that are not project management specialists.
The following sections explore in more detail the sensitive aspects of customer projects, notably the proposal phase and the contractual agreement.

External Customers Have a Requirement or a Problem

An external project normally begins with an invitation to tender or a request for proposal. Putting together a proposal or bid is a time-consuming exercise. Over half of all proposals end up in the bin. So the first thing that the person putting together the proposal should do is check that the customer has a genuine intention to buy, and that the customer actually knows what they really want. An initial discussion with the potential customer should address the following issues (if they are not clearly answered in the request for proposal):

- When does the customer need to receive the proposal?
- Who will take the decision on placing the order?
- When will that decision be taken? (Making a phone call shortly before the deadline to answer any questions can help improve a project's chances)
- What criteria will the decision be based on? What are the essential results? What are the most important desired results? What else is important?
- Does the customer want to buy? If the organisation's tender shows that it fulfils all of the relevant criteria, will the customer award the project to the organisation?

The Bid Manager

Bid management demonstrates that a highly organised formal approach is the key to success for critical projects. That doesn't mean that an organisation will be successful in every customer project that it bids for. But it does optimise the use of resources, help concentrate on profitable (beneficial) projects and on high-quality proposals that the company can be proud of.
Many organisations have adopted a formal process for bid management, consisting of several milestones with formal Go/No Go decisions. The first step in this process is always qualification of a potential project. This involves developing a basic plan and a rough business case. The result of the qualification is evaluated by a team of decision-makers who take a joint decision on whether to start the project or not. This first hurdle is generally a very significant one. And so it should be. Once the green light has been given, resources will be put in place to enable a proposal to be put together in good time. A bid manager will also be appointed, who will then act

as a project manager to coordinate all the contributions for the proposal according to a strict time frame.

Bid managers work in a pool and concentrate exclusively on drawing up proposals using a specific methodology. They are specially trained, and often have no detailed knowledge of the solution that they are responsible for putting together. But they are extremely capable when it comes to assigning work packages, setting deadlines and rapidly escalating something to management when there is a bottleneck. They are measured by whether the proposals are complete, correct in terms of content, and reach the customer by the deadline.

Many bids are invalid if they are submitted too late, so the whole success of the project can depend upon meeting this one objective (Fig. 14.1).

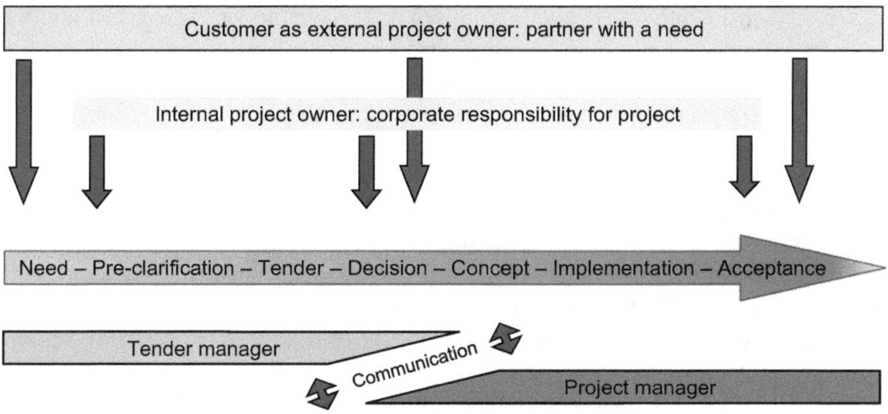

Fig. 14.1 Synchronising all of the functions involved in the customer project

If the project bid is successful, everything continues in much the same way. A market project manager takes over responsibility for managing the project until the agreed product or service has been delivered to the customer. In this phase, the running of the project is closely controlled and managed. This is where it will become clear whether the margins that were assumed at the beginning can actually be realised. Tight risk management and rigorous claim management must be used to continuously assess potential risks and to initiate measures that will minimise any risk. Bid managers and market project managers are given full support by a project office, which looks after administration and methodology. Here are the points that the bid manager must pay special attention to.

The Bid

For companies that have to give binding price quotes to their customers (in the engineering and architecture fields, for example), accurately estimating the costs is essential to avoid losses, since in many sectors the market does not allow for large reserves. This makes the cost estimate and the associated uncertainty all the more important. Drawing up a meaningful cost estimate needs experience, attention to detail and an experienced group of potential project staff. The costing units will also be external, rather than internal.

In order to produce a full cost estimate, a comprehensive situation analysis and a good solution plan will need to be drawn up. That takes a lot of time. But if the organisation does not go into the right level of detail, the risk of incorrect estimates increases. That leads to cost overruns, which cannot usually be charged to the customer. How much time each organisation will have to invest in drawing up the bid will vary, depending on its experience in similar projects and on the likelihood of the customer placing the order.

From the perspective of the customer and the user, the purchase cost is not the only important consideration. They will also look closely at the operating cost and the costs for repairs, maintenance, training, disposal and other on-costs that they will incur whilst using the solution. More enlightened customers will be interested in low life cycle costs, which reflect the total costs that the investment will incur over its full lifetime. Depending on how far-sighted the customer is, the project manager may choose to emphasise solutions that focus on low overall costs, provided this doesn't push up the production costs and project costs too much (or unless the customer agrees to cover these additional costs).

Before submitting a bid, the contract should be checked internally. A responsible person, often a Board director or senior manager (internal project owner), checks the contract. The most important questions during this process are whether the organisation is in a position to fulfil the contract in terms of content (available expertise), time frame (availability of resources) and legal aspects. Checklists can be used for this purpose. Companies that frequently produce bids have this process covered by a separate internal process, and have it documented.

Project managers or bid managers give the draft bid to the internal project owner, who will approve its release to the external customer. If the draft is rejected and does not receive internal approval, ways must be found to reduce the costs or to increase the development speed. The most significant consideration is selecting the most effective members for the project team. The project manager will need the help of the internal project owner in order to do this.

The aim of the bid is to sign a contract with the customer. During the tendering phase, bidders and potential customers should share information about the risks associated with the project in a factual, formal manner and in financial terms, and should get to know each other as loyal partners. Both partners should endeavour to negotiate a balanced and mutually fair contract. Even in this preliminary contract phase, both parties should negotiate in good faith. The bidder has a lot of in-depth expertise, and they are usually communicating with a non-expert. So it is particularly important that the bidder puts any important but non-obvious points or issues on the table. Designs and written topics can be taken into consideration later when drawing up the contract, if the intentions of the parties are not already clear from the contractual document. Providing incorrect information during the tendering phase gives the customer good grounds to withdraw from the contract. More importantly: If the project management organisation knowingly makes false statements they can be held liable, despite any waiver clauses.

The following considerations during the tendering phase may have legal implications:

- The contractual deadline that bids must be submitted by, or when the decision not to submit a bid must be announced.
- The legal conditions under which the organisation will carry out the project.
- Observation of strict confidentiality of all information contained in the tender specification documents.
- Returning the tender specification documents if the organisation does not get the order.
- The fee charged for submitting a bid (normally free).
- The purchaser's right to make free use of the bid documents.

For fixed price tenders, the risks and the scope for changes should be taken into account when thinking about the project. Project planning should be realistic (or even better, pessimistic!) as far as the market and the competitive situation allow. If a decision is taken for strategic reasons to submit a bid even if the order will not cover the costs, then to protect the project manager the internal calculations and the resulting commercial decision must be set out in writing. Special attention should be paid to the critical path, and to the likelihood of meeting the required deadlines. Any contractual penalties should be proportional to the project's potential profitability.

The Contract with External Customers

If the bid is successful then the project must be contractually agreed in writing, just as with an internal project. The contractual objects are therefore the project objectives, goods or services to be provided, costs, deadlines, contractual conditions, liability, etc. Contract management includes managing the contract negotiations between the client or project owner and the contractor, implementation of the contracts and carrying out any changes to the contract. The list below highlights a few important points for the contractual relationship:

- The signature block and the project title identify the project and the contractual parties. This may seem self-evident, but it has important implications if any aspect of the contract is later disputed.
- The specification describes the technical content of the product or service.
- Information and plans for the product, with delivery deadlines.
- Definition of the contractual conditions and the scope summarise the applicable delivery conditions and payment terms, and should describe the framework conditions of the work in sufficient detail.

- Information and services to be supplied by the customer. These list the client's additional contractual obligations. This may include concerns such as access to certain areas during the work.
- Project agreements will be needed at various different levels throughout the project life cycle. The customer may be asked to sign off each status report before the project team moves on to the next phase of the project. In some cases, approval may be required from government authorities.
- Payment terms are normally included in the standard terms of the contract. Stage payments can be agreed monthly on the basis of work done, or at the start of the project, on submission of the project plan and after implementation. These sums are then paid to the contractor as payments on account. The final payment is made after the work has been checked and accepted. It is important that the acceptance criteria are clearly defined.
- General terms and conditions are standard elements of a contract. They are sector-specific and cover the main rights and obligations. This would include, for example, how changes are handled, applicable law and jurisdiction.
- Supplementary conditions specified by the customer govern specific delivery conditions such as noise limits, working hours and access. "Small print" doesn't work well for project contracts.
- Provisions relating to changes and variations are normally included in the general terms and conditions. For large and complex projects, these should be covered by a separate clause in the contract. It covers the arrangement and design of any variations, and describes the process for changing the project fee.
- Dealing with differences of opinion and with conflict. Most contracts specify arbitration as the first line of intervention. If arbitration does not resolve the issue, the aggrieved party can take the matter to court, which triggers an expensive and time-consuming process. Arbitration or mediation offers a faster and cheaper alternative, provided that it is specified as the first escalation level by the parties to the contract.
- Deposits and guarantees indicate which precautionary measures are to be taken and how. The client can require guarantees right up to the point of readiness and hand-over. The guarantee relates to the quality and reliability of the finished product when the project is handed over and during part of the use period. The guarantee may also need to be backed by an insurance policy.

In establishing a project contract, the following points should be noted:

- Tendering for a contract and having the tender accepted is very different to buying something over the counter. When the customer accepts the bid, the contract is established.
- Remuneration is the exchange for something of value, usually money, in return for providing some sort of service. Will the full price be paid at the end of the project, or will remuneration be made in stages, beginning with a down payment?

- The ability of the parties to fulfil their contractual obligations is a key element of the contract. For example, in Great Britain the contract would be null and void if one of the parties conceals the fact that they do not have the capacity to deliver their part of the contract.
- The risks increase if a contract is entered into in a different cultural or legal environment, and early assistance should be sought from people with relevant experience.
- Legality: The contract must not be either unlawful or inappropriate. The contract must state that the intention is to establish lawful relationships.
- The acceptance of the contract must be clearly communicated, so that both parties know the other party's wishes and intentions.

The Project Organisation 15

15.1 Basic Points

The project organisation is a special organisational structure that is set up for the duration of the project. A special team is needed because the existing line organisation is designed to fulfil its regular specialist work, rather than to manage and carry out new, one-off initiatives that involve different areas and departments. The line organisation also lacks the flexibility to react quickly enough when problems are encountered or changes are needed.

Key requirements for a well-functioning project organisation are:

- Clear project agreement: challenging goals and framework conditions or guidelines that define the scope for change.
- Clear decision-making authority and management responsibility by project managers.
- Project team made up of interdisciplinary representatives from different specialist areas and different departments.
- Actively involved users and stakeholders, in order to achieve maximum acceptance.
- A working culture that encourages communication, commitment and creativity.
- Good links back into the core organisation, if possible "connected" to the relevant decision-makers.
- Availability of resources.
- Methodical support from the "project office".

J. Kuster et al., *Project Management Handbook*, Management for Professionals,
DOI 10.1007/978-3-662-45373-5_15

15.2 The Line and the Project: Two Different Worlds

If a company sets up project teams to resolve specific problems, that means that a new work group will be set up alongside the existing line organisation. The new team will be different from the existing line organisation in terms of its responsibilities, in its teamwork, in how it communicates and handles conflict, in terms of taboos, etc. These differences should be "designed in". Depending on the project, the differences may be barely noticeable, or they may be very marked. For example, when introducing new product structures that need new forms of cooperation, the development process can be helped by using the future culture as part of the project itself. That would be a significant difference between the project environment and the core organisation's culture. But in standard projects or implementation projects the project culture will be closer to the main organisation's hierarchy, with only minimal differences between the two worlds (Fig. 15.1).

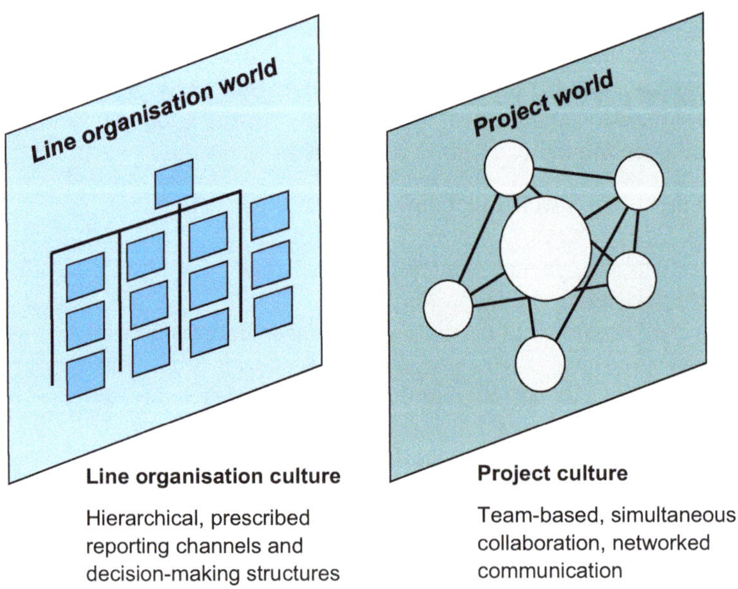

Fig. 15.1 Two worlds: The line organisation and the project environment

15.3 Roles and Committees

People are often brought into project organisations without a clear definition of their role. Consequently, roles are not clearly defined or delimited, or they may even not exist. For example, a project owner who enjoys working as part of the team takes

"top down" decisions by virtue of his or her position of authority (as is expected). But they also take "bottom up" decisions within the team, as a team player. This dual role by a decision-maker puts the project management team in a difficult position.
Multiple roles are dangerous. In small projects, it is clearly not possible to separate out every single role. But in those circumstances, a person should hold "neighbouring roles". So the project manager could work on the content of the project, the manager of a sub-project could also be part of the core team, etc.

To summarise: Roles must be clearly defined, agreed and transparent.

The committees and roles should be set up to suit the project and corporate culture (and sector culture), so that all levels of authority are represented and to enable the project goal to be achieved as effectively as possible. In many projects, it is not really clear who takes decisions. Or the project management role is seen as a necessary evil, so little attention is paid to it.
The project must have clear connections to the decision-making level. The decision-making process must be transparent (Fig. 15.2).

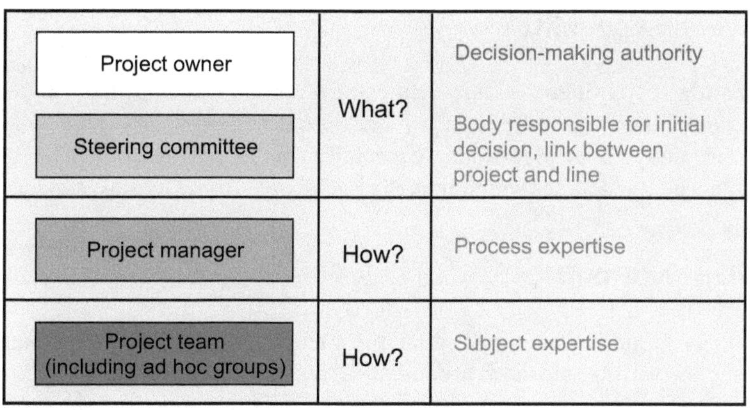

Fig. 15.2 Institutional project bodies and levels of authority

The Project Owner

In most cases, the project owner is also the decision-maker. There can also be situations where the decision-making function is split; e.g. between management and the board, or between the executive and legislative teams. The key aspects of the project owner's role are:

- Setting the strategic framework conditions.
- Specifying priorities, which projects are important or urgent.
- Reaching contractual agreement for the project order.

- Taking milestone decisions.
- Supporting the project managers, backing them.
- Securing resources. Project managers often do not have the authority to simply take resources. This is where the project owner must give them support.
- Opening doors, such as providing access to people with important information.
- Informing: It is often important for the project owner's office to inform external parties (representation).
- Motivate people (at the kick-off meeting, for example).

The Project Portfolio Manager

This is often the central project controller or project officer in a permanent role. It is beneficial if the project portfolio manager is a member of the senior management team, a central staff function in large corporations. This central controlling role supports senior management by prioritising the various different projects and running the project portfolio.

The Steering Committee

The steering committee is also called the project committee or top-level management. It is an extension of the project owner. Its role is general monitoring and initial decisions, especially in large projects. The steering committee usually includes senior managers or people from key stakeholder groups.

The Advisory Group

The advisory group is sometimes called the sounding board. It supports the project by sharing views on substantive content. This can be useful for projects where different stakeholder groups are represented. Its purpose is to get issues relating to the project onto the table for open discussion, in order to ensure wide acceptance.

Project Management

Project management is responsible for the operational running of the project. In other words, they are process designers. The project management role is generally fulfilled by one person. But there could be a management team, in which case the various complementary roles would need to be very clearly defined. For large customer projects or construction projects, a project manager is generally appointed in each of the two companies (contractor and customer or developer).

The Project Team

The project team is generally responsible for delivering the content of the project. For large projects, it may be beneficial to structure the team into a core team and an extended team. This sort of structure makes meetings and workshops more efficient, since they do not need to involve every team member at each meeting.

Sub-project Managers, Sub-project Teams

These represent a normal structure in large projects that can be broken down into relatively autonomous sub-projects.

Temporary Groups

Working groups can be set up to handle specific aspects. These are temporary, with a shorter lifetime than the overall project.

15.4 The Project Management's Responsibilities

The project management runs the operational management of the project (Fig. 15.3). They are responsible for the execution of the project. They bear responsibility, have a management function and are also front-line workers in the project. They are responsible for the following tasks:

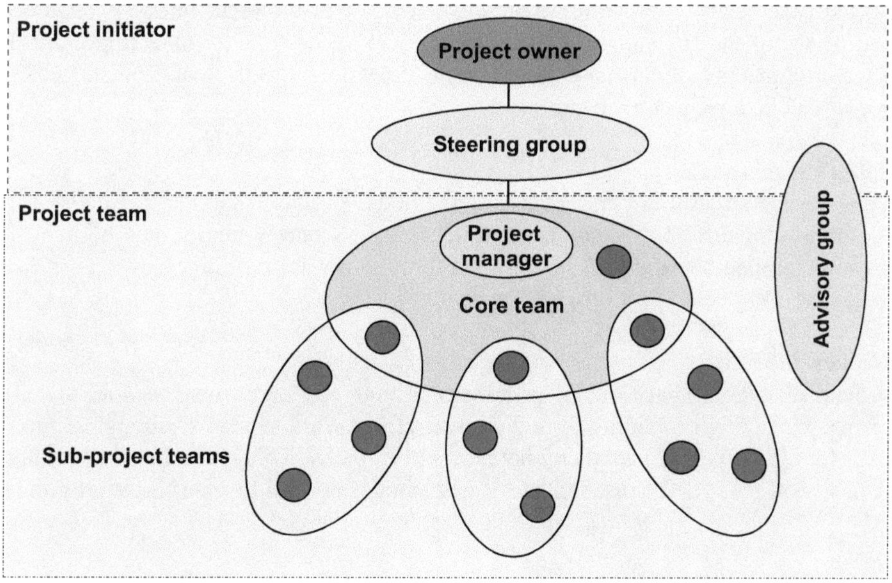

Fig. 15.3 Ideal project organisation

- Negotiating the project agreement with the project owner and setting it out in writing.
- Obtaining the necessary resources.
- Setting up the project team, together with the project owner.
- Carrying out the project start and kick-off meetings.
- Structuring the project into phases and milestones, and possibly into sub-projects and work packages where the order doesn't specify this in detail.
- Coordinating the different sub-projects and activities.
- Planning, monitoring and managing the schedules, people, costs and quality.
- Leading the team and coordinating the stakeholders.
- Carrying out risk analyses.
- Assessing the project's cost effectiveness.
- Preparing the milestone decisions.
- Putting information, communication and project documentation in place.
- Moderating meetings and workshops.

These tasks can also be carried out by the team, or some of them may be delegated to individual team members. But the project manager is responsible for ensuring that they are done.
The project manager should be able to offer subject knowledge, familiarity with different methods and leadership skills. Of course, he can't be an expert in all of these fields. That sort of Superman doesn't exist. But he should be able to contribute relevant expertise in all project management disciplines that are appropriate for the project and its goals. That means:

Subject knowledge
Specialist expertise is less important here than being aware of the interdependencies and knowing the specialist language(s). Subject knowledge is more important in standard projects or recurring projects than in potential projects. So a "non-specialist" could well be a project manager there.

Different methods
This covers structuring and planning methods, problem solving methods, etc. Awareness of different methods of doing things is barely taught at school, so it must be learned "on the job" or acquired through further training. However, it is a skill that is expected of a project manager.

Leadership skills
Project managers shape social processes. Above all, they must be able to lead teams. Managing or leading, in the context of a service, is very challenging and it calls for high levels of social competence and empathy. This is another skill set that depends on the type of project, and it is generally needed in inverse proportion to

subject knowledge: The lower the importance of subject knowledge, the greater the need for leadership skills, and vice versa. The project manager doesn't have administrative authority over the people working on the project, so he has to be able to lead without formal authority.

The following criteria can be helpful when looking for a suitable project manager:

- A team player: Has a strong preference for group work, is flexible in terms of role.
- Determination: Able to persuade and convince, never loses sight of the objective.
- High frustration tolerance: Does not respond negatively or destructively to failures.
- Networked thinking: Maintains an overview, makes connections, allows different views, and sees indirect consequences.
- A negotiator: Willing to take risks, not afraid of taking decisions. Dynamic, energetic. Focuses on finding solutions rather than placing blame.
- Forward-looking: Acts rather than reacts. Anticipates the future.
- Generalist: Has a wide range of interests and an inquiring mind.
- Commercial: Thinks in terms of cost effectiveness.
- Self-confident, self-aware: Able to think for themselves, confident.
- Comes across well: Can sell ideas and the project, good communicator.
- Good judge of character: Can see strengths and weaknesses in others, and can reflect on his own performance and learn from it.

15.5 The Role of Deputies in the Project

Any proper management structure needs clearly identified deputies. This can take a number of different forms:

- The deputy's role can be a placeholder function, a function that is used in the absence of the main post-holder, with major decisions being held over until the main post-holder is present again. Limited delegation of authority.
- Deputy as co-leader. Again, the function is only delegated in case of absence, but there is a constant sharing of information so that the deputy can react faster if they need to step up. Limited delegation of authority.
- Executive deputising. There is an ongoing division of work in order to ease the load on the main leader, or to build up the deputy's knowledge and experience. This enables responsibility and authority for various aspects of the project to be fully delegated. In the event of absence, all duties and authority can be transferred to the deputy.

Deputies are normally appointed from within the team, which is often problematic in terms of team dynamics. In that context, it is a good idea for both sides to clearly express and explain their expectations and wishes. It helps if there is good understanding between the relevant people (similar leadership philosophies) and if the relationship with the deputy works on the basis of loyalty rather than rivalry. The support of an effective deputy can take a lot of strain off a project manager. Similarly, taking on the deputy's role can be a highly effective development tool that helps give "on the job" training to prepare committed staff for management roles. Deputising for decision-makers can be problematic if not everyone is comfortable with the decisions that are taken, and if there are attempts to go behind people's backs or to use the organisation's hierarchy to overturn decisions. That is where real delegation of authority comes in, and the decision-maker must loyally support their deputy.

15.6 Basic Project Management Organisational Structures

Any temporary organisation is faced with the question of how it should be linked to the main organisation. For project organisations, there are three basic structures: project coordination, pure project organisation and matrix organisation. Criteria to be considered when selecting the appropriate structure would include:

- The type of project, project goals, project scope
- The company's management structure and culture
- The project culture in the organisational units

These organisational structures differ in both the manner and the extent to which the management responsibility and the decision-making authority are divided between the line organisation and the project organisation, or the amount of freedom given to the project manager and his team (Fig. 15.4).

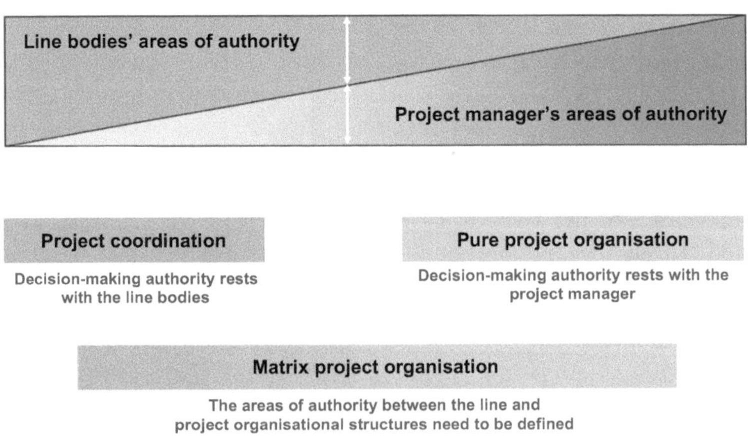

Fig. 15.4 Project management organisational structures and lines of authority

Project Coordination

The project coordination is the most minimal type of project organisation. A staff role for the project coordinator is simply added on to the primary organisation. The existing functional hierarchy remains unchanged. The coordinator (staff project manager) has no authority. However, he is responsible for the material content of the project, for ensuring that it remains on schedule, for distributing information on time to the relevant line bodies, and for a qualitatively correct approach.
As part of the coordination function, he suggests measures and the next steps to the line organisation. This type of project management needs the line organisation to cooperate constructively and to make the necessary information available to the project manager. With this project structure, the project manager should not carry sole responsibility for achieving the goals, because he has no management authority in the project and he is dependent on the goodwill of the line organisation (Fig. 15.5).

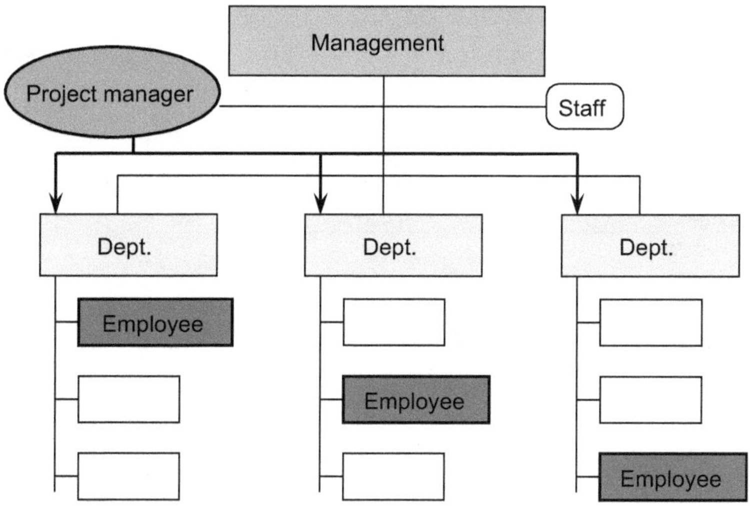

Fig. 15.5 Project coordination structure

Benefits
- Greater flexibility in use of personnel.
- Makes it easy to collect and share information about experiences across different projects.
- The organisation does not need changing.
- Responsibility for the project remains largely with the line organisation.

Drawbacks
- Nobody feels responsible for the project.
- Low reaction speed.

- Difficult to get a cross-organisational perspective.
- No real project team.

This project coordination structure is best suited to projects that do not massively exceed the normal type of work (e.g. customer orders, simple product development, etc.).

Pure Project Organisational Structure

With this structure, the project team is set up as a new, independent organisational unit. The project manager and the project team members work full-time on the project. The project manager has full management responsibility, and also full decision-making authority (with the exception of the milestone decisions). This results in an efficient, independent task force. This structure can be considerably more expensive, as new people have to be recruited into the existing posts. And if the project does not make full use of everyone's time, there can be a cost for unused staff time. This structure is best suited to complex projects, for time-critical activities or for projects that will produce results that are crucial and decisive for the company (Fig. 15.6).

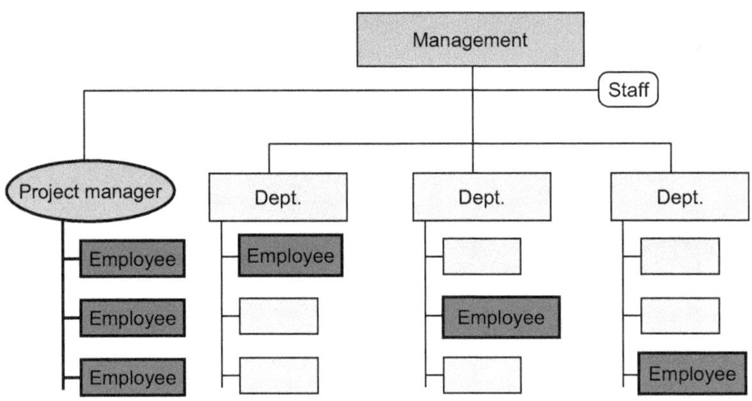

Fig. 15.6 Pure project organisational structure

Benefits

- An efficient structure for large projects.
- The project manager has clear responsibility and decision-making authority.
- Fast reaction to any problems.
- The project team identify closely with the project.
- Independence from the influence and vagaries of the line organisation.

Drawbacks

- Limited flexibility of personnel, particularly where specialists are only needed for part of the time.
- Recruitment of project team members and reintegration after completion of the project can be difficult.
- Slightly increased risk of authoritarian or non-team focused leadership by the project manager, since they are in a special temporary leadership role.

The pure project organisation is clearly separate from the line organisation and has a high degree of autonomy. It is well suited to very large initiatives that have relatively little contact with day-to-day tasks (e.g. developing a completely new product line, new construction), for high-risk projects (spinning the project off from the company) or where the results have to make a real impact. As a task force, they are able to quickly handle important and urgent projects.

Matrix Organisational Structure

This structure is a combination of a pure project organisation and a project coordination. Responsibility and authority are shared between the project manager and the line bodies. Quite how they are split will depend on the project in question, and can vary significantly. A matrix organisational structure needs clear agreement about what the line team will do and what the project team will do, and it is important that this agreement is adhered to. There must also be high awareness of roles. This all increases the likelihood of conflict, so there must be good communication to resolve issues and reach agreements (Fig. 15.7).

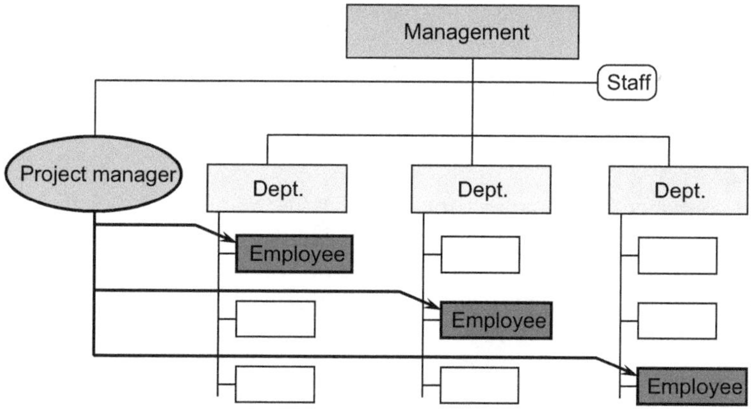

Fig. 15.7 Matrix organisational structure

Benefits

- The project manager and the team feel responsible for the project.
- The project manager has clear responsibility and decision-making authority.
- Flexible use of personnel, no capacity utilisation problems.
- Continuity of ongoing specialist training, no loss of contact with the line organisation.
- Goal-oriented coordination of different interests.
- Encourages a holistic, interdisciplinary perspective.

Drawbacks

- Risk of conflicts of authority between the line and project manager.
- Possible uncertainty on the part of managers (no exclusivity) and personnel ("servants of two masters").
- Needs a lot of communication and information sharing.

In practice, the matrix structure is by far the most common form of project organisation. For many companies, limited resources mean that there is almost no other choice. Because of the interdependencies and the differences arising from the hierarchical line organisation on the one hand and the project organisation's team culture on the other, this is the most challenging and difficult structure in terms of organisational psychology. It only works well if:

- The tasks and responsibilities are clearly defined (in the same way as there are rules about who has right of way at a road junction).
- The line organisation can optimally coordinate the resource utilisation of the personnel between line work and project work (capacity planning).
- Problems and conflicts can be identified and discussed. In other words, if there is a suitable conflict management culture.
- The different parts of the company want to work together and do not compete over the project.

15.7 The Project Management Handbook

For any company that is project-focused, it is useful to produce a set of project management guidelines in the form of a handbook. The document should contain rules for running projects. The rules become binding when the document is published. Staff who are working on the project will absorb and live the project culture more quickly if they can easily understand the rules, and if they view this document as a helpful tool. The document is most effective if it is short and simple to follow, when the staff are involved in its writing and when helpful support tools (e.g. templates, instruments and checklists) are supplied.
For complex projects, these guidelines should be described in full, and should be followed to the letter. For smaller, routine projects, the steps can be adapted to suit the company's needs and can be applied in a scaled manner. If the guidelines

document gets bigger, it becomes known as the company's project management handbook. For important projects, it may even be adapted to a form that suits the project's requirements, the project handbook (example: Fig. 18.6). The project management handbook applies throughout the entire company. The project handbook applies just to one specific project. Publishing these documents on the company's intranet makes them more accessible, and enables them to be linked. It helps make them more user-friendly, and offers a better overview.
Topics that might be covered in a project management handbook are:

- Scope
- Types of project (organisation, ICT, product development, infrastructure)
- Project categories (strategic, international, small projects)
- Overview of process landscapes, value chain, phase plan, milestones, reviews, approval
- Description of activities, associated documents
- The project organisation, team composition, steering committees
- Authority, responsibilities, competences
- Stakeholder analysis, organisational chart, escalation procedures
- Scheduling and cost planning, multi-project management
- Information, communication
- Change request management, handling risks
- Project controlling, monitoring of costs and schedules, reporting
- Ending the project
- Other applicable documents, tools, templates, checklists
- Transfer of expertise and experience, continuous improvement process
- Cooperation between line organisation/project organisation
- Authority, responsibilities, project manager career planning
- Definitions, glossary

15.8 Areas of Responsibility

It is recommended that a general description of the goals, tasks and responsibilities of each individual project body should be drawn up (specification document). However, it is not worth trying to regulate every last detail of areas of responsibility. Bureaucracy should not destroy project management's flexibility. How any problems that occur are discussed and resolved in a manner that is appropriate to the situation is actually more important. In other words, the important things are a culture of dialogue, the ability to deal with conflict, and a willingness to work together. Matrix organisational structures are particularly challenging in terms of clarifying who does which tasks and identifying areas of responsibility, since both line managers and project managers have responsibilities. Decision-making powers and areas of authority could be divided using the following criteria, for example:

What are the tasks or activities that need doing? In projects, it is normally the project manager who plans and takes decisions about this.
When (or by when) do the activities need completing? Again, this is normally something the project manager decides. But in most cases, it is advisable to agree and coordinate activities and scheduling with the line organisation, or at the very least to inform them.
Who is being delegated to the project from the line organisation? All the project managers can do here is make recommendations. The decision is taken by the line organisation. However, it is crucial that the project manager should be involved in the selection of the project staff. Firstly, the project manager will know the composition criteria for the project team (what is required of each individual team member). And secondly, the project manager will know the people that this staff member will subsequently need to work with and to achieve results with.
What is the technical solution? In projects that are closely associated with technical departments (for example, product development), the technical decision is generally taken by the line organisation. In other projects with higher-level overall objectives (e.g. new salary system), the line organisation can take decisions about parts that affect them. Innovation projects are increasingly being run by so-called "task forces", in order to deliberately involve a different perspective to the line organisation's thinking. The technical expertise and experience is often "bought in" (for example, by using technical consultants) and the technical decisions are taken by the project team or by the project owner.
Where/how will the activity be carried out? In which specialist area? Internally or externally? What equipment or resources will be needed?

At the end of the day, the project manager is responsible for planning and managing the running of the project and for the approach used. The line organisation is responsible for solving the technical problem. The project manager intervenes if, for example, the selected technical solution does not meet the agreed goals (Fig. 15.8).

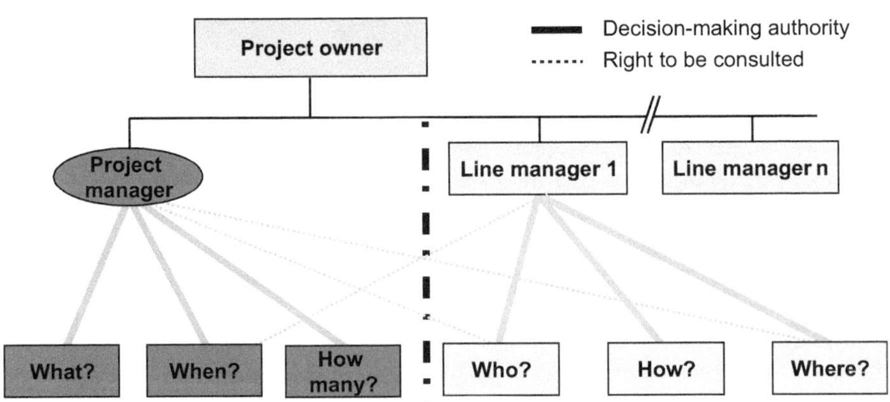

Fig. 15.8 Areas of responsibility for a matrix organisational structure

Areas of responsibility can also be shown as an organisational chart (Fig. 15.9).

Phases / Activities \ Involved parties	Project owner	Project manager	Sub-project manager	Line manager	Other participants
Initialisation					
• Project order	D			A	
• Organisation / Areas of authority	D			A	
• Project manager	D			A	
Pre-study					
• Goals	D	A		C	
• Project structure		D	C		
• Planning (activities, deadlines costs)		D	C	C	
• Preparation of information		C, A	C		
Milestone decision	D	A			
Concept					
• Comparison of alternatives	D	A	C		

Key: D = Decision (final) P = preparation, drawing up, planning
A = Making Application (initial decision) I = Implementation
C = Consultation, having a say G = Giving approval (yes / no)

Fig. 15.9 Example of an organisational chart

15.9 The Management Continuum with Changing Responsibilities

Companies generally set up their structures and procedures on a decentralised basis. This means that tasks are normally carried out by organisational units. So projects are often run on the same principle. That can mean that the business management and the project management work independently of each other. A silo mentality takes hold in the centres of power. Everyone just looks out for their own interests. With the left hand not knowing what the right hand is doing, in practice the exchanges between the business and the project tend to be brief and sporadic. And that represents a serious risk to the success of the project.

Examples of possible combinations include:

Business management	Project management
Banking specialists	ICT specialists
Sales	Engineering
Planning	Development
Development	Production
Production	Support/After-sales

This process is well suited to more complex projects. Complex projects frequently affect different areas of the company. They need a well-integrated organisation and overarching processes that can cope with these contradictions. And that is why an ongoing management process with a gradual transfer of responsibilities must be put in place (Fig. 15.10).

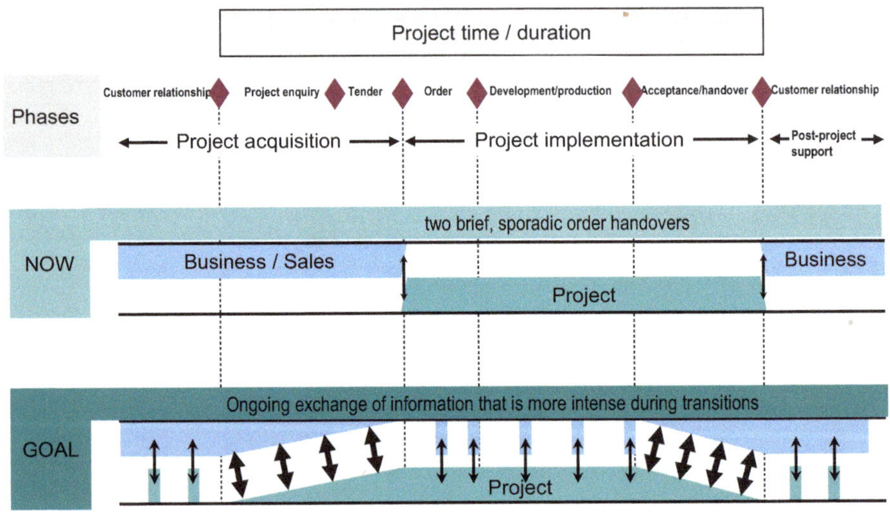

Fig. 15.10 Sporadic project handover vs. continuous management

Close cooperation with a constant, ongoing and open exchange of information between the management teams and the specialists is essential. It is also needed before the project starts, and after it ends. In particular, a lot more attention should be paid to the transition phases.

Before any promises are made to the customer in terms of products, product components, scheduling, costs, etc., agreement must be reached on what is realistically achievable, and this should be formally agreed between all stakeholders. An especially important point is that the two project managers should always work together and talk together, with a shared aim of achieving the objectives. It is useful to involve the next project manager at the start and the previous project manager at the end. It is helpful to hold interdepartmental workshops, rather than bring in expert consultants.

In practice, this handover of information and responsibility occurs in the context of a hand-over meeting (HOM). And to make sure that for example the developer clearly understands what the customer is looking for, and how hard it can be to agree realistic goals, they should be present during the commercial negotiations.

Another form of project management that is becoming more common in practice is "co-leadership". This means a "with one another" or even a "next to one another" approach to managing a project involving two (to three) people with different backgrounds and experience. As well as intensive technical and procedural coordination, this form of joint project leadership also calls for "good chemistry", excellent team skills and a high awareness of the importance of working together cooperatively.

For this sort of "multi-person leadership", the following aspects are important:

- Proper agreement about roles.
- Not allowing themselves to be played off one against the other.
- They must work as a "core team", with clearly agreed tasks, areas of authority and responsibilities.
- There must be one designated contact person for external contacts, both incoming and outgoing.
- Regular differentiated two-way consultation.
- Selecting people with the maturity to have a positive approach to working together.

For projects with co-leadership(s), the company's senior management has a particularly important role and function:

- They bear corporate responsibility.
- They provide structures and processes that encourage the exchange of information.
- The carefully define and assign roles.
- They assign clear and unambiguous responsibilities.
- They agree clear goals and orders, and also goal-oriented, performance-related and resource-based areas of authority.

15.10 Setting Up the Project Organisation

Generally speaking, most organisations are not critical enough or careful enough when putting project organisations together. The reasons why include seeing things from only one perspective (the project owner's, or the technical department's, etc.), a culture of constant staff changes, and fear of having sensitive and conflict-prone discussions about personnel needed for the team (commitment, interests, availability, expertise, sympathy/antipathy).

Here are a few tips to bear in mind when setting up a project organisation:

- For projects with a wide range of divergent interests, the contribution from stakeholder groups becomes more important. That makes the project organisation larger. But the project organisation should still be kept as lean as possible. Wider involvement can also be achieved through the use of advisory

groups, special workshops for people who are affected and for users (possibly even events for large groups).
- Pay attention to defined roles and avoid multiple roles. Aim to achieve separation of powers between the different advisory boards.
- Don't leave selection of the project team members to chance. As well as thinking about who should be on which advisory board, clarify how they should be nominated: By senior management, by the project owner, by the project manager, on the basis of popular support from stakeholder groups, etc.
- Make sure that project staff have the resources that they need. Overworked project staff will not be motivated, or will set their own priorities according to their own interests.
- During the course of the project, the project organisation may change from one phase to another. But within each phase, the organisation should be kept as constant as possible, in order to avoid disrupting team processes.
- It is better to run fewer projects, but with technically excellent and highly motivated people. And management must be committed to the project, too!

15.11 Virtual Teams

Project teams that work together on an international basis have become an increasing reality. They may rarely or never actually meet in person, and communicate primarily using technical tools such as e-mail or workgroup applications.
Quite how much need there is to meet physically, at least for a project kick-off meeting or for occasional meetings, will depend on the type of project. For a standard project in a global company, the degree of virtual working will certainly be higher than for a pioneer project involving various different firms or colleges that have not worked together before.

Here are a few tips and things to think about when working virtually:

- The formal and technical requirements must be agreed (data formats, which tools to use, hardware and software for all participants, etc.).
- There should be an agreed project methodology and project language.
- There must be a shared understanding of the project. In most cases, this makes it worth holding a meeting with everyone in the same physical room, especially if the project participants don't know each other.
- Rules about information and communications must be very clearly agreed, and must be adhered to.
- The division of work (work packages and responsibilities) must be defined, and must be understood by everyone. A workflow plan is essential, with milestones and hand-over points. It is used both for orientation, and for controlling.
- In between the synchronisation points, the team members have full responsibility for working on their work packages. That demands a lot of the

team members. They have to accept responsibility, work largely independently, make decisions, and flexibly adapt to working in a virtual team.

15.12 Working in a Multicultural Environment

With globalisation and increasing cooperation in international teams, project managers are also increasingly faced with the reality that their project environment involves different cultures and ethnic origins. That can lead to problems, since communication is made more difficult by different value sets, different perspectives, behaviours, languages, etc. On the other hand, many project managers now recognise that having different cultures in a project team should not necessarily be seen as a problem. Indeed, the range of cultures might even be a resource in itself. There is a lot of good literature available these days on this varied and complex topic (see bibliography). To summarise, to work well in a multicultural project environment, a project manager should ideally have the following skills:

Attitude and conduct
The project manager should develop an open approach to other cultures. This should enable him to demonstrate respect for his partners, even if their behaviour might appear unusual to him. He must develop his awareness of the external impact of his way of working. He must be capable of adapting his behaviour to suit the situation in a flexible manner.

Cultural awareness
The project manager needs to have a basic awareness of the culture that he will encounter. The key thing is to be aware of expected gestures and of the significance of rituals and ceremonies, to avoid any taboos and to have a basic understanding of religious sensitivities. Without some form of appropriate preparation, there is a high risk of tripping up along the way, and of losing credibility (loss of face).

Language and communication skills
It is absolutely crucial that the project manager has the relevant language skills. That means having the correct vocabulary to enable precise formulation of his views, and a range of expression that is appropriate to the culture. An awareness of linguistic traps is also extremely important. This also includes being able to use a metalanguage to develop team communication. And of course, the other culture will have implications in terms of body language, some of which may be quite alien to the project manager's own culture.

Leadership skills
A knowledge of process characteristics such as dominance, understanding and familiarity with hierarchies, acceptance and power is extremely important for the project manager in all leadership situations. An awareness of which management

styles are appropriate or usable in a particular culture helps overcome many potential problem situations in a multicultural project team. How can I persuade people? How should meetings be planned? (Expectations, preparations, punctuality, language, who to invite, agenda, leadership, reaching a consensus, next steps, etc.) What are the implications of the work ethic, qualifications and networking in a different cultural environment?

Negotiating skills
In Asia, there are very different rules about negotiations. The Asian ways of decision-making are sometimes unfamiliar for Europeans, and using an overly direct approach can quickly lead to a loss of face. The project manager must rely on precise knowledge of which negotiations methods can be used when (win-win, win-lose, poker or chess strategies, which opening or ending is most likely to succeed, who should take the lead).

Conflict resolution skills
In order to deal with work conflict in multicultural project teams, the project manager must develop both the relevant cultural awareness (what is allowed? what is not appropriate?) and excellent resolution skills. This helps him to act authentically and convincingly in conflict situations. Finding the right words or the appropriate degree of sensitivity whilst also developing the ability to drive things to a resolution is usually a challenging and delicate balancing act.

Example

The Western communication style is characterised by directness, "getting to the point", adopting a clear position, and is predominantly verbal. The Asian communication style is more subtle, circular, about throwing light on the context, and is predominantly non-verbal. We Westerners have our own prejudices, and interpret this as meaning that we are direct, open and honest, and that others beat about the bush. But positive interpretation means that the Asian style is currently in vogue. The communication style outlined above is recommended to modern management these days. Somehow, these differences have to be made transparent in a project team. If that happens, a multicultural team can work through complex problems as if they all came from the same cultural background.
So in multicultural project teams, it is even more important not to rely purely on virtual cooperation. Making full, effective use of these resources needs occasional "face to face" sessions. A physical meeting is particularly useful at the start of the project.

Project Planning 16

There are two steps to planning complex projects:

Rough Planning
This is done as early as possible in the project, generally towards the end of the preliminary study. If the project is highly innovative, rough planning may not be possible until the basic concept has been drawn up or towards the end of the concept phase.

Detailed Planning
This is done as early as possible, generally towards the end of the concept phase when the possible variants have been decided (Fig. 16.1).

J. Kuster et al., *Project Management Handbook*, Management for Professionals,
DOI 10.1007/978-3-662-45373-5_16

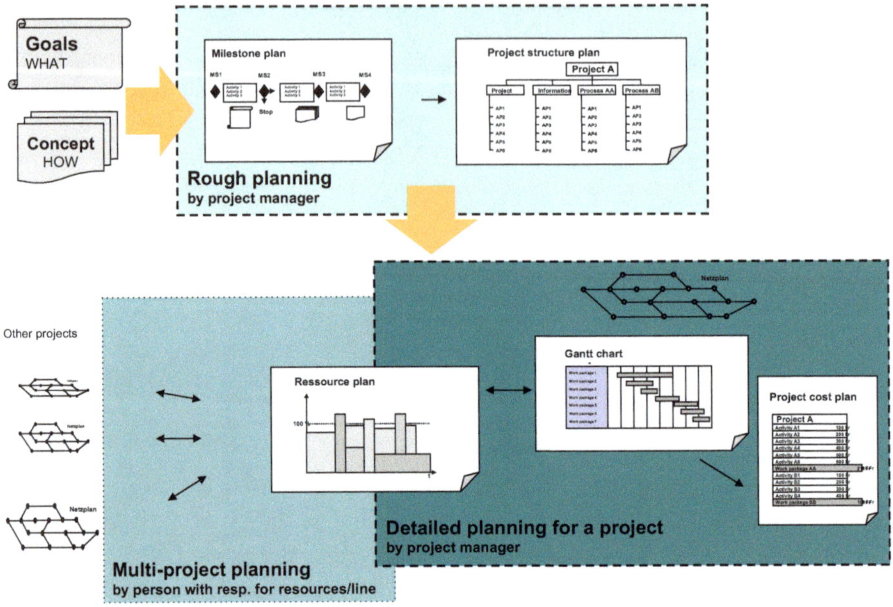

Fig. 16.1 An overview of the project planning process

The Purpose of Planning

The project manager and line managers work to achieve the following goals with the planning:

- Check that the project owner's specifications are realistic.
- Structure work packages and identify the boundaries between them so that responsibilities can be clearly and usefully assigned.
- Identify which subject experts (with defined expertise and experience) will be available for what percentage of the project.
- Recognise any bottlenecks or conflicts with respect to resources at an early stage and take appropriate action in good time (people, money, other critical resources, machinery, etc.).
- Ensure that everyone involved knows who has to do or deliver what and when.
- Provide a target figure for the purposes of checking the project's actual status (compared to the planned status).

What Will Be Planned?

The project manager is responsible for three variables: achieving goals (quality and quantity), time and costs. He therefore plans these three variables so that they can be checked in the course of the project. They are also interdependent. If the planning shows that the desired results (goals, deadlines, costs) cannot be achieved, the project owner must be consulted to establish priorities. The project manager recommends alternatives to the project owner (Fig. 16.2).

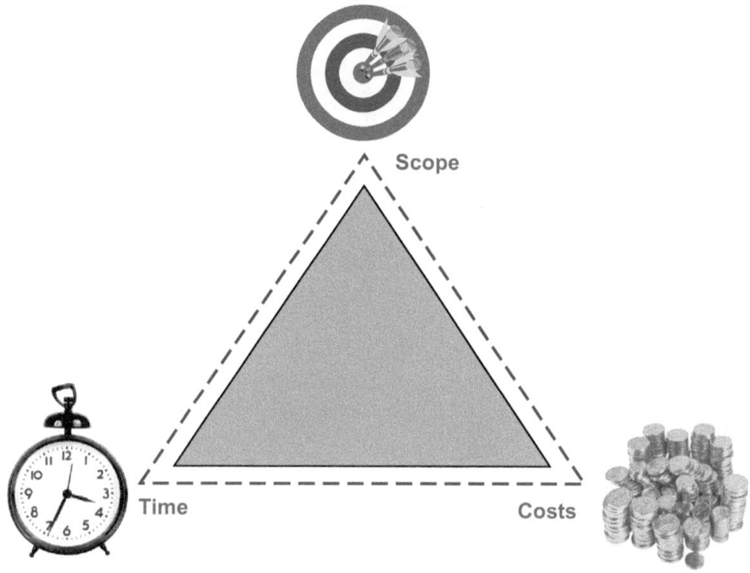

Fig. 16.2 Interdependence of goals, time and costs

Requirements for Meaningful Planning

- The set goals must be known
- The approaches to achieving those goals must be roughly defined
- It must be possible to assess the risks (problems, feasibility, acceptance)
- It must be possible to assess the expertise that is required and available
- The availability of critical resources must be known

16.1 Rough Planning

Rough planning involves clearly structuring projects according to their size and complexity, and thus making them manageable (Fig. 16.3).

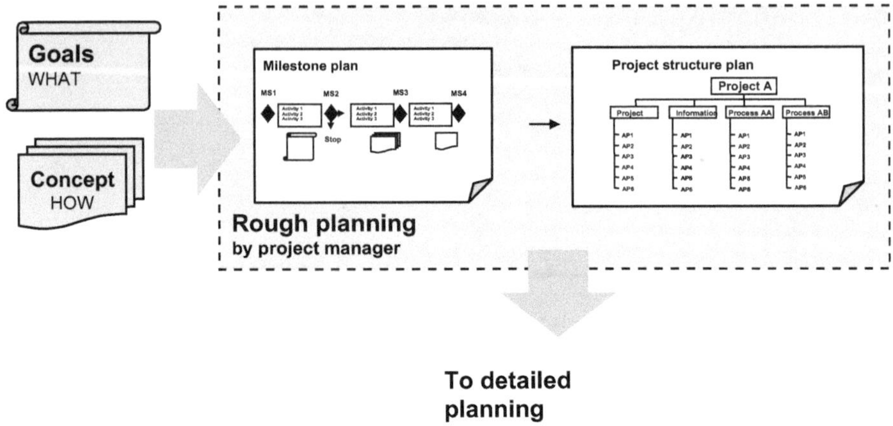

Fig. 16.3 Rough planning

Establishing Milestones in the Project: The Milestone Plan

The milestone plan sets out a rough project schedule and should clarify the following questions:

- Into how many phases should the project schedule be broken down?
- Which work packages will take place in which phase?
- How many milestones will be needed?
- Which interim goals, documents and results must be achieved or produced and checked at each milestone?
- Which decisions will be taken at these milestones?
- At which milestone will a review be needed?

Which Criteria Are Applied to Break a Project Down into Phases?

It would be risky to carry out a large project in one go and wait until the end to check whether the intended goal was achieved. For this reason projects are broken down into individual phases with verifiable intermediate results. These are the interim goals. There are some that have to be carried out at the start and others that can wait until later. Thus the project manager considers which tasks need to be done in which phase or for what a concept needs to be drawn up and in which phase. A decision will be taken about that concept at the milestone.

A project or project phase needs to be broken down further if an error is subsequently discovered, a wrong direction is taken or there is lack of acceptance, and this would have disproportionately large consequences. In such cases it is a good idea to set a milestone at this point. This means a scheduled time at which to check what has been achieved and consider how to proceed. Two phases may be sufficient for a small, non-problematic project for which there is already plenty of experience, specifically a definition and concept phase and an implementation phase. For larger projects or if there is greater uncertainty within the project, more phases and milestones at critical points are a good idea (Fig. 16.4).

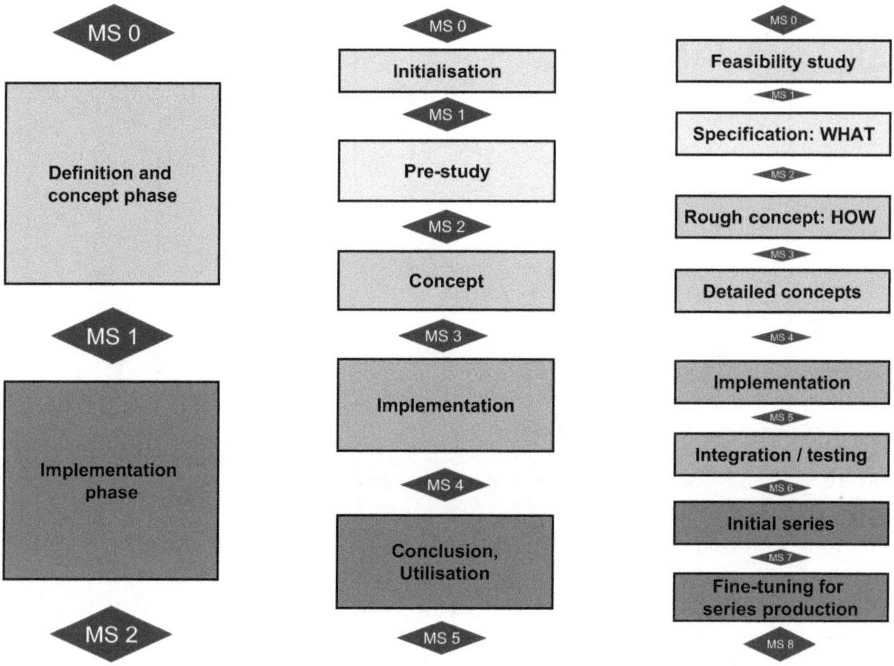

Fig. 16.4 Milestone plans for simple and more complex projects

Example

If the customer or project owner does not know exactly what he wants or if the people involved are unable to agree which goals are important and how important they are, then it is a good idea to add a milestone between the definition phase (What is to be achieved?) and the concept phase (How should it be achieved?).

If the effort and risk involved in introducing a new procedure into a company are very great, it is helpful to introduce the new solution first in a pilot project in a small, manageable area. It should not be introduced across the board in all of the company's subsidiaries and countries until the results have undergone critical assessment.

The number of phases introduced should be limited to those actually needed to effectively control the project. Phases that last too long, over many months or years, should be avoided (Fig. 16.5).

Phase	00 Pre-study	0 Study			1 Development					2 Production Market launch
Subphase / Milestone	Feasi-bility 0.0	Require-ment 0.1	Solution concept 0.2	Plann ing 0.3	Detail spec. 1.1	De-sign 1.2	Verifi-cation 1.3	Inte-gration 1.4	Quali-fication 1.6	Initial series Adjustment for series production 2.0
Review	R		R	R					R	R
Software develop.										
- System	Idea	Analysis	Design					Int. test	Acceptance	
Hardware dev.										
- System	TS	Outline	Solution concept		PT	Design	Testing	Settlement		IS testing
Documents										
Syst. analysis	D	A								
System design			A							
Module spec.					P			A		
Prog. listing						P	A			
Test plan			D			A				
Accept. plan			D					A		
Marketing										
Product plan	D	P A								

Key: D = Draft, P = Provisional, A = Approval,
TS = Tender specification, PT = Prototype, IS = Initial series

Fig. 16.5 Example of a milestone plan for complex innovation projects

The term "milestone plan" is used particularly to illustrate the structured schedule and the time planned for checking the interim results. The term "phase plan" is used when the emphasis is placed on representing the work packages and activities in detail.

If similar projects are frequently carried out within a company, the company defines one or more processes describing how the projects should be run, and then introduces them as mandatory standard project procedures or standard project processes with a process owner authorised to approve the process. The following points are defined as mandatory in such processes: phases, milestones, reviews, activities, documents to be created and their current status (draft, approved), documents to be considered, decision-making situations, (person responsible for: initialisation, implementation, decision, right to be consulted, etc.).

If such a standard procedure exists, the project manager will apply it. If the project requires moving away from this procedure for sound business reasons, the project manager will make, justify and document the necessary adaptations. Examples: Merger of two phases and introduction of an additional review (Fig. 16.6).

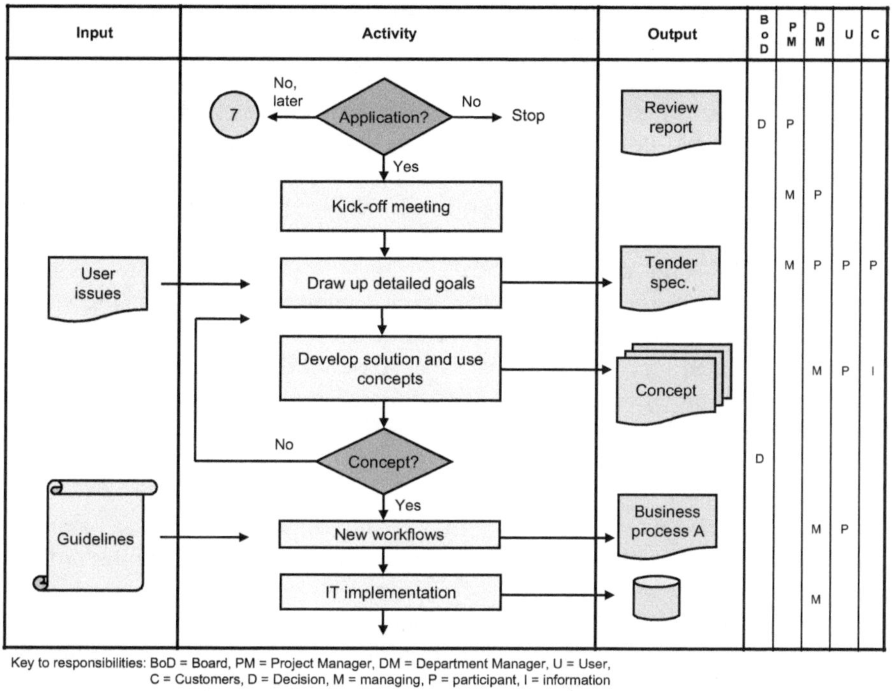

Fig. 16.6 Extract from a standard process for running a project

When Will a Review Be Needed?

After a milestone, if the further course of action is particularly likely to trigger significant consequences or financial risks, if the company's image might be affected or if the project breaks new ground with correspondingly greater uncertainty, then it is always sound business practice to carry out the milestone check with a particularly careful and critical eye. Then the project manager applies an established method and carries out a review (critical check).

The purpose of the review is to cast a critical eye over the interim status achieved and the further course of action from a different, independent viewpoint. It should be ensured that the level of maturity of the documents created and the basis on which decisions are taken are appropriate to trigger the next phase. Any errors that might subsequently have an impact should still be identified in good time. Other possible topics include assessment of the risks, do the earlier assumptions still apply, are the framework parameters still appropriate and is the adopted course of action still sensible? The project owner often demands such critical examination before he signs the approval and thus releases the resources for the next phase. The project manager and project owner discuss where a review is needed.

Breaking a Project Down into Work Packages: The Project Structure Plan

The overall project is broken down into clear work packages (task packages) with simple interfaces and assignment of subject experts to technical tasks, ensuring that responsibilities can be clearly allocated. It is important to ensure that boundaries are clearly identified and the persons responsible for specific tasks maintain their responsibility for the entire duration of the various project phases. The project should be described in full.

The term "work package" means the totality of a number of activities. These must be complete in themselves in order to achieve a verifiable result. A responsible person is appointed for every work package. A well thought-out work breakdown structure (WBS) provides clarity and allows responsibilities to be easily allocated. Work packages can be put together in two ways:

Top-down

The entire project is disassembled step by step into smaller units, with meaningful boundaries between work packages. This requires experience with similar projects and a good overview of the project.

Bottom-up

This method is used where there is little experience available. Every activity that comes to mind is compiled by the team. Then activities that belong together are grouped to create work packages. Finally, the listed activities are checked to ensure they are complete.

There are various criteria that can be used to break down the project structure plan. It may be object-oriented (oriented by content, goal or product) or procedure-oriented (process, activity or function). Mixed orientation is also possible (both object and procedure-oriented). How do we identify the best criteria by which to structure the project?

In production-oriented companies, projects are often structured by objects or contents, e.g. assemblies. This provides a good breakdown for management and processing purposes (Figs. 16.7 and 16.8).

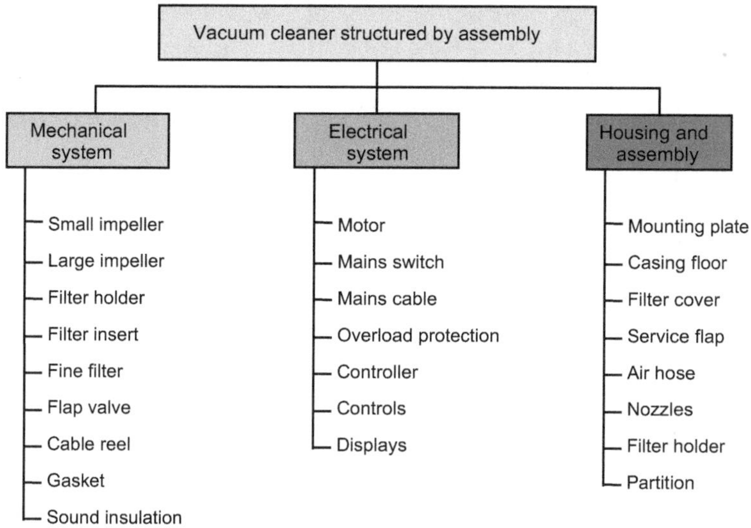

Fig. 16.7 Example of an object-oriented PSP structure

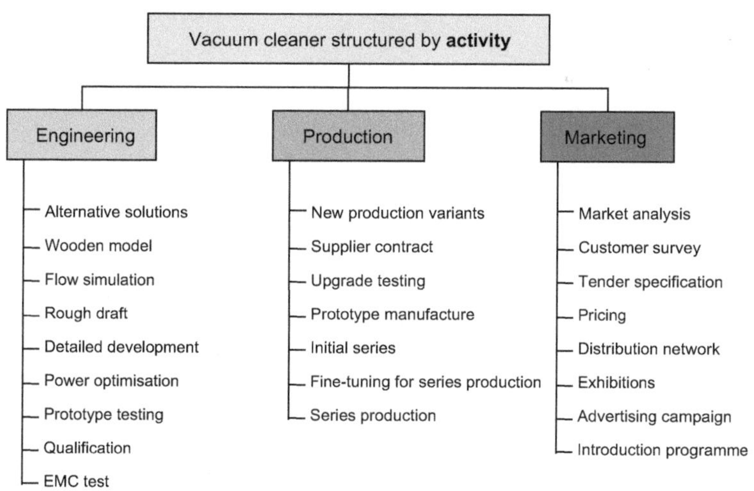

Fig. 16.8 Example of a procedure-oriented PSP structure

In this case, however, it is more important for responsibilities to be clearly allocated to the responsible persons. This is done by structuring by function, which often corresponds to the organisation's specialist areas such as Sales, Marketing, Customer Service, Development, Design, Production, etc. This type of demarcation is also useful in service-oriented companies.

In large or complex projects, a description is created for every work package. This details the requirements (input), activities to be carried out in this work package, results (output), generally in the form of documents, and the time, responsibilities and framework parameters needed for this work package. "Requirements" is a definition of the information and documents that must be available in order to start the work.

"Activities" defines what is contained in the work package and what is not, if necessary (demarcation). "Results" stipulates the results or documents to be drawn up and made available to other work packages. All "requirements" must be supplied by another work package. All results must be used in another work package, otherwise they may be omitted. The project manager checks the work package descriptions for completeness and for superfluous results and documents (Fig. 16.9).

Work package description **Project: Mobile PDA**

Work package: Transmission module No. AX12

Input, requirements

Tender specification documents, including ergonomic requirements
Approach at the system level
Interfaces to subsystem AX4
Draft testing concept

Activities

Draw up and add to detailed concept
Rough design with documentation
Detailed design
Simulation with APX
Dimensioning of 3 alternatives
Pilot trial in the field
Software development
Produce the sample for testing, produce testing instructions and test it

Output

Schematic, parts lists
All test reports with evaluation
Draft customer documentation
Full software documentation
Sample for testing and draft testing instructions

Effort (internal): 3200 hours
Investment: € 120,000
Person responsible: Markus Müller
Date: 23.2.2005

Effort (external): 1550 hours
Follow-on costs: 500,000 €
Dept.: TDE
Initials: MM

Services not included

Customer documentation suitable for publication
Adaptation for USA

Authority

Access for PPS SAP module

Notes

The effort is only realistic if project A21 is implemented as well since they share work packages.

Fig. 16.9 Example of a work package description

Will Sub-projects Be Needed?

For small to medium-sized projects, project management falls under the responsibility of a single person. For very large or complex projects, if the scope of management is too great for a single project manager, it is a good idea to introduce sub-projects and to split the project management activities (planning, assessment and control of the process) between two or more people. This will, however, require additional discussion and coordination between the sub-project managers and the overall project manager. Creating sub-projects only makes sense if the advantages (manageability and independence) outweigh the disadvantages. Only people with appropriate authority and sufficient time should be appointed as sub-project managers. Breaking down the content into work packages or activities is not of itself reason to introduce sub-projects. In practice, the project manager also manages a sub-project (Fig. 16.10).

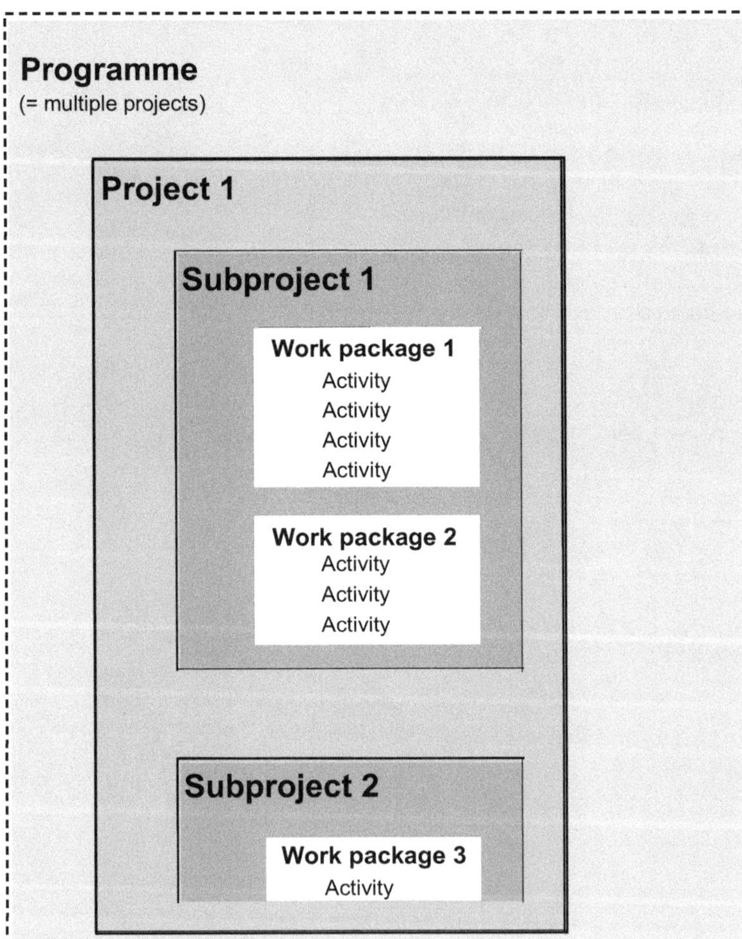

Fig. 16.10 Hierarchies within the project

The main project manager and sub-project managers must discuss the interfaces between the sub-projects and define consistent milestones at which meaningful decisions are taken for the project as a whole. At a milestone, all sub-projects must have reached a state at which the appropriate decisions can be taken, enabling a go/no go decision to be taken for the entire project (Fig. 16.11).

Phase \ Sub-project	**Subproject 1 Engineering** SPM 1: Developer nn	**Subproject 2 Production** SPM 2: Production assistant	**Subproject 3 Marketing** SPM 3: Head of marketing
MS 0	**Decision on tender specification and product development**		
Development	Development trials Prototype design Produce prototype Test prototype Revise documents	Prelim. clarification for production Consultation on design Work preparation Prelim. costing analysis	Advertising concept Aftersales concept Sales concept Contractual negotiations Cost effectiveness
MS 1	**Decision on initial series**		
Transition to production	Documents initial series Test initial series Fine-tuning for series production	Procure production resources Produce initial series Optimise production	Advertising materials Aftersales and sales organisation Initial series field test
MS 2	**Decision on series production and market launch**		
Market launch	Revise documentation	Optimisation of series production	Advertising campaign
MS 3	**Decision on follow-up projects**		

Fig. 16.11 Example of a project structure plan

The project risk increases if the sub-projects are not all ready at the same time. If the schedule and timing for the work packages is roughly agreed between the sub-projects, scheduling in the next steps (detailed planning) will yield good results sooner.

16.2 Detailed Planning

Once large and complex projects have been clearly structured and thus made manageable at the rough planning stage, the detailed planning involves defining who must have done what and by when, ascertaining the associated costs and identifying and resolving any resource conflicts (Fig. 16.12).

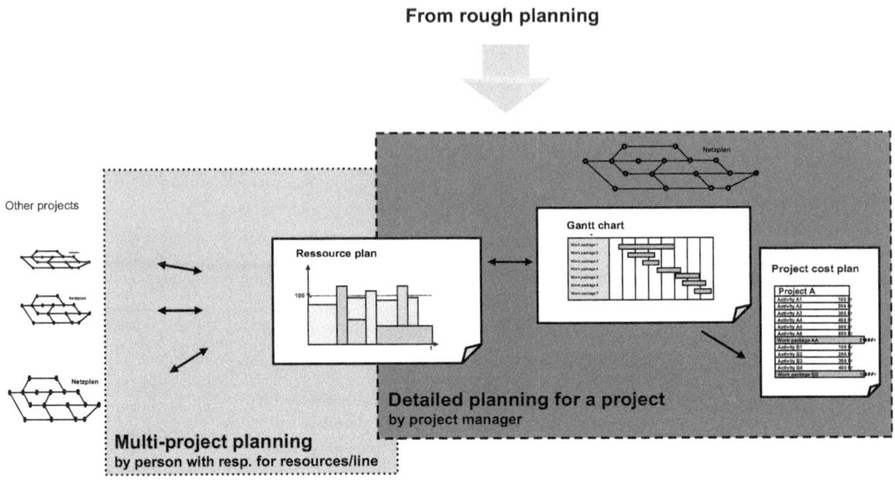

Fig. 16.12 Detailed planning

Create List of Activities and Schedule

The content of the work packages is now described in greater detail and recorded in full in a list of activities. Every activity to be carried out, specifically everything that needs time or money, including checking of milestones, is listed in full and ideally in chronological order. Every activity is given a unique ID number. This can be made up from the project ID, the work package ID and a consecutive number for the corresponding activity, e.g. 07-12-22 (Fig. 16.13).

Activities and deadlines Date:

No.	Activities / Measure / Event	Resp-onsible	Pre-conditions	Effort in weeks	Duration in weeks	Scheduling / Gantt chart 1 2 3 4 5 6 7 8 9 0 1 2 3 4 5
1	**Project**					
2	***Phase 1***					
3	Activity A	MM	-		6	
4	Activity B	MM	3		1	
5	Activity D	PP	-		2	
6	Activity E	PP	5		3	
7	Activity C	MM	3;6		2	
8	Activity F	TT	7		3	
9	Activity G	MM	8		1	
10	Milestone Preparation	PL	4;9		1	
11	Milestone Decision	AG	10		0	
12	***Phase 2***					
13	Work package H	MM				
14	:					

Fig. 16.13 List of activities

The unit that will do the work or the relevant department is then entered for every activity, if known. The unit that will do the work must have the necessary subject expertise for the activity and sufficient free capacity at the specified time. For critical resources, the availability should be taken into account as soon as it is roughly known.

At the very least a period should be roughly estimated, as the precise time of deployment is not yet known. The subject specialist estimates the effort needed to solve the problem and the necessary processing time. The time is the time required by the responsible person (expressed in full-time equivalent FTE months, weeks, days or hours) to fully resolve all the tasks associated with the work package, with due regard to the resources that will likely be used.

The processing time (duration) is the shortest period needed to achieve the previously defined effort, taking into account the availability of critical resources, unavoidable waiting times and other framework parameters. Example: A painter takes 4 h to prepare and spray a bath. The bath then takes 72 h to dry before the next work step is possible. The painter needs another 4 h to sand the bath. The time required for the painter is thus 8 h, while the processing time is 4 days.

The activities are then set out in the order, which is most sensible or indeed necessary with respect to the work itself. For every activity the results to be obtained must be considered, along with the framework parameters to be fulfilled so that the activity can start. These chronological dependencies are entered as preconditions in the list of activities. The scheduling can take place once this information has been gathered.

Scheduling

The scheduling can be done using Gantt charts or precedence diagrams. The scheduling provides the end date and interim dates for the project, taking account of any dependencies. The planning process provides an earliest and a latest start and end date for every activity. The earliest end date indicates the earliest point at which an activity can be finished if all the conditions are fulfilled as satisfactorily as possible. The latest end date indicates when the activity must completed at the latest to ensure that the project will not be delayed.

If the latest and earliest dates are different, the activity has a degree of leeway (slack or buffer). The slack is the time by which an activity may be delayed without affecting the end date. The critical path connects all those activities that have no slack. It also determines the end date. If an activity on this path is delayed, the end date for the project is put back by the extent of the delay. To allow the project manager to take early action if required, he must check the activities on the critical path regularly and at useful intervals. Just like a helmsman whose boat responds to corrections with a slight delay.

The network diagram highlights, in particular, the dependencies between activities very effectively, although it does have the disadvantage of being large and rather cumbersome. It is for this reason that the Gantt chart is normally used in practice.

The duration of the activities is represented by a thick horizontal bar, while the slack is a thin line.
The Gantt chart is merely a different representation method to the precedence diagram; they contain the same information. The Gantt chart shows the earliest possible start, along with the earliest possible and latest possible end date for each activity in a clear and compact view. The duration, slack, chronological relationship between the individual activities and the interdependencies are clearly shown (Fig. 16.14).

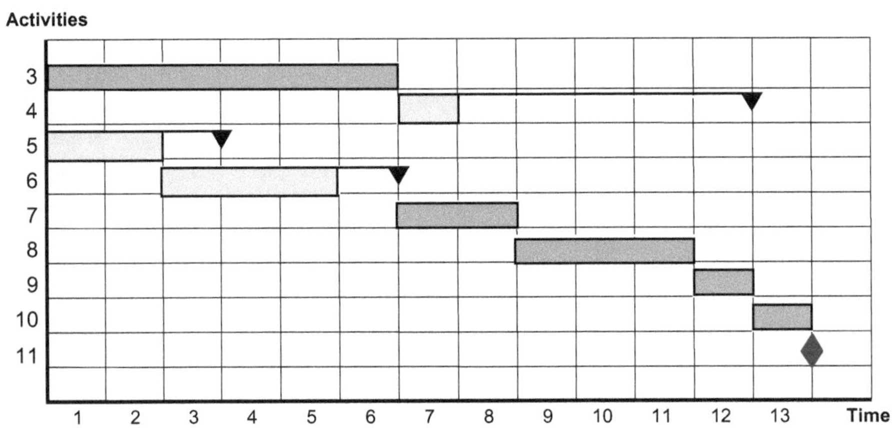

Fig. 16.14 Gantt chart with critical path and slack

Planning the Use of Resources: Line Management and Project Manager Working Together

The project planning is done by the project manager. Scheduling allows to plan what is to be done and when. The availability of resources is not yet certain, however, there is a need to plan the use of resources in detail. The resource planning records at least all the resources that will be available to a limited extent at a defined moment.

Examples

Subject experts, external participants, machines, equipment, special areas and all other irreplaceable resources. The resource usage plan (resource usage profile) for a project contains at the very least the planned data for all critical resources involved in the project (Fig. 16.15).

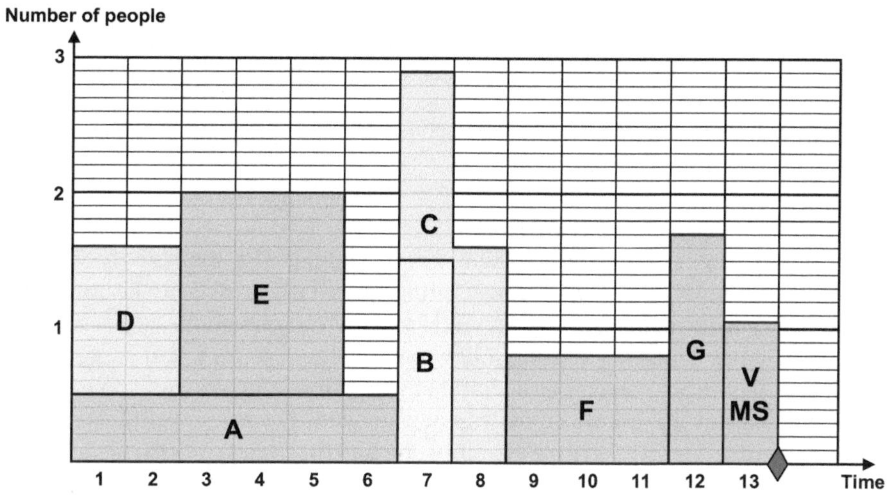

Fig. 16.15 Resource histogram for a project

However, the project manager generally does not have access to the specialists who will need to contribute to his project. Specialists from different areas are often needed at quite specific times to differing extents. To this end, the project manager needs to cooperate with the line managers who have these specialists in their areas. The project manager considers the knowledge, skills or experience that is needed to efficiently achieve the defined goals. He may already know with which specialist he would like to work. He does not know, however, whether the desired subject specialist is actually available at the desired time. The first draft of a resource usage plan allows him to specify in detail and schedule his needs for resources or his wishes for those resources.

The project manager and line managers are dependent on one another for resource planning. The project manager lists his needs and wishes in terms of resources to the line managers. If the necessary resources are available at the desired time, they will be promised and confirmed as reserved for him. A confirmed promise from the line manager requires the resource usage planning to be updated across all projects (see Multiple project planning). If resource conflicts occur because multiple projects wish to access the same resource at the same time, the project managers and people responsible for the resources must jointly discuss the problem and make suitable arrangements. It is advantageous if the line manager acknowledges that he is responsible for providing services within his specialist area on behalf of the projects.

But it is also helpful for the project manager to be aware that he cannot always have exactly the specialists he wants at the time that is best for him. He must therefore offer a little flexibility with respect to the date. This will provide a favourable basis for collaborative negotiations between project managers and line managers.

If the resulting date is not as specified, the project manager has the task of examining suitable measures and making the necessary corrections. Several different measures are often needed to meet the desired deadlines.

Possible ways of fine-tuning when the planned result gives an end date that is too late:

- Deferrals within the buffer period
- Increased parallelisation: look along the critical path for the activity with the greatest potential to be broken down further (generally an activity lasting a long time). Check the dependencies with follow-on activities with the aim of bringing forward those dependencies (the follow-on activities can then start earlier). This requires the activity under consideration to be split into two or more parts, which in turn requires the list of activities to be extended. The new schedule will show the dependencies that were brought forward, and the project processing time will have shortened.
- Use of additional resources available from the project manager's own or adjacent areas of the business
- Project priorities stipulated by senior management, giving preference to certain projects to the detriment of others.
- Outsourcing of suitable activities or work packages
- Use of intelligent resource usage profiles (using resources within the project at the point at which they will bring the most benefit to the project).
- Separation of basic functions and options to be completed later (update)

If the specified time or the effort proves to be unrealistic, the issue should be discussed and agreement reached with the project owner.

16.3 Multi-project Planning by the Line Organisation

If several projects are carried out within a company at the same time and they wholly or partly access the same resources, conflicts may arise with human resources (specialists) or special equipment. In this case multi-project planning will be required by the people responsible for the resources, generally the senior management or managers of specialist areas (Fig. 16.16).

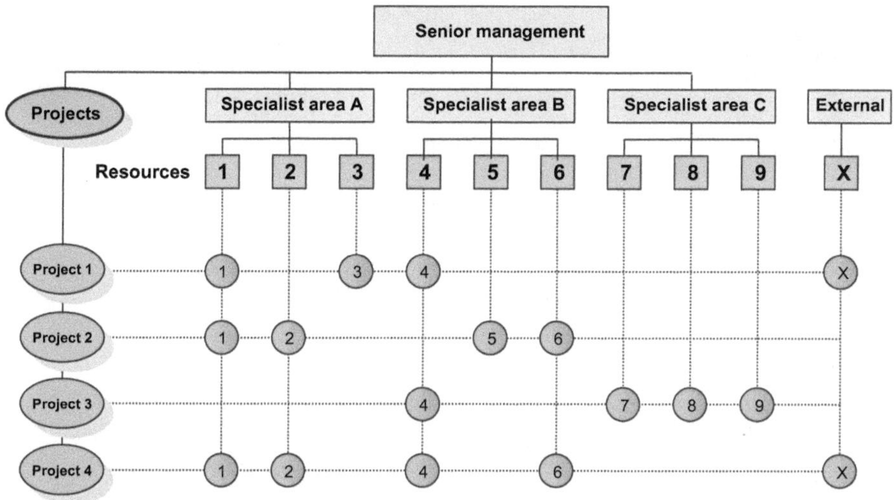

Fig. 16.16 Multi-project structure

The project managers plan the activities and deadlines for their projects (project view) and pass on to the line managers their need for resources and their wishes with respect to the resources they would prefer at a given time. The line manager, who has the resources, must incorporate the needs of all projects into a resource usage plan for all projects and all the resources for which he is responsible. He also takes account of all the basic workloads of every employee (line organisation tasks, absent colleagues, ancillary tasks, etc.). The usage plan that the line manager creates for each employee represents all of that employee's tasks across all the projects in which he is involved.

If an employee is likely to be overloaded for a long period when an additional, new project is scheduled, for example because he is involved in various projects, the conflict must be discussed and resolved between the line managers and the affected project managers. The simplest way is to make use of any slack available in one of the projects. If the problem cannot be fully resolved in this way, additional resources may be used, provided they are available in the line manager's own area or other areas and the necessary requirements are fulfilled (expertise and experience). If the task does not fall within the core business, it is a good idea to examine whether it should be subcontracted to external suppliers or service-providers. If neither alternative is possible, the end date is put back. Once the resources have been reconciled between line management and project manager, the project manager corrects his project plan. This also provides an amended resource usage profile for his project (Fig. 16.17).

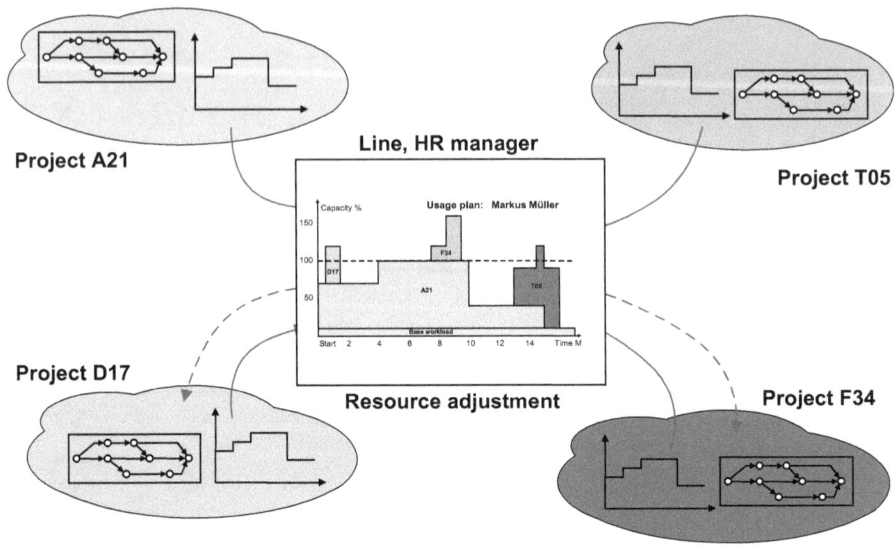

Fig. 16.17 Multi-project planning

The line managers are responsible for giving a firm commitment on the allocation of suitable subject experts. This is one of the most important human resource management tasks. To be able to do this, the managers must keep an informative and periodically updated overview of all reserved resources in all projects in which their subject experts are used. It must be complete, specifically it must contain all projects and all line tasks and other ancillary tasks. It should only be detailed enough to reliably draw attention to medium-term under-loading or overloading and to identify whether a subject expert still has sufficient free capacity for this project. It should not be too detailed, however, as the effort required increases disproportionately with the level of detail and the reality of the projects rapidly overtakes the planning. It is the line manager who decides whether another suitable specialist or additional staff need to be appointed as he has responsibility for employing them after the project has ended. If no acceptable compromise can be found, the problem is put to the internal project owner or senior management for a decision (e.g. change of priorities). The project manager then incorporates the agreed corrections into his project plan.

The resource usage plan created and updated by the person with responsibility for resources does not require the same resolution terms of timings and content as the project plan. Resolution to the level of work package and month is sufficient to be able to answer whether a certain resource will still have sufficient free capacity at a given time to be scheduled into a new project (Figs. 16.18 and 16.19).

Outline resource planning **Resource: Markus Müller, Dept. TDE**

In hours Status: dd.mm.yy Sig: KE

	YYYY												YYYY+1					
Activity	1	2	3	4	5	6	7	8	9	10	11	12	1Q	2Q	3Q	4Q	later	total
Project A21																		0
Work package A10		10	50	50	130	120	50											410
Work package A11				10	10		90	120	120									350
Work package A12											50							50
Project D17																		0
Work package D30	120	130	50	50														350
Work package D31										120	100	100	150	50				520
Project C19					10	30												40
Project R08												50	300	400	400	500	800	2450
Outside project	20	20	20	20	20	20	20	20	20	20	20	20						240
Total workload	140	160	120	130	170	170	160	140	140	140	170	170	450	450	400	500	800	
Capacity	160	160	160	160	160	160	160	160	160	160	160	160	480	480	480	480		

Fig. 16.18 Resource usage plan

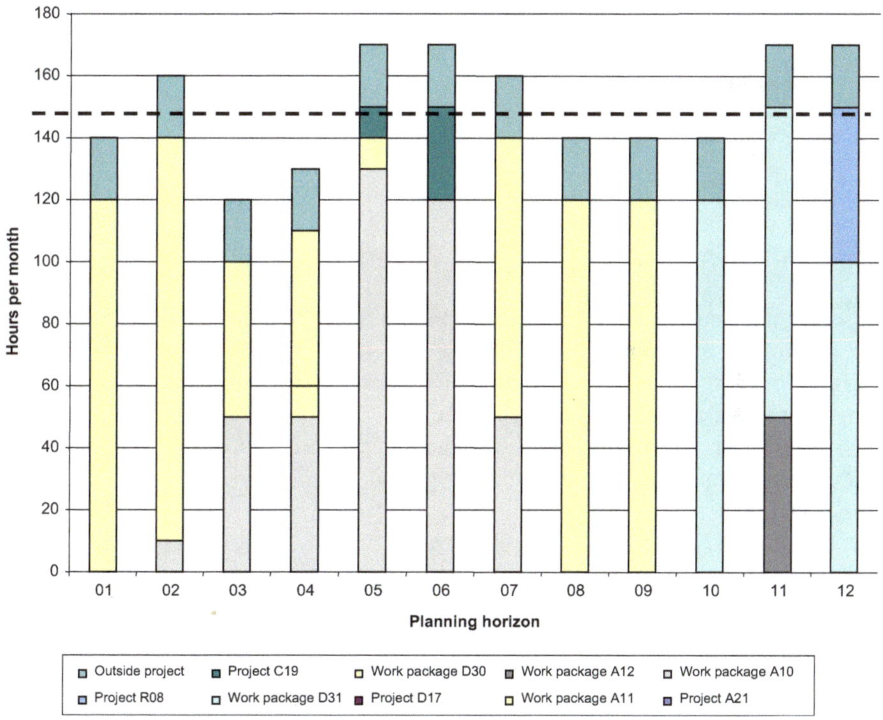

Fig. 16.19 Example of a resource usage graph

If the available resources are no longer sufficient to be able to implement all the projects, the relative priorities of the projects need to be assessed and adapted by senior managers. It should not be left to each project or line manager to determine his own priorities as he sees fit and which project he will tackle first or which he will leave and thus delay.

The priorities between all ongoing projects should be clearly defined according to the company's overriding interests and communicated to managers and everyone involved. Senior management should monitor the list of priorities on an ongoing basis and update it as required.

16.4 Cost Forecast and Cost Curve

The cost forecast records all the resources used in the project that give rise to a cost or direct expenditure, such as:

- All internal and external personnel working on the project, including the project manager
- Temporary use or hiring of special facilities (rooms, machinery, instrument, ICT equipment, etc.)
- External investment (purchases)
- Other direct costs (expenses, fees, insurance, etc.)

In view of the costs incurred over time, it is normally assumed for the sake of simplicity that resources are used in one of three ways and their costs can be assigned accordingly:

- Usage and costs incurred evenly distributed over the period of activity (offsetting with cost rate)
- Usage or costs incurred at the start of the activity (e.g. upfront investment by making necessary purchases in advance)
- Usage or costs incurred at the end of the activity (e.g. invoicing for a service or procurement at the end of the delivery period)

Cost forecast and budgeting within the project fulfil the following purposes:

- For the company (investor): as the basis for providing the financial resources (liquidity planning)
- For the project owner: as the basis for decision-making at milestones within the project
- For the project manager: as an instrument for operational monitoring and controlling

The costs of each individual activity (or individual resource item) are calculated and totalled across the entire duration of the project in order to determine the project costs (Figs. 16.20 and 16.21).

No.	Resource and cost planning informatio Activities	Duration in days	Number of people per activity (int / ext) Investment	Average staff use as %	Effort in days	Costs per person and per day	Planned overall costs (per activity)	When costs are incurred At start	Spread	At end
3	Activity A	30	1 person	50	15	800	12'000		X	
4	Activity B	5	2 people	70	7	1'000	7'000		X	
5	Activity D	10	1 person	100	10	800	8'000		X	
6	Activity E	15	2 people	70	21	1'000	21'000		X	
			Investment				89'000			X
7	Activity C	10	3 people	50	15	800	12'000		X	
8	Activity F	15	1 person	80	12	1'000	12'000		X	
9	Activity G	5	2 people	80	8	1'000	8'000		X	
10	Milestone preparation	5	5 people	20	5	1'200	6'000		X	
11	Milestone decision	0								
						Total	175'000			

Fig. 16.20 Details for the cost planning

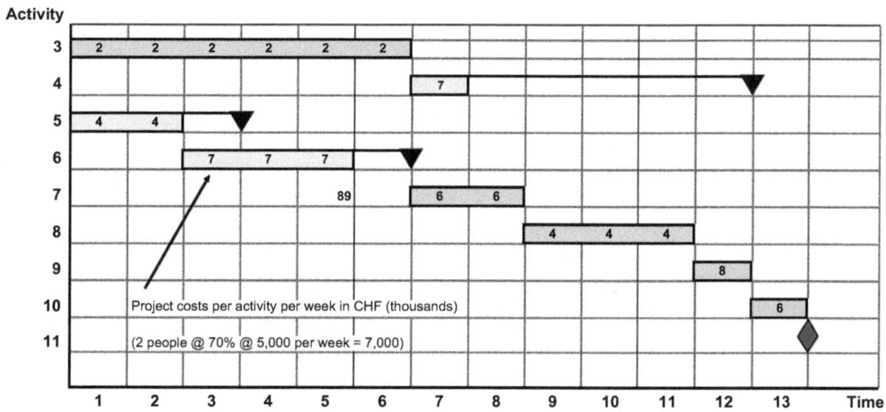

Project costs	1	2	3	4	5	6	7	8	9	10	11	12	13	
per month	6	6	9	9	98	2	13	6	4	4	4	8	6	kCHF
cumulative	6	12	21	30	128	130	143	149	153	157	161	169	175	kCHF

Fig. 16.21 Cost planning

If the budgeted data from the resources and cost forecast are overlaid onto the schedule, this gives the cost curve. It represents the changes in the cumulative total costs of the project over the term of the project (Fig. 16.22).

The cost curve is a useful tool for assessing:

- Period costs when the individual project phases are approved
- Budgeted amounts for the individual financial years (provision of the financial resources)
- Liquidity result from investment
- Basis for comparing planned/actual costs

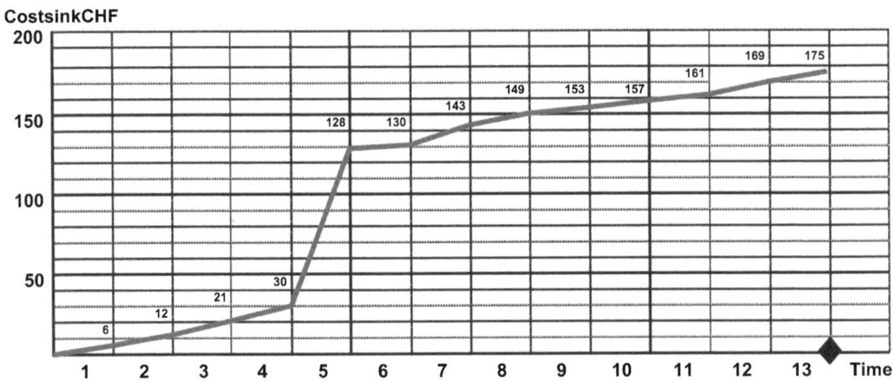

Fig. 16.22 Cost curve

16.5 Overview of the Project Planning Procedure

The most important requirement is that the order be clarified in terms of goals, costs and deadlines. The following illustration is a textbook procedure for planning demanding projects with a high level of innovation. For routine projects with values based on past experience, the detailed planning may have already been done during the initialisation phase (Fig. 16.23).

	Actions	Results
Initialisation	**Rough estimate:** Estimate the processing time (project end and interim stages), financial and human resources and follow-on costs. Obtain commitment from the people responsible for the resources.	Draft milestone plan, effort estimation, and rough plan for use of critical resources.
	Check plausibility with key data or an experienced project manager.	
	Plan the preliminary study phase in detail.	Gantt chart
Pre-study	**Rough planning:** Define the work breakdown structure: define the content, responsible persons, demarcation and total effort required. For large projects, indicate the interaction between the work packages (WP).	Work package description Project structure plan
	Define the chronological order for working through the WPs and times for checking the interim goals. Schedule reviews.	Milestone plan
	For large-scale projects or special situations the project management can be broken down into sub-projects, if need be.	Sub-projects
	Plan the concept phase in detail.	Gantt chart
Concept	**Detailed planning:** Break work packages down further into activities. Estimate the effort and processing time, clarify dependencies and determine the expertise and experience required and the posts that will do the work.	A procedural plan and a schedule in the form of a Gantt chart.
	Discuss the availability of resources with the responsible persons. Resolve any resource conflicts	Resource usage plan
	Compile all financial expenditure for resources, investment, etc. Break down as per the project structure plan and by time.	Cost forecast
	Communicate the planning results to all affected parties.	
Implementation	Adapt the planning from the introduction phase.	Gantt chart
Introduction	Consolidate empirical values into key data.	Close-off report

Fig. 16.23 Project planning in the individual phases

16.6 The Planning Procedure

The planning often starts with specifications (e.g. completion date) and the work steps are broken down into ever-increasing detail as far as individual activities, the effort of which can be assessed. The detailed deadline, resource and cost planning are carried out at this level. The results are shown in consolidated form and compared against the specifications. If there are any differences, the project manager then attempts to find solutions that allow the specifications to be fulfilled. The project manager arranges a meeting with the project owner if the differences are so great that the specifications are essentially unrealistic or priorities need to be established between deadline, resources and costs.
Today's most popular management concept is management by objectives (MbO). This requires the process to be carried out both from the specification (top-down) and from the adjustment (bottom-up); otherwise it will not conform to the principle of "tasks, authority and responsibility" (T-A-R). If the person who performs the task does not have a say (bottom-up), it will also be difficult to create a feeling of shared responsibility for the implementation.

Three methods are commonly used to plan a project:

- At the start of the project, the entire duration of the project is planned in detail
- At the start of the project, the entire project is planned in outline, and at the end of one phase, the next phase is planned in detail
- At the start of the project, the entire project is planned in outline, and every month the next phase is planned in great detail

If considerable experience is available from similar projects or if a fixed price quotation has to be provided, the full scope of the project must be planned in detail across the entire duration. This takes a great deal of effort (Fig. 16.24).

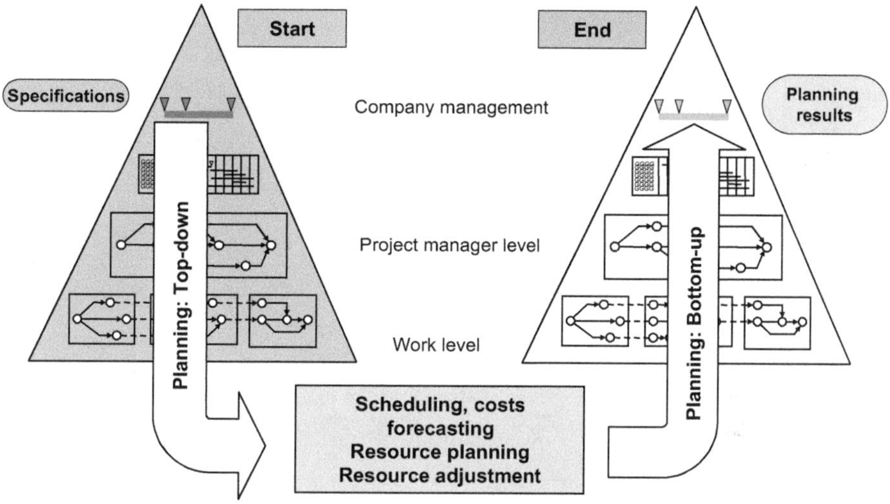

Fig. 16.24 Top-down planning and bottom-up consolidation

If there are no usable empirical values or the order is open and entirely new, the initialisation phase is limited to rough planning for the entire project with rough estimates of the effort and resources required. One phase must be completed before the next phase is planned in detail. Rough estimates should always be given with a margin for error, e.g. ± 10 %. This method is very efficient and often used.

It is a good idea to plan large-scale projects in two stages so as to make the project manageable. First level of detail: procedural planning and scheduling for work packages in the form of a project overview. The level of detail is increased in the second step, breaking the work down into activities, optionally by the sub-project managers. The depth of planning should be matched to the available knowledge and to the consequences of any deviations from the plan.

If the level of innovation is high and the deadline is critical, then the level of detail in the work packages is increased step by step as the available knowledge improves. At the start, the project is broken down into a few, rough work packages, and the entire planning horizon is defined for it. As the available knowledge increases, a manageable part of the project is planned in greater detail. The third step is to plan the immediate future, e.g. the next 2 months, in great detail (a few FTE days, matched to the controlling interval). This step is repeated every month (Fig. 16.25).

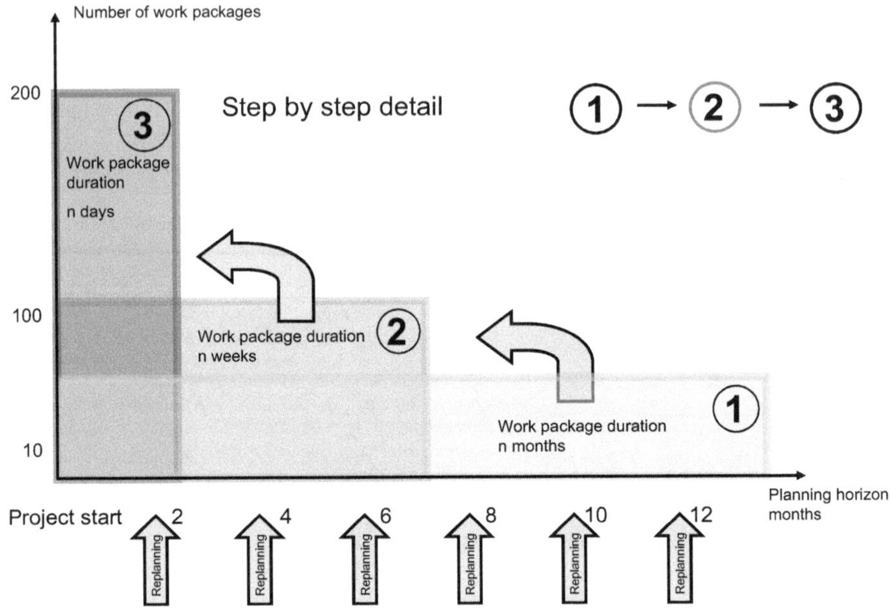

Fig. 16.25 Step by step planning

With large-scale projects, the planning involves a number of steps. The project manager creates a first draft. If the processing times are too long or are incompatible with the specifications, he optimises the plan for the first time. Resource conflicts are negotiated, agreed and confirmed with the people with responsibility for those resources. The initial planning is now frozen and must be approved by the project owner. If the framework parameters (e.g. the priorities of senior management) change, then it will have to be adapted. For significant changes, it may be necessary to replan the remaining tasks (time-to-complete and cost-to-complete).

The planning within the project will emphasise adherence to deadlines or available capacity, depending on the goals: If the end date is very important, then planning will emphasise adherence to deadlines. What do we need to do and which resources do we need to use, and what measures must be taken to ensure we meet this deadline? How overloaded will we be as a result, and how long will this last? (Fig. 16.26).

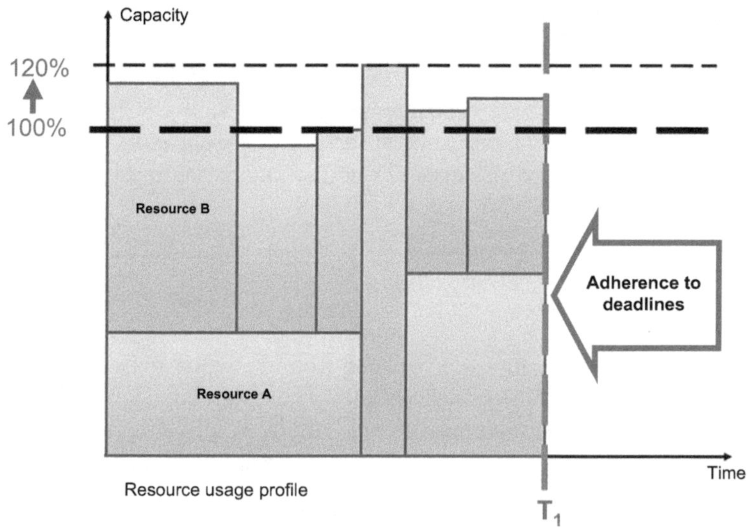

Fig. 16.26 Planning for adherence to deadlines

If the costs or other projects have higher priority, then planning will emphasise the available capacity. What deadline is possible with the available resources? (Fig. 16.27).

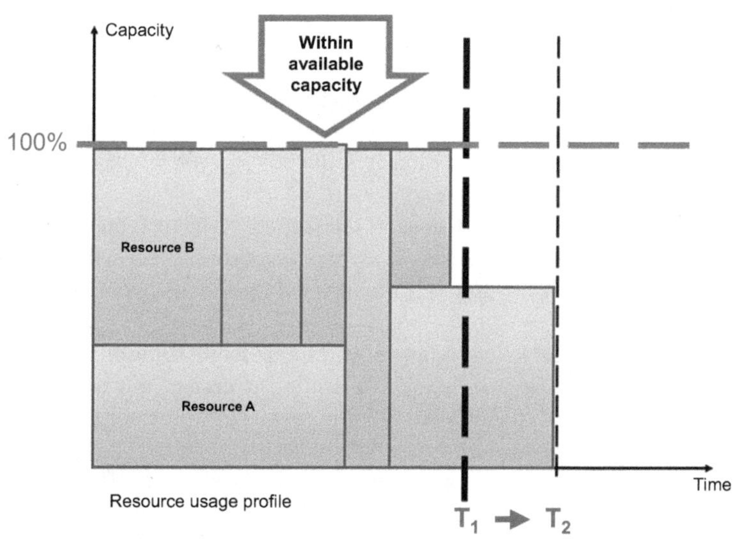

Fig. 16.27 Planning within the available capacity

Forward Scheduling

Scheduling from the start of the project to its completion. If the specified deadline is exceeded, measures must be considered (more resources, parallel working, improved infrastructure, external subcontracting of partial tasks, checking goals, etc.). Forward scheduling with necessary corrections is the most commonly used method.

Backward Scheduling

The project is scheduled backwards, starting from the requested end date: when do we need to start to achieve the scheduled goal? If the start date would be in the past, the time up to the planned date may be broken down in relation to the processing times that are ideally needed, and this used as the specification.

16.7 In Which Phase Does the Planning Take Place?

The planning does not represent a phase on its own. Rather it is an accompanying process that is carried out at the earliest possible stage as soon as the WHAT and the HOW are known. Once the WHAT and the HOW are known, the HOW MANY and with WHOM may be defined.

The effort is roughly estimated at a very early stage. Further detail is added on an ongoing basis and the accuracy is improved. The "Initialisation" phase is a rough estimate with a correspondingly greater tolerance, which may be +100/−50 % or even more, depending on the type of project. In the preliminary study phase, an rough plan is created on the basis of the milestone plan and the project structure plan. The tolerance of this rough plan is reduced, e.g. +40/−20 %. The detailed planning should aim to be accurate to +/−10 %.

If there is experience with similar projects (sequence, problems) and the approaches or concepts for a solution are largely known, the entire project will be planned in detail very early, e.g. during the preliminary study phase or even the initialisation phase.

If the approaches are new and require comprehensive solution concepts to be devised during the concept phase, rough planning is carried out before the end of the preliminary study and the detailed planning waits until the solution concept has been selected, before the concept phase is concluded.

If a binding bid is given to a customer, e.g. for fixed-price offers, the detailed planning must be carried out before the bid is submitted in order to determine whether it is possible to deliver the promised services, with the available resources, by the promised deadline.

16.8 Who Will Plan?

The project manager is responsible for the planning (procedures, methods, tools, checking plausibility). He will include the project team in the scheduling process (demarcation between work packages, estimating costs) as a mental test:

1. The whole project team working together has more of the necessary interdisciplinary technical knowledge than the project manager, who is a generalist.
2. The sub-project managers will be better informed if they are involved in the planning. This will also increase their motivation and willingness to accept responsibility for their activities.

16.9 How Detailed Should the Planning Be?

The level of detail will be determined by the purpose. It must be detailed enough so that the necessary actions can be derived from it and the project can be reliably monitored and controlled. The planning must be at least detailed enough to allow subsequent controlling. The project manager will distinguish between 10 and around 200 work packages, depending on the project's size and complexity. If the checks are to be carried out at shorter intervals to allow problems to be tackled at an early stage, then the tasks must be broken down with even greater resolution. The quality (accuracy and completeness) of the data is more important than the level of detail. The level of detail should not be greater than necessary as the effort involved in planning and controlling will quickly become very great.

16.10 Planning Large and Small Projects

The methods and level of detail for the planning should be matched to the project's complexity. For large and complex projects, those that break new ground or are crucial for the company's survival, it is revealing that all the measures involved in good project management are necessary and helpful for the successful conclusion of the project. The basic considerations for small projects are the same, albeit in a reduced and simplified form. For smaller, manageable projects, particularly where similar projects have been completed in the past and for which there is broad experience, it is not always necessary to apply every measure to the same extent and level of detail.

For a large project with many dependencies and risks, scheduling with a network diagram is a good idea. The project is divided into perhaps 5, 6 or 7 phases and multiple reviews are carried out. A small, routine project may have 2 or 3 phases and the schedule will be created directly as a Gantt chart.

16.11 Effort Estimation

The effort estimation is the basis for calculating how long the project will last (scheduling) and the project costs. The company needs to know whether the project should be implemented, whether the resources are available within the intended time frame, whether the project is cost-effective or whether investment in various other options would be the better alternative for safeguarding the company's future. The quality of the decision therefore depends on the accuracy of this estimation of the effort required. For companies that have to give binding price quotes to their customers (in the engineering and architecture fields, for example), a very accurate effort estimation is both necessary and crucial to avoid losses, since in many sectors the market does not allow for large reserves.
This makes the effort estimation and the associated uncertainty all the more important. Drawing up a meaningful estimate needs experience, attention to detail and preparation.

Basic approaches for estimating the effort required are:

- Based on experience and analogous situations by comparing work packages with similar tasks that have already been completed (empirical values from similar projects).
- Analytical: the task is broken down into individual manageable activities, and the effort required for each activity is estimated.
- A combination of the two approaches.

Many estimating methods are based on empirical values, which need to be built up systematically over a long period after analysing a large number of projects upon completion. The work packages used to do this must be clearly structured and demarcated. Analysis of the actual values obtained from completed projects provides adjusted estimates of the effort required and other influencing factors for the future. Possible influencing factors include the scope and complexity of the task, experience of the person performing the task, acceptance by the affected parties, available infrastructure, applicable regulations, etc. Only the most important influencing factors should be selected, and this key data applies only to the company and type of project for which it was analysed.
A company must select an estimating method and adapt it to conditions within that company. To allow the method to be applied to an entire organisational unit within the company, the definitions and demarcation between work packages and their contents must be defined, interpreted and handled in a uniform manner. Empirical values cannot simply be accepted from other units without critical examination (Fig. 16.28).

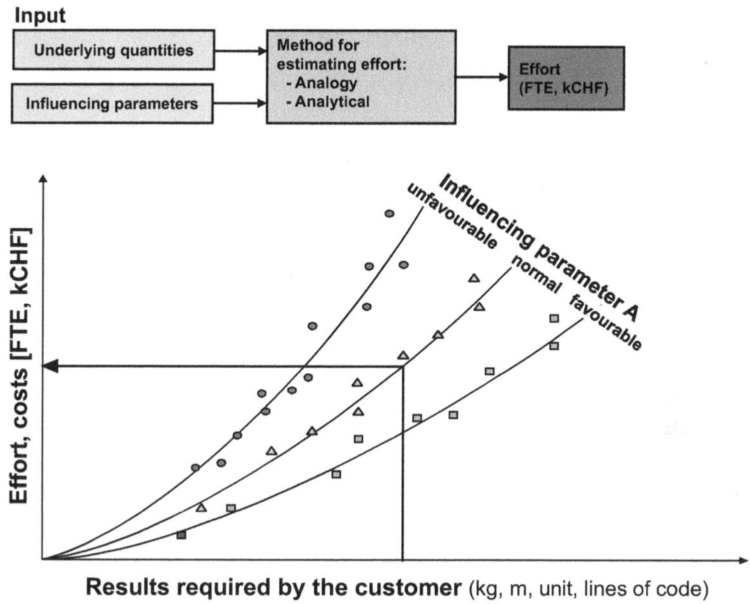

Fig. 16.28 Effort estimation with empirical values or key data

In many methods, an attempt is made to correlate one or more variables (e.g. result variables, customer requirements) in the project with the time required for them in FTE months. The correlation may be represented as a formula or graph. Examples of variables: Performance or accuracy of a system, volume of a structure or number of processing steps for software. The customer's requirements are used as independent variables. This method can be used in the early project phases (e.g. tendering) before the precise approach is known in detail.

Function Point Analysis

One proven and commonly used estimating method is function point analysis as developed by IBM for software projects. It is based on the analogy and weighting method. The method starts from the requirements from the users' viewpoint, without breaking the project down. The number of business transactions and their complexity, such as inputs, outputs, data functions, reference data and inquiries, are identified and analysed. This gives the number of function points. The function points multiplied by those influencing factors that have a decisive impact on the effort in this type of project give the weighted function points. Evaluating completed projects identifies the association between weighted function points and FTE months.

Multiplier Method

The task to be performed is broken down into small, manageable units where the effort required is known or has been tested on an example (number of modules, number of pages, number of graphics). The effort per unit multiplied by the number of units, summed across all types of unit, gives the total effort.

Percentage Method

This gives empirical values for the percentage shares of the total effort for each phase. The percentages per phase depend very much on the type of project. A phase is estimated in detail and implemented. When this phase is completed, conclusions are drawn from it for the entire project. Caution is needed if conclusions are to be drawn for the whole project from an initial small project phase involving little effort. The method is very suitable, however, as a plausibility test for checking estimated values that were derived in another way. The empirical values or key data must originate from projects that were carried out under comparable framework conditions (e.g. same infrastructure or corporate culture).

16.12 Planning for Great Uncertainty

The project manager may select another method, depending on the size of the risks and thus the variation in the effort of a project.
For pioneer or highly innovative projects that are being tackled for the first time or are suspected to be barely feasible, the project manager carries out an extended preliminary study with well-founded risk analyses. If necessary, he will also commission a feasibility study before starting a project. The project will not be started until it is more assessable and the risks are easier to detect. Risk management must be carried out if there are still significant risks.
If the uncertainty is reduced to two alternatives with greatly differing effort, both scenarios are included as alternatives in the planning. Example: if the goal can be achieved with the introduced methods, the effort will be 2 FTE months. After this time has elapsed, if it emerges that the goal cannot be reached in this way, a new method with an effort of 8 FTE months must be devised. Both scenarios are scheduled and a milestone is set as the latest point for deciding whether to abandon the first alternative (deadline).
If there is considerable uncertainty with respect to the effort estimation, questionnaires are sent to a panel of experts. The Delphi method questions multiple experts independently of one another.
The forecasts from the first questionnaire are sent to all those who submitted their own arguments and are still able to modify their estimates. The mean value is then used. Another useful procedure is the estimation session in which all the subject experts come together prepared and set out their own estimates and arguments. This

creates a welcome group dynamic within the process, which is moderated by the project manager.
For projects with considerable uncertainty (such as pioneer projects, acceptance projects and research projects), the variation can also be taken into account. In the three-point estimation technique, a most likely estimate of the time m (most frequent case) is estimated along with a minimum time estimate a (optimistic, best-case estimate) and a maximum time estimate b (pessimistic, worst-case estimate). The planning is continued with a single planned value:

$$\textbf{Planned value} = \mathbf{1/6\,(a + 4\,m + b)}$$

Even with medium-sized projects (number of work packages), the statistical adjustment favourably affects the deviation in the total effort, provided there are no systematic errors in the individual estimates. However, even with the most careful planning there are unforeseen events. Effort items are forgotten, minor changes occur or the effort required is underestimated. This is why a project needs reserves, which may be accounted for as cash reserves. It is the project manager who holds these project reserves. If a member of the project team is running late with one of his work packages and is unable to recover the time within his own work package, he must inform the project manager. The project manager checks how much of the project reserve has been used and can adapt his willingness to accept this in return for desirable changes to the situation.

16.13 Special Situations

Simultaneous or Concurrent Engineering

In many companies or dynamic markets it is absolutely essential to launch new products for sale in good time. The time to market is a critical factor in the success of such projects. Whoever is first on the market will make money and gain market share. This requires companies to drastically reduce the processing times for projects, which is possible by running activities in parallel as far as possible and developing products and processes at the same time. For this to be done successfully in practice, other measures must be carried out to accompany the project planning, controlling, communication and quality assurance. This technique is known as simultaneous engineering or concurrent engineering.
If two activities are parallelised, i.e. overlapped in terms of time, there is a risk that the effort involved will be wasted as the preconditions will change or a milestone decision will go a different way. The risk of having to do the work twice should be weighed up against the time saved by an early start. Activities should only be parallelised if the time saved is greater than the risk of time lost. It is useful to prepare for work or orders with high follow-on costs and then not to release them definitively until the milestone decision has been taken and approval given.

The greater the parallelisation, the more attention must be paid. Various accompanying measures are needed to prevent this technique turning into a stumbling block:

- More intense and improved communication between the persons responsible for work packages. Creation of a pure project organisation and release of staff to work 100 % on this project. This can be done by bringing together the subject experts from the various departments such as Marketing, Development, Production, Purchasing, etc. into a single room or close proximity.
- Accompanying quality assurance measures determined by the content of the specific project, the environment and the risks in the project. These measures will be defined in the quality planning.
- Increasing the level of detail when planning the work packages, at the same time clearly defining the criteria for when the work package is complete (doneness criteria), and checking the project at short intervals (e.g. short project meeting every week). If this check identifies delays or problems, remedial action should be sought and initiated immediately to avoid wasting valuable time (Fig. 16.29).

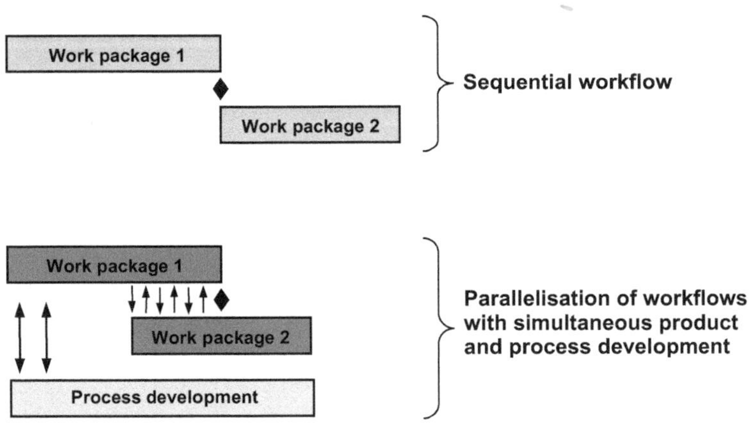

Fig. 16.29 Parallelisation of activities

16.14 Further Planning as Required

Target Costing, Design to Cost

In a price-sensitive environment, the costs of the result of the project, service (e.g. work sequences) or product (manufacturing costs) are more important than the "one-off" costs of the project. Eighty percent of the product costs for new products and services are determined in the definition and development phase. They can be changed very little in subsequent project phases. This gives the project manager and the developer a special responsibility for achieving the cost goals agreed in the list of goals.

A maximum value is set in the target cost definition. This places a ceiling on the product manufacturing costs (target costing) and is included as mandatory in the list of goals. The criteria to consider are:

- How much is the market prepared to pay for this service? How many units can we sell? Under which conditions?
- Where do our competitors stand (prices, market shares)?
- On what target costs and number of units is our profitability calculation based?
- What is our cost structure, with and without actions?

This target cost value is itemised against the individual partial systems, subassemblies or the posts involved in the service so that every participant knows the proportion of the product costs that may be attributable to his added value or partial system. The project manager monitors compliance with the manufacturing costs budget.

From the customer's and user's perspectives, the purchase cost is not the only important consideration. They will also look closely at the operating cost and the costs for repairs, maintenance, training, disposal and other on-costs that they will incur whilst using the solution. They are interested in low life cycle costs over the full lifetime of the investment. The project manager should choose to emphasise solutions that focus on low overall costs, provided this does not push up the production costs and project costs too far (or unless the customer agrees to cover these additional costs).

For products that are manufactured (duplicated), the project manager and developer should find the optimum ratio of project costs (one-off costs) to product costs that results in the lowest overall costs for the forecast number of units. Project costs are one-off costs. Manufacturing costs are incurred for every copy of the product that is made.

If the intention is to manufacture large quantities of a product because the market is able to absorb it, greater effort can be invested in development during the project so as to find more cost-effective solutions and further rationalise production. This will increase the project costs, however. In contrast, the proportion of the project costs per product will be higher for one-off systems or small quantities. In this case, the one-off costs of the project need to be kept low. These considerations apply just as well to services or work sequences. It helps to find the optimum ratio between the one-off costs of investment and the recurring costs of work. Examples: stock management and nursing care.

The break-even analysis helps to clearly show this association and to identify the effects if the planned number of units cannot be sold on the market (Fig. 16.30).

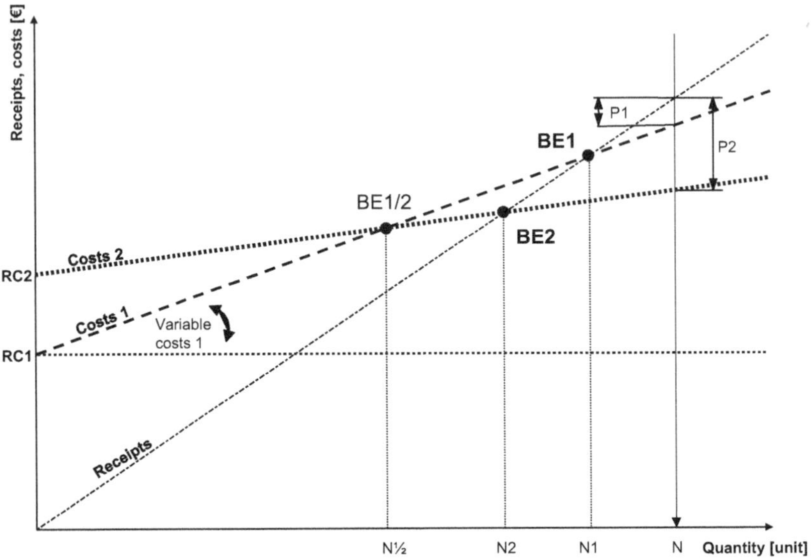

Fig. 16.30 Break-even analysis

Make-or-Buy Decisions

The current economic environment demands careful handling of resources. Greater diversification of demand, the need for greater flexibility and pressure to quickly adapt to market changes have made it essential to focus strengths on the core business.

A marketplace full of suppliers and service companies, technical progress that demands greater specialisation and increased standardisation are making it easier to cooperate with partners.

If the project manager has sufficient in-house resources available to him, then he will give preference to those resources. If resources are short, however, he will examine the options for outsourcing work packages. For highly specialised activities or those for which there is only sporadic demand, it may be sensible to essentially build long-term cooperation with another company.

Reasons for including external suppliers in projects:

- Greater flexibility and faster response in the event of market changes
- Everyone specialises in his strengths, bringing top results
- A partner with more experience can greatly increase efficiency
- Human and financial resources are freed up for investment in the core business
- There is no risk of underutilising specialist knowledge or special facilities that are only needed occasionally

- Insufficient in-house capacity for sporadic large-scale projects that need to reach the market quickly

Important Points
- Careful selection of suitable partners with the same level of quality
- The partner's market position and future prospects (risk)
- The partner's dependency relationships
- Comprehensive contractual arrangements
- Periodic evaluation by auditing
- Incorporation into own project management and support
- Active collaboration in the event of problems
- Dividing line for subcontracting: expertise and experience specific to the core business

Quality Planning

Quality planning means defining all the actions that are needed on the basis of regulations, good practice and the company's own quality requirements to ensure that the quality goals for the project or product are met. The quality plan lists all defined and cost-effective actions, the deadline for completion and who is responsible for them. In large-scale projects, these actions are set out in a standalone quality plan, while in small projects the information may be incorporated into another planning document.

Examples of Actions in the Quality Plan
- Testing the finished product, interim tests
- Trial in the operating environment, on the customer's premises, qualification test, conformity test, homologation, clinical trials, approval tests.
- Defining special quality goals, quality inspections
- Configuration management
- Design reviews, design verification, design validation
- Value engineering, value analysis, reliability analysis
- Special tests: integration test, safety tests
- Risk analyses, FMEA, sensitivity analysis
- Preparatory treatments (endurance test, burn-in, etc.)
- Environmental test, climate test, vibration test, drop test
- Corrosion test, chemical residues, flammability
- Electromagnetic compatibility, radiated interference
- Prototype, pilot series

For cost reasons, the actions that are stipulated and carried out must be needed for the matter at hand, financially justified, of proven effectiveness or requested and paid for by the customer.

16.15 Use of the Computer as a Planning Tool

The choice of tools must be suitable for the task. For tiny projects or simple sequential workflows it may be entirely sensible to carry out and document the planning with simple means such as pen and paper. Planning is often an iterative process, so simple and descriptive tools such as Post-It notes, adhesive tape, small cards or Lego blocks may be helpful for making the planning process flexible. Even with complex projects, these tools may be used in the first phase, e.g. at the start of a major planning exercise when everything is still unclear.
If projects are more complex, however, the use of computers and project planning software becomes inevitable.

Criteria for the Use of Computers for Project Planning

- Frequency of the projects
- Similarity of the projects
- Scope and complexity of the project
- Frequency of planning changes
- Own needs are covered by the software
- ICT experience

What Are the Benefits of Using IT Tools?

- Changes are easy to make
- Cross-project analyses
- Impressive presentations
- Need for rigorous planning and structuring
- Efficient use of standards

What Are the Disadvantages of Using IT Tools?

- Initial effort involved in their introduction
- Setting up a complete database
- Investment in training, hardware and software
- Pseudo-accuracy, interpretation errors
- Danger of overuse

For many important planning steps, the computer does not help substantially with the planning process, e.g. defining project goals, defining the project structure, defining activities and work packages and their boundaries, identifying dependencies, stipulating responsibilities, estimating effort, defining expertise and experience and resources. In these cases it merely helps administratively with breakdown functions, ease of correction or clear documentation.

The following main planning steps can be carried out using integrated project planning software. The cost/benefit ratio becomes less favourable from top to bottom:

- Scheduling
- Allocating resources and identifying resource overload
- Adjusting resources in the event of overloading
- Project progress monitoring

Integrated project planning software is just one tool among many, however. Under certain circumstances, simple and commonly available IT tools such as spreadsheets, databases for an overview of resources, word processing, e-mail, prioritisation programs, etc., can be very helpful or even entirely sufficient.
Within a company, project data is often recorded in other systems (e.g. actual values in the financial management software). It can be beneficial in some cases to be able to exchange or update data automatically. The benefits of greater integration and networking are often small in comparison to the disadvantages, however. The effort of entering data is in fact small compared to the effort of interpreting that data, communicating it, holding meetings or deriving actions from it (Fig. 16.31).

PM functions / Software category	Integrated PM software	Word processing	Document management	Spreadsheet	Graphics / presentation	Calendar management	Database management	Email	Performance accounting	Prioritisation programme
Project structure plan	■				□		□		□	■
Project running plan	■				□					□
Project organisation	■				□		□			
Deadline management	■			□		□	□			
Resource management	■			□			□		□	
Cost management	■			□	□		□		■	
Pending issues management				□		■	■			
Project applications / orders		■	□					□	□	
Kickoff / milestone files		■	□			□		□		
Project reports		■	□		□					
Presentations		■	□		■					
Correspondence		■	□					□		
Memos / logs		■	□					□		
Performance accounting	□			□			□		■	

Key:
■ = OK, suitable software available
□ = Can be used under certain conditions

Fig. 16.31 IT tools for various project management tasks

An important guiding principle for using computers in project planning is that the planning is only as good as the quality of the data.

Project Controlling 17

17.1 Overview

Project controlling has now become a concept of its own. It describes the "processes and rules that are used in project management to ensure that the project goals are achieved" (Fig. 17.1).
Nowadays, project controlling covers far more than just economic auditing of an initiative. Quality controlling is a key component of this role, along with the identification and assessment of possible risks (risk controlling). In an ideal situation, it is used as a tool for corporate planning or for strategic development (portfolio controlling).

Why Have Project Controlling?

Increasing cost pressures and increased competition force companies to use flexible tools to optimise decision-making and to improve transparency about the resources they are using. The results of project controlling should help the company's senior management to stay on top of the various projects. This is particularly important when projects are becoming more complex, and of greater significance, and when the environment includes constantly changing demands and conditions. In so doing, project controlling helps to secure the company's long-term competitiveness.
So project controlling is a necessary management task for the project manager, working with the project owner, steering committee, decision-makers or the company's board. Distinctions are made between the following terms, depending on the amount of external influence:

- Strategic controlling (by company management)
- Multi-project controlling (by the steering committee)
- Individual project controlling (by the project manager, steering committee)

J. Kuster et al., *Project Management Handbook*, Management for Professionals,
DOI 10.1007/978-3-662-45373-5_17

Project monitoring	Continuous monitoring that the goals are being met in terms of schedule, costs and quality.
Project assessment	At regular intervals, but at the very least at the end of each project phase, the project should be evaluated against predefined criteria, and the anticipated risks should be reassessed.
Project reporting	Reporting includes documentation and communication of the results achieved by the project to the relevant offices and decision-makers.
Project steering	Corrective measures should be drawn up on the basis of the results identified during project controlling.
Changes to the project	Document changes to the ongoing project (requirements, technology, market, etc.), and formulate and implement appropriate measures.

Fig. 17.1 Aspects of project controlling

17.2 Project Monitoring

The current project plan forms the basis for this task. The project must be checked (or controlled) at regular intervals to ensure that it is on track in terms of schedule and costs. The more complex or time-critical the project, the more frequent the controlling intervals should be.

Cost and Schedule Monitoring

For cost and schedule monitoring, a comparison against the following key data is normally sufficient. However, assembling the necessary up-to-date data is often fiddly and time-intensive (e.g. extracts from the reporting systems, statements from project staff, etc.). The planned data and the actual data must be available in compatible form.

- Planned and actual duration
- Planned and actual costs
- Degree of compliance as a percentage

The status of the project should be assessed at regular intervals, including cost control. The intervals should be appropriate to the overall project duration (weekly, monthly, etc.). However, it is not always easy to identify whether the costs incurred thus far match what has been delivered so far (Fig. 17.2).

Fig. 17.2 Assessing the status of the project and cost controlling

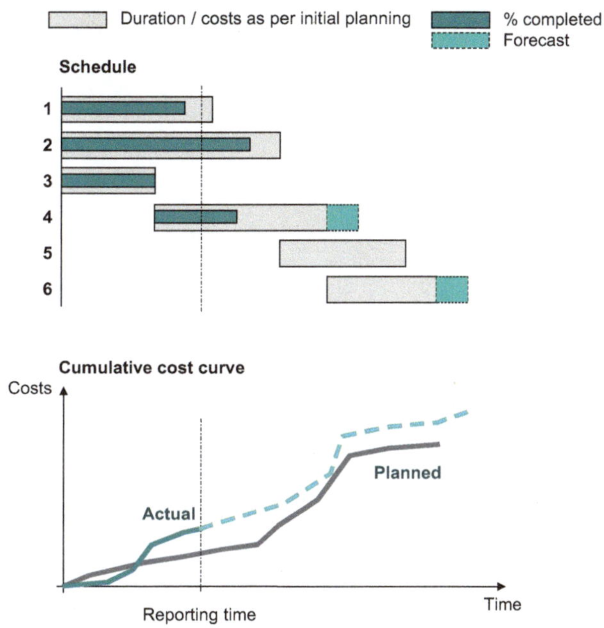

In general, effective cost controlling needs the following conditions to be met:

- Transparent cost planning
- Ready availability of the cost status
- Periodic review of the predicted final costs

If there are any significant changes to the goals during the course of the project, or if there are big differences between the planned values and the actual values, then the original plan should be revised. The remaining time (time-to-complete) and the anticipated remaining costs (cost-to-complete) must be determined for any remaining activities, or for any activities that are not yet fully completed (Fig. 17.3).

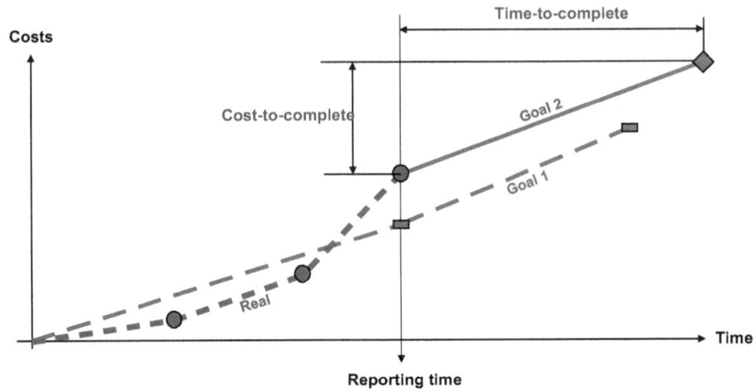

Fig. 17.3 Time-to-complete and cost-to-complete

Resource Monitoring

The use of personnel resources is difficult to plan, and even more difficult to implement. If no values based on past experience were available when the project started, or if the estimates of the resources needed were unrealistic or based on incorrect assumptions or requirements, then there may be big differences.
In practice, the problem is usually made more difficult by the planned resources not being available to start at the agreed time, or not being available for the full agreed length of time. Possible causes include:

- The company does not have an established project management philosophy.
- The project managers received only half-hearted support from senior management.
- Senior management do not have a clear overview of the resources they have agreed to provide, or do not stick to what they have agreed.
- More projects are started than resources are available for.
- There is no resource planning, or inadequate resource planning and coordination of the available resources.
- Projects are not centrally monitored and prioritised.

17.3 Project Assessment

In most cases, the most important sections (milestones) of a project are agreed in the project order. Milestones are ideally suited to project evaluation, because clearly delimited work packages have to be fully completed by each milestone.
Milestone meetings are a perfect opportunity for a critical review of how the project has gone that far, of how it should progress and of the maturity of the available intermediate results. The project manager must work together with the project owner or the steering committee to clearly answer the following questions:

- Will the project continue to achieve the project's goals?
- Does the project remain commercially viable?
- What risks might endanger the project (including measures)?
- Are the initial assumptions still valid, or have new framework conditions emerged?
- Are there any unresolved problems that prevent the project being continued?
- Has approval been given for the next project phase? What experience do they have? On what conditions?
- Should the project be abandoned?

At the end of the meeting a report should be drawn up with all conditions, lines of responsibility and schedules, and should be signed by both sides.

Feasibility

When a project first starts, there is limited knowledge of the project content and the possible solutions. But that knowledge increases as the project progresses. The risks are greatest at the start of the project. Reducing this risk as quickly as possible and as much as possible is a clear objective in any project.
If the requirements (goals) push the limits of what is possible, or if the possibilities are not precisely known (technological limits, politically sensitive goals), then it is useful to carry out a preliminary study (feasibility study) before starting the main project. At a limited cost, this approach offers targeted and detailed insight into the critical points that decide the project's feasibility (Fig. 17.4).

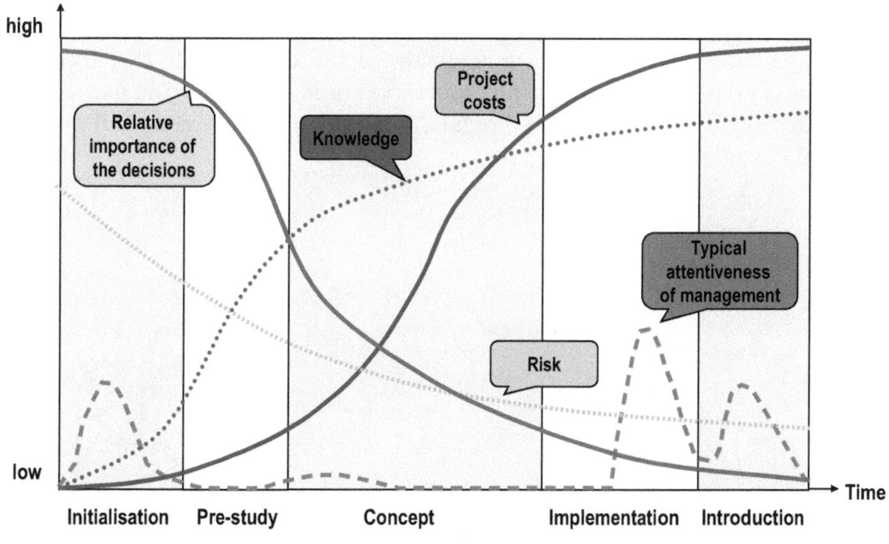

Fig. 17.4 Opportunities to influence the project

If it becomes clear that there are not enough current or available resources available to realistically achieve the goals, a decision to abandon the project can be taken when this milestone is reached. That avoids committing precious limited resources to a project that has no hope of success.

Cost Effectiveness

The project request provides the initial decision-making basis for approving the project. The decision-makers have to be guided by one very simple question: "What will this project cost, and what benefits will it offer?" It is important to keep coming back to this question at regular intervals throughout the entire lifetime of the project. If the answer is no longer a clear "yes", the option of abandoning the project should be discussed, and a conscious decision about whether to continue should be made.
A simple cost-benefit analysis is often used when calculating the cost effectiveness of an initiative. For complex projects, classical processes and figures from capital budgeting are often used, e.g.:

- Profitability calculation
- Return on investment (RoI)
- Net present value method
- Break-even analysis
- Dynamic payback method

Risk Management

A number of risks, which cannot be anticipated, might occur in any project. So there needs to be some way of assessing potential project risks. New risks can crop up in each project phase, and appropriate measures will need to be drawn up in response. This risk process must be reviewed regularly throughout the project, and must be constantly expanded (Fig. 17.5).

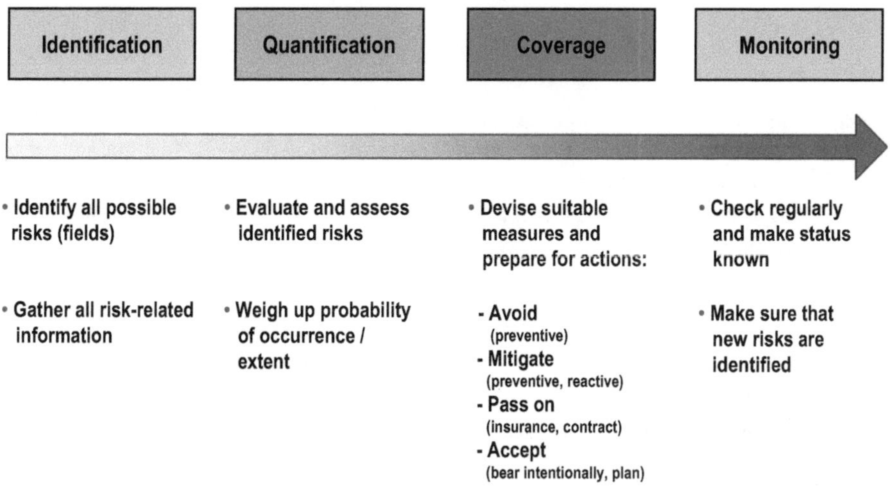

Fig. 17.5 The risk process

1. Analysing the risks

A range of different tools are available to help identify potential risks, depending on the project phase. An analysis of the project environment (e.g., using a context diagram) is often enough to give an initial rough risk analysis. The key thing is to identify the most important risk categories. Examples of risk categories include:

- Methodological risks (complexity, approach)
- Technological risks (new products, material characteristics, etc.)
- Commercial risks (cost ceiling, creditworthiness of business partners, etc.
- Personnel risks (sickness, resignations, etc.)
- Political risks (changes in corporate strategy, legislation)
- Competitive risks and market risks (competitor's product is better or cheaper)
- Legal risks (product liability, contracts, etc.)
- Environment risks (politics, strategy, weather, etc.)

The outcome of a project (the product or service) can also represent a risk for the company, the product risk. Bringing a newly developed product onto the market can be a considerable financial risk (product liability, image). Since resources and tools are finite, it is not possible to take every conceivable measure to head off the full range of possible effects or possible material or non-material damages.

2. Quantifying the risks

When it comes to defining the extent of the risks, there are both qualitative and quantitative methods available. This section seeks mainly to describe the "extent" of the risk. The following two variables enable us to describe the risk in quantitative terms:

Risk = P × I	P = Probability of the risk occurring (in %)
	I = Maximum impact (damage or loss) of the risk (€)

A sensitivity analysis can be used to show the influence of each individual parameter (e.g. observation period, interest rates, annual earnings, etc.) on the factors that influence the decision-making. This can help to show up any sensitive assumptions or critical parameters, so that they can be closely monitored during implementation. A simulation enables several parameters to be played out using different scenarios (Fig. 17.6).

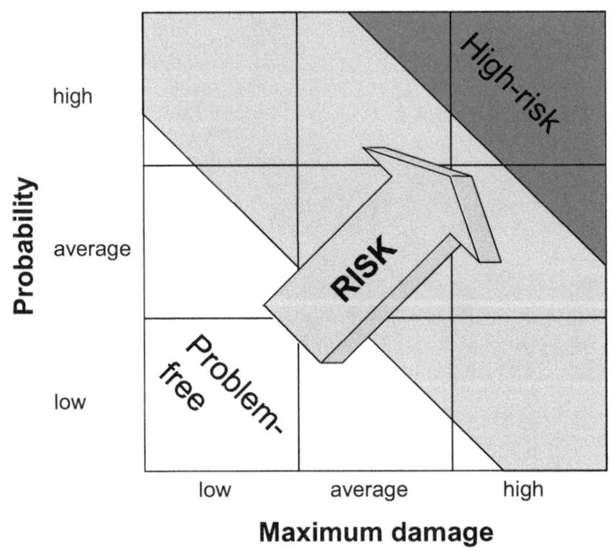

Fig. 17.6 Identifying the extent of risks

One method that is based on the underlying concepts of sensitivity analysis and which is used by a lot of companies in many different sectors is Failure Mode and Effects Analysis, or FMEA. In some sectors, its use is compulsory for all new systems and newly designed components, including those from subcontractors (automotive, medical technology) (Fig. 17.7).

Risks	Assessment before measure			Actions		Assessment after measure			Decision
	P	S				P	S		
Project owner's changed goals	7	6	42	Clearly word orders	Establish changeover procedure	3	5	15	yes
Supplier failure	3	9	27	Contractual coverage	Include alternative suppliers	2	5	10	Prevent-ive only

Key: P = Probability of occurrence
S = Scope of consequences

Fig. 17.7 Example of a risk assessment

Other sectors adapt this method to suit their needs, such as in the foodstuffs sector (HACCP, Hazard Analysis and Critical Control Points).

3. Countering the risks

To ensure that a project can still be carried out, or to make the risk "bearable" for the company, targeted preventive measures should be developed in order to reduce the risk to a reasonable level. The following questions may be helpful when developing risk management strategies:

Avoid	How can a risk be avoided or eliminated? Develop solutions that do not contain the risk
Reduce	How can a risk be reduced or minimised? Identify potential for improvement, explore alternative scenarios (e.g. second supplier)
Pass on	How can the risk be passed on to other parties? Agree changes to the contract (e.g. limiting the warranty), insurance, penalty clauses
Accept	Is there a way to simply bear the risk? Build a budget for it into the project plan, make provisions, document and communicate (customer, decision-makers)

But there are some risks that the project manager cannot directly influence, such as external changes or events that influence the project. Depending on how well or poorly the project manager can predict such events, and on how quickly or slowly he can react to the new situation, the project manager can implement various different strategies (Fig. 17.8):

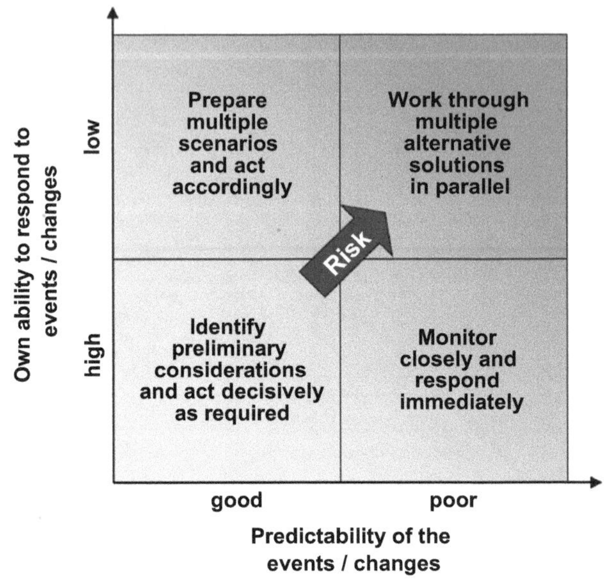

Fig. 17.8 Risk management strategies

- Observe carefully and act quickly, e.g. temporarily introduce shorter control cycles, review critical path.
- Prepare scenarios for alternative courses of action in good time, and keep them ready so they can be implemented without delay if the event occurs.
- Draw up several alternative solutions in parallel, so that a replacement is always available (redundancy). As the project progresses, the cost increases significantly, so this is only justified where there are major consequences.

17.4 Project Reporting

Reporting to all stakeholder groups is one of a project manager's most important duties and one of the least popular. A regular exchange of interim results with the relevant offices and decision-makers helps make sure that management are aware of the project, and that the project manager can count on their quick support if problems occur. Reporting is often (and incorrectly) seen by many project managers as "jumping through hoops", whereas in actual fact it is an ideal instrument for keeping project owners and decision-making bodies involved with the project.

It should be possible to submit a status report (progress report) containing the following information to the decision-makers (customer, project owner, steering committee, etc.) at regular intervals (often monthly):

- Which work or work packages have been started or completed?
- A comparison of planned and actual progress in terms of time, costs and resources.

- Can the remaining milestones be achieved as planned, including results?
- What problems have arisen since the last status report?
- What measures have been taken? Who is solving the problem? And by when?
- What new risks have been identified since the last report?
- Where is management support needed?

17.5 Project Steering

A project must be managed in a flexible manner that is appropriate to the circumstances. Each project is different, and the range of problems encountered is so wide and varied that there is no single magic formula for managing projects. Measures for managing and guiding the project should be drawn up on the basis of the results identified during project controlling.

The project manager's responsibilities include:

- Updating and monitoring the agreed key project figures
- Updating the project plan to achieve the original project goals
- Drawing up alternative scenarios if the project strays away from the project plan
- Requesting additional resources and financial support
- Changing what the project team members are working on
- Placing sub-contracts
- Negotiating with the project owner
- Implementing the steering committee's decisions
- Initiating project reviews and project audits
- Seeking authorisation to abandon a project

The steering committee's responsibilities include:

- Authorising or refusing additional financial support
- Authorising and implementing additional resources (from the line organisation)
- Giving approval for the next project phase
- Taking decisions about abandoning a project
- Changing the project priorities

The Magic Triangle

The magic triangle is used in project management to show the interrelationships between scope (or project goals), costs and time. None of these factors can be changed without affecting the other two. These are also the key variables, which need to be discussed in detail and agreed with the project owner or client (Fig. 17.9).

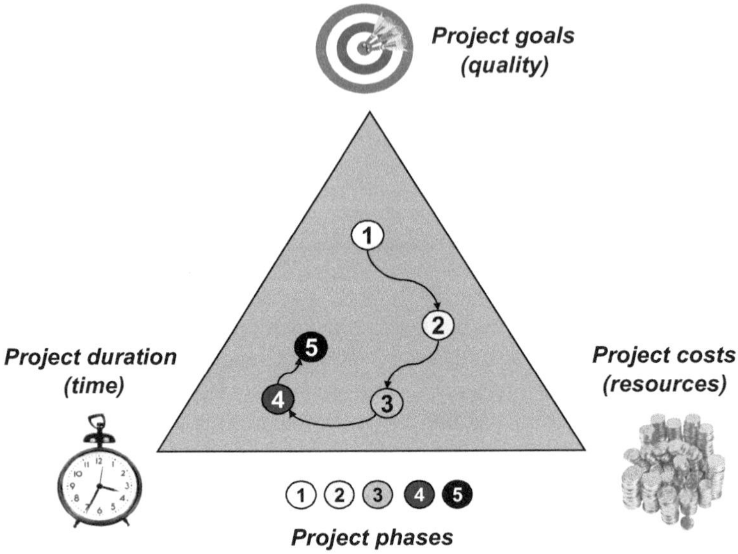

Fig. 17.9 The magic triangle

- Project goals: The required results of the project: scope, quality, functionality, ease of use, service levels, etc.
- Project schedule: The required time frame for achieving the agreed project goals.
- Project costs: The maximum costs including labour and other resources that can be used for the project.

These three constraints are shown at the three corners of an equilateral triangle, forming a symbol that represents project management. The triangle is intended to show that a project's goals, time or cost cannot be changed in isolation. Sometimes, a meaning is given to each connecting line: The line between goals and costs represents profitability, the line between goals and time represents effectiveness, and the line between costs and time represents productivity.
It is also important to get the project owner to clarify how important each of these three variables are to them. Different strategies can lead to different weightings. For example:

Best-in-Market
Quality and functionality have top priority here. The attractiveness of the product is intended to appeal to as many new customers as possible.

Time-to-Market
Speed is of the essence here, so the top priority is to minimise the duration of the project in order to get on the market as quickly as possible (e.g. product development). Cost and quality are less important in this case.

Design-to-Cost
The top priority here is meeting the target cost, in other words, how much the result that is developed will cost. The project manager will break the target cost down into smaller units, and monitor these individual cost elements. Also known as target costing.

The 90 % Syndrome

Relatively soon after the project gets properly underway, some of the people involved in the project will believe that most of the project work is completed (90 %). This phenomenon occurs because some achievable solution ideas can be seen by that stage. However, this view underestimates the obstacles and the as yet unknown problems that are bound to crop up during the implementation phase. Progress monitoring in a project is typically viewed from an optimistic perspective during the first half of the project, meaning that a massive correction is needed during the second half (Fig. 17.10).

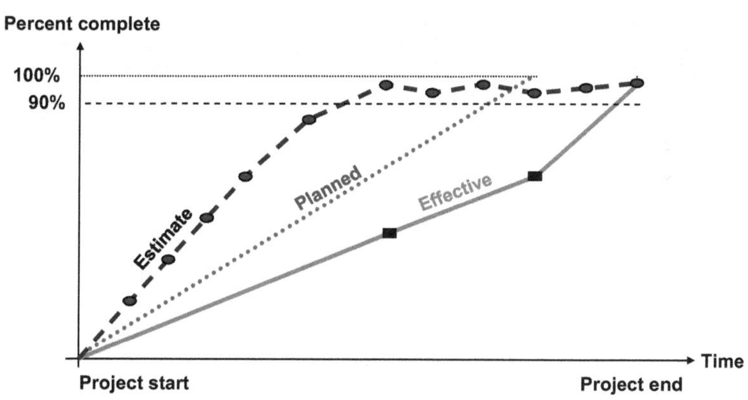

Fig. 17.10 The 90 % syndrome

To avoid the 90 % syndrome, active control methods are needed. The most reliable predictions are obtained using the 0/100 method, under which no credit is earned for an element of work until it is finished More complex processes need a metric for monitoring the progress of ongoing delivered results in relation to the overall planned results (e.g. the 50/50 method).

Preventive measures that support efficient management of projects include (Fig. 17.11):

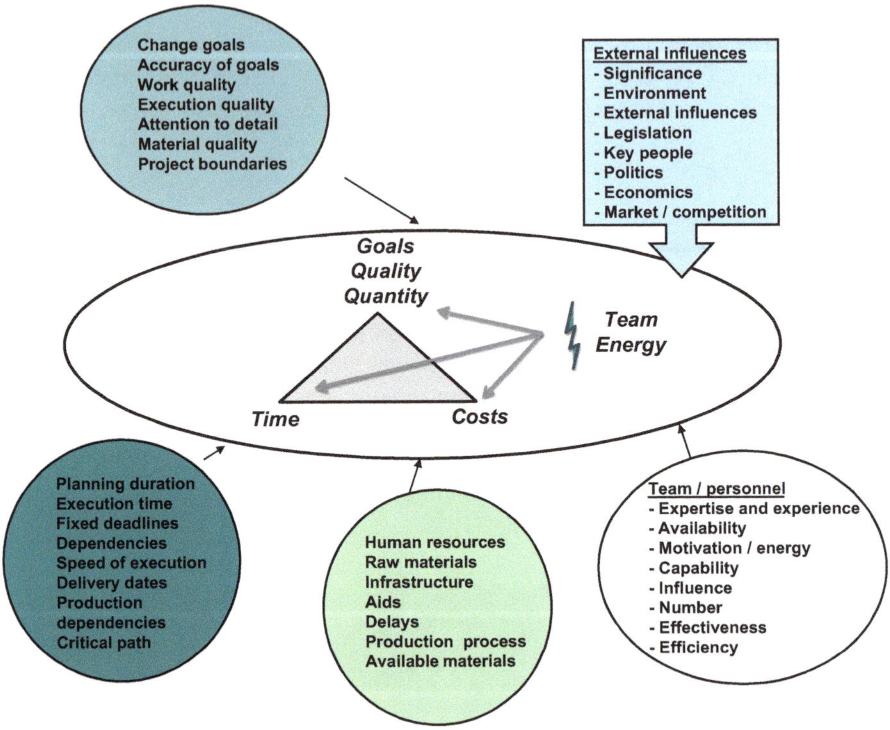

Fig. 17.11 Internal and external influences on managing a project

- Clear, measurable goals.
- Responsive rolling project planning.
- Periodic coordination meetings for the project team allow the current situation to be identified, anticipated problems to be discussed, preventive measures to be agreed and decisions to be taken.
- Any additions or changes to the detailed goals must be confirmed in writing by the project owner immediately.
- The client or project owner must be consistent in how they take key decisions.
- Periodic team meetings.
- Contextual or situational analysis.
- Reviews and milestone meetings.

Cost Transparency and Realistic Assessment of the Project's Economic Situation

In order for project management measures to work well and to be appropriately structured, an accurate assessment of the project situation is essential. As well as progress made in the project and the use of resources, the cost situation must also be assessed on an up to date, complete and realistic basis. This is important both for planning and for implementation (Fig. 17.12).

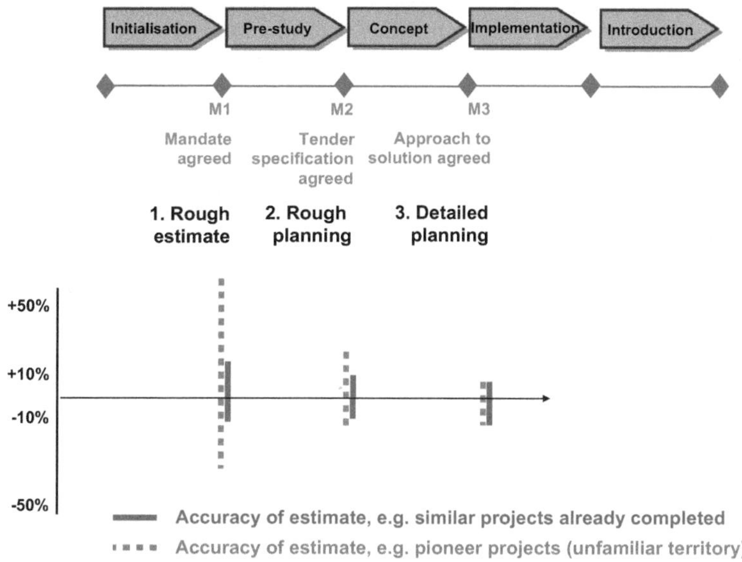

Fig. 17.12 Accuracy of estimates

Full cost accounting should be used for planning and for project requests, i.e. both internal and external costs must be shown. As well as listing the direct project costs, the follow-on costs that could be incurred as a result of this project should also be shown (operation, follow-on projects, maintenance, decommissioning). The project involves one-off costs, and then later after it has been introduced it involves operating costs. Provided the client or project owner agrees, it is better to use life cycle costing (the sum of the one-off costs and the operating costs) as the assessment criteria. A break-even analysis can be used for optimisation.
At the start of the project, or in the case of a fixed-price quote, the project costs are planned to the best of the teams knowledge and belief. Under certain market conditions, or if there is spare and unused capacity, senior management may sign an order at a sale price that doesn't actually cover the costs. That is a commercial decision taken at company level. The project manager's responsibility is still measured by the planning values. Any plan contains elements of uncertainty, such as changing external influences impacting on the result, or management decisions being delayed. The project manager sets out the situation as realistically as possible, outlines the alternatives by way of scenarios, quantifies the degree of planning uncertainty and shows the consequences if a decision is not taken as planned.
For planning and controlling, all costs and cost types should be listed and separated out: Work for different work packages, investments, external orders, modification costs, work done on a goodwill basis, etc. It is important to define which cost types will need figures producing at the end of the project in order to develop values based on past experience, how much was planned for and how much was actually used.

Controlling can only go into as much detail as the planning. So even at the planning stage, the project manager should think about what he needs to check in detail. The work breakdown structure during controlling must match the work breakdown structure used during initial planning. If subsequence changes mean that the discrepancies are too large, then new structuring and new planning will be necessary. The actual values that should be recorded must be entered in a financial tracking system at the start of the project.

The values used during initial planning are "planned values", and the values used during project controlling are "actual values". The important thing here for the company's senior managers is not primarily the arithmetic difference between the two, but the "predicted values" estimated by the subject experts at the time and in the light of the actual project situation. If discrepancies from the expected values are identified, the responsible people should inform the project owner or senior management immediately.

The actual values should be recorded as realistically as possible. Orders that have been placed should be recorded from that time and follow-on costs of decisions that have been made should also be taken into account. Changes during the project are inevitable. The project manager should ensure that any additional expense and costs brought about by changes are clearly identified so that they can be used to show the project owner or during negotiations with customers and suppliers (claim management).

When a project is finished, the project team is disbanded. Whether or not the saving objectives are actually realised after implementation will usually not be clear until later. So it is important to agree a date when the project manager will assess the results and also undertake a final costing analysis and communicate the results to the project owner and the former team members.

17.6 Changes to the Project

Projects exist in a heavily interlinked and dynamic environment. Events, which can have major effects on the project, can occur at any time.

But it is very difficult to control external events or influences such as sudden changes in the market, new legislation, new competitor products, etc. They usually occur suddenly, and without warning.

Events or influencing factors that are internal to the project can generally be anticipated, since the project is under the control of the project manager and the developments can be followed.

Any anticipated event, or event that has occurred, should be investigated by the project manager to check what effects it has on the project, especially in terms of its impact on the scheduling, performance and cost goals.

The impact on organisational and content changes is generally easier to sort out than the interpersonal aspects. Project managers and members of the project team need specific skills and expertise in order to address these aspects.

Configuration Management

In the case of complex products and some services, there may be many changes during their life cycle. Or there may be different versions and variants. In these cases, it is important that project managers are able to keep an overview of all the changes or product versions.
Problems that are frequently encountered include:

- Consistency problems: programmes, components, modules, etc. in different versions and different states.
- Change problems: Corrections to products or parts.
- Lifetime problems: changes to staff during the lifetime of the project and the product life cycle.

Configuration management covers all the technical, organisational and decision-making measures and structures that relate to the configuration of a product. It therefore forms a link between a company's product portfolio and its project portfolio, at least in the area of product development projects. Outside of project management, configuration management is generally considered to come under quality management.

Change Request Management

Change request management covers the organisation, management and implementation of change requests (project goals and project processes) whilst the project is ongoing. A clear distinction should be made between general change request management and systemic change management.

Project changes can occur due to:

- Customers' wishes, customer complaints
- Development errors
- Components or materials becoming unavailable
- Regulatory changes
- General product improvements
- Improved cost effectiveness

Regardless of how important or urgent they are (can they be implemented, are they necessary, what are the risks, what are the benefits, what is the cost), changes should be handled, summarised as a To Do list and then be processed together and approved as versions.
The subsequent costs must be bearable for the company. For this reason, change requests must also be assessed from a commercial perspective when decisions are made. For customer projects, clarify whether the change request is a goodwill gesture, or whether it should be treated as an additional requirement.

In many companies that have a high proportion of development work, the change request management process is clearly set out and must be followed. For change request management, the project manager has to go through the committee, which approves all major changes. If there is no change management process in place, the project manager should ensure that the decision-making body and the responsibilities are clearly defined. The relevant stakeholder groups should be represented on this team of experts. In practice, the project manager has the authority to decide whether a change request should be accepted as a goodwill gesture, or whether it should be chargeable work.

Change requests contain the following items:

- Type of change
- Reasons for the change request
- Work packages affected by the change
- The impact on project costs, duration and other parts of the project
- The effect of turning down a change request
- The impact on safety, new risks
- Links to earlier change requests that are now making themselves felt
- Body responsible for taking the decision
- Customer approval
- Placing of order (from design change request to design change order)

Claim Management

If authorising a change request causes additional costs, changes the project outcome or delays the completion date, the project partners who are disadvantaged by this generally make a claim against the party that caused the change. Handling these claims is part of claim management, which is positioned at the interface between change request management and contract administration.
Systematic monitoring and evaluation of discrepancies and changes, and of their financial consequences, is an important factor for the success of a project, and especially so for complex projects. Identifying the additional services and the related claims serves as a basis for compensation for them. It is up to the project manager to find an appropriate balance between the claims of stakeholders involved in the project, and the optimum progression of the project.
To achieve this, the warranty claims and time frames should be clearly set out in the contract, and acceptance of part performance should also be agreed to. The contract should therefore state which factors cannot be influenced and, consequently, for which no claims can be accepted. The whole subject of claims management is particularly delicate in projects, because during the start phase it is not normally possible to set out in crystal clear terms what the project should deliver.
There is an art to drafting the project contract so that the expected outcomes are defined so tightly that anything, which goes beyond, will count as an additional

service, and be paid for as such. On the other hand, the service procurer should be able to use the contract to demonstrate any deficiencies. This shows that the documentation should be suitable for providing any evidence that is needed in terms of the project work. Backed up by appropriate quality management, this enables defects to be identified quickly and unjustified claims to be limited.
But rather than relying on legal recourse, the interests of long-term cooperation and good image are better served by developing a relationship where the client and the contractor work together in a spirit of partnership.

17.7 Reviews as a Higher-Level Controlling Mechanism

Reviews are critical checks on the project, both while it is ongoing and afterwards. There are several different types of review (Fig. 17.13).

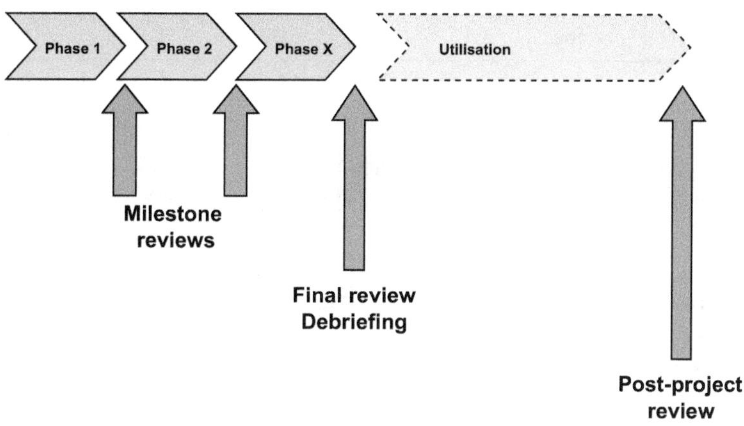

Fig. 17.13 Different types of project review

The reviews look at the intermediate results or overall results, and also at the processes used: general approach, the selected tools and methodology, cooperation, communication, organisational effectiveness, etc. (Fig. 17.14).

	Main aim	Likely participants
Milestone review	Management of the project process Assessing how well the phase goals (deadlines, finances) have been achieved, the methodology used, the project organisation, information, communication and teamwork, identifying consequences for future approach in the next phase.	Project manager, project team, steering committee, customers, possibly other stakeholder groups
End review	Assessing how well the project's goals have been achieved, critical review of the project process, identifying lessons learned, i.e. identifying measures for the organisation's future project management.	Project manager, project team, steering committee, internal project owner, external moderation
Post-project review	Assessing how well the long-term goals were achieved, and the effects of the project.	External team

Fig. 17.14 A project review's aims and participants

In general, the potential offered by reviews is not used anywhere near enough, either for ongoing phase controlling or for evaluating a complete project.

17.8 Coaching for Project Managers

Coaching is about supporting the project manager in terms of things to think about in the future such as planning or implementing the next steps, and for their conduct in critical situations or in dealing with conflict. But reflection is also useful in order to better understand our reactions, behaviour and outcomes from past situations, and to analyse them. This helps project managers in dealing with their own disappointments and anxieties, but more importantly, it helps them to manage future situations in better ways.
Using the three-level model (relationships, content and organisation), it is good if the project manager can reflect on whether the structures they set up and their management interventions:

- Encourage a spirit of trust and acceptance at the relationship level
- Facilitate knowledge-based and results-oriented ways of working at content level
- Provide order and effective structures at the organisational level

In coaching, the emphasis should therefore be on supporting the project manager so that he is able to correctly assess where the project is at, identify where the current risks are, and to develop scenarios to shape future processes so that he can lead the project proactively. Coaching a project manager is often extremely challenging, because it is very difficult to stick to the role of a coach in the strict sense. Rather, what the coachee often needs is someone with whom they can talk through strategy issues and reflect on their experience so far in the project, and discuss their planning for the next work stages, conflict management, someone who can suggest methods, someone they can vent their frustrations to, and someone who can help them stay focused on their role.

This wide range of requirements places high demands on the coach. The coach must therefore skilled in the following areas:

- Different theoretical approaches such as Systems Engineering, systemic thinking, simultaneous engineering, organisational development, change management
- Techniques such as time management, moderation, crisis intervention
- The dynamics of projects in organisations
- Familiarity with pressure situations brought about by the various different interest groups

The more experience the coach brings with them, the higher the acceptance on the part of the coachee. It also makes it all the more important to specify in the contract what consultancy or guidance the coachee is looking for (coaching in the sense of process consultancy of expert advice relating to content and methodology).

Coaching should be something entered into voluntarily. The coachee should actively want to be coached, and not view it as interference. When reviewing situations, it is not about making mistakes sound better, or about glossing over issues in order to show the project manager in a better light.

Coaching is not about judgements in the sense of a test or exam. It is an opportunity to take advantage of expert help in order to review the quality of your own work, and to ask yourself critical questions about it. It helps with identifying and qualifying risks, with understanding our own motivations and the motivation of others, in expanding our perception through external feedback, and above all, with developing alternatives for a wide range of problems that can be encountered when managing a project.

17.9 Crisis Management

Crises can occur in any project. A project crisis is a situation that prevents or limits project progress, and endangers the reaching of the project goals. A crisis is not a catastrophe. It does not need to lead to the project's failure. It can generally be managed.

A crisis can develop gradually, or it can occur unexpectedly. There are several warning signals or indicators of an impending crisis. These include, for example, mounting cost overruns, incomplete partial results, missing decisions, dwindling

motivation and lack of engagement by people involved in the project, calls for help from within the project team, etc. If handled properly, a crisis can even bring benefits for the ongoing project and for the project management team.
Fear or loss of face often result in a crisis being swept under the carpet, where they continue to simmer until they flare up again. By that stage, it is often very difficult to deal with the crisis without incurring substantial losses. A crisis is an exceptional situation. It needs special handling, and suitable organisation. There are various different phases to resolving a crisis (Fig. 17.15).

Recognise the crisis, crisis definition	Identify the crisis through project controlling. Decide if the situation counts as a crisis (possible with decision-makers).
Crisis organisation	Special organisation for crisis management: Project owner, senior management representatives, consultants, auditors, etc. Pay attention to information and communication
Take immediate steps if necessary	Damage limitation
Plan crisis management measures	Analyse the crisis situation, draw up crisis scenarios, plan appropriate measures, e.g. new resources, new contractual agreements, re-start the project, communication guidelines, etc.
End the crisis, review lessons learned	Consciously end the crisis, hand it back to the "normal" process, review the crisis and make clear notes of any important lessons learned from it

Fig. 17.15 Crisis management process

The key to early detection of a crisis that is starting to develop, and to initiating appropriate management measures, is careful and regular controlling. They key thing is to look for any developing trends. For example, if the mountain of unfinished items is growing bigger and bigger, or if the gap between deadlines and costs is constantly increasing, etc. It is also important to address any resistance as it begins to develop, and to deal with emerging demotivation at an early stage. Managers often wait too long, in the belief that the problem can still be solved, or that it will resolve itself.
One very important preventive measure is risk management. Generally speaking, it simply does not take place. And if it is carried out, then it is left to one side and not taken into account or updated during ongoing controlling activities. Thorough, comprehensive risk management is the very best form of crisis prevention.

Information, Communication and Documentation

18

Information and communication are always found in projects in one form or another. The question is whether they "happen" by chance or are intentionally designed. Allowing them to happen by chances involves high risks: The people involved in the project do not receive the information they need or when they need it. Documents are not filed, let alone found. Stakeholder groups feel excluded and offer resistance. And so on.

Failing to create an information and communication plan also opens up the door to abuse: information can be actively withheld, or rumours can be started in order to sabotage the project. And without a plan, very little time and money (for software, for example) is set aside for communication.

Information and communication are handled in different ways within the project and in the line organisation. Communication within the line hierarchy uses predefined reporting channels. Direct channels of information that are as short as possible are better in projects with smallish teams as they allow decisions to be taken quickly and flexibly, and improve coordination.

Information and communication in projects are still viewed as a necessary evil or simply overlooked altogether as the project management structure is unwieldy or the project team is occupied with content issues. However, project managers increasingly need to be aware that good information and communication are fundamental contributors to the project's success. Indeed they are essential (Fig. 18.1).

J. Kuster et al., *Project Management Handbook*, Management for Professionals,
DOI 10.1007/978-3-662-45373-5_18

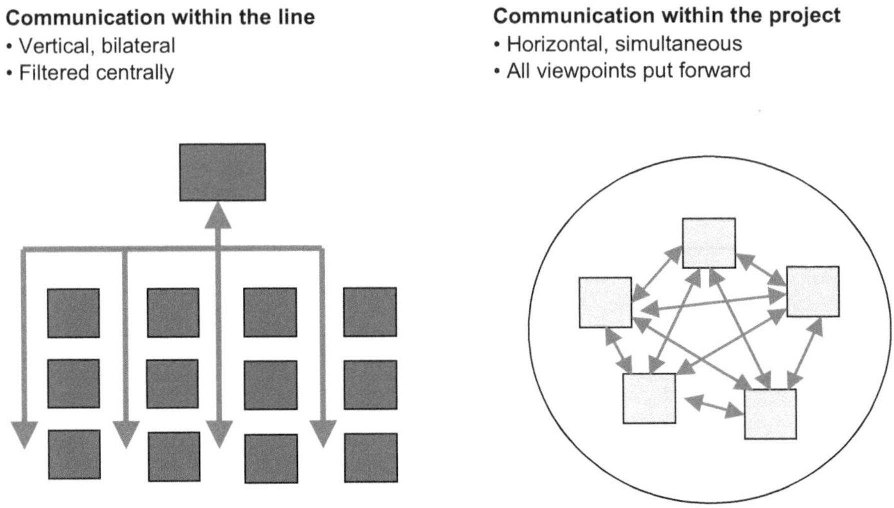

Fig. 18.1 Communication within the line and project

18.1 Goals of Information and Communication

We distinguish between: Information and communication directed inwards, i.e. within the circle of those involved in the project, and directed outwards, i.e. between project and users, customers, interest groups, the general public, etc.

- **Directed inwards**: goal-oriented action requires a high level of transparency and comprehensive information for the people involved in the project. The facts and documents, such as contracts, tender specification documents, schedules, etc., must be accessible to every team member.
- **Directed outwards**: it is vital to establish trust and obtain acceptance and support during the project. Rumours are replaced by facts, thus reducing the negative effects on the project.
- **Post-project phase**: operating information and project documentation often needs to be available during the utilisation phase.

18.2 Principles of Information and Communication

Information is normally insufficient or provided too late. Even in sensitive projects, the people who are likely to be affected notice very quickly that "something is going on". This often leads to rumours that have an unsettling effect, arouse vague fears and can greatly damage the project. For all these reasons, it is generally sensible to provide information as early as possible. Of course, it is not possible to

reveal any results at this stage, but it is possible to say what is going on, who is involved, what is the object, vision or goal of the project and when the first results can be expected.
The project is not a glasshouse: conceptual experiments have to be carried out and alternative solutions derived from them. In this respect, the project needs to have a specific, but delimited space and must offer protection, particularly in its development phase.
It is important within the project and within the team to agree both organisational and interpersonal rules for information and communication, e.g. at the kick-off meeting.
Information may also be provided about processes, and not simply the content and solutions. Indeed delays and problems can be made transparent to the outside world, but this demands a lot more confidence than when simply couching information in positive tones or not giving any information at all.
Incorrectly prepared information can do more harm than good. Recipients generally have different areas of interest, questions, understandings or ways of looking than the project participants who rarely make the effort to put themselves in the shoes of the customer, user, affected party, etc. It is not by chance that large projects often appoint communication officers who have a certain outsider's view and can better understand the language and interests of the recipients. Projects bring with them changes, fears, imaginings and frustration, which can often be "worked around" with open communication.

18.3 Scope of an Information and Communication System

The entire project information and communication system can be structured as follows:

- Verbal communication: e.g. discussions, meetings, problem-solving sessions and cooperation on the content of workshops
- Reporting: minutes, progress reports, revisions, etc.
- Project documentation: project handbook (folder), project-specific files
- Project marketing: e.g. lobbying, creating trust and acceptance
- Exchange of data and collaboration over the intranet or Internet

The components do not all need to carry the same weight within a project. In a research project, for example, the logbook for recording data on an ongoing basis may be important, while almost no marketing will be needed. Or in a change project, it is the verbal communication and marketing that are important. Greater significance will be placed on the documentation in an ICT project.

The following thoughts will help to review the aims of providing information:

- What will be more successful in the project if appropriate information is given?
- What would fail if this information were not given?
- Who could obstruct or delay the project if not given the necessary information?

The communication concept has to be practically reinvented for every new project, so it is worthwhile in standard projects to standardise the relevant reporting channels, checklists and documentation principles in project management guidelines and to provide the appropriate templates, as is already the case within the line organisation.
All the information and communication can be illustrated in a communication matrix (Fig. 18.2).

Characteristic Information type	Responsible for writing report	Recipients / Participants	Deadline / Frequency	Notes
Verbal information				
Project status (presentation)	Project manager	Steering group and relevant department	As appropriate	Minutes
Steering group meeting	Project manager	Steering group	Monthly	Minutes
Reviews	Project manager	Steering group and project manager	As appropriate	Minutes
Project meeting	Project manager	Project team	Weekly	Minutes
Written information				
Project status report	Project manager	Steering group	Monthly	
Intermediate report	Project manager	Steering group	Milestone	
Final report	Project manager	Steering group and relevant department	Projectend	
Work report	Project team	Project manager	Weekly	

Fig. 18.2 Communication matrix

The "W" questions may be helpful when drawing up the communication matrix:

Who is the sender, who provides the information? It can often make a considerable difference whether it is the project manager or the CEO who provides the information!
Who needs it for the project? Who are the recipients?
What is the object of the information? What goal is it working towards? What is the message?
When and at what intervals is the information provided? Will the information be announced?
How is the information given, in what form and with which media? For example, paper, e-mail, blackboard, in-house magazine, etc. What methods are used to pass on the information? How should feedback be gathered?
Where and within which framework should the information be passed on and discussed?
To what extent should it be accessible and serve as a working platform for project team members in different locations, e.g. website or intranet?

18.4 Making Communication Visible

Here the question is which social force field is used to network the project?

Even during the "preliminary study" phase, it is a good idea, at least for innovation projects, to get a picture of the environments that are of relevance for the project and the relationships between them. In other words, who would support or reject the project (if they were aware of it), and who could have what expectations and fears. This is known as a project environment analysis. Risks, information and communication strategies and organisational inclusions can be easily derived from it (Fig. 18.3).

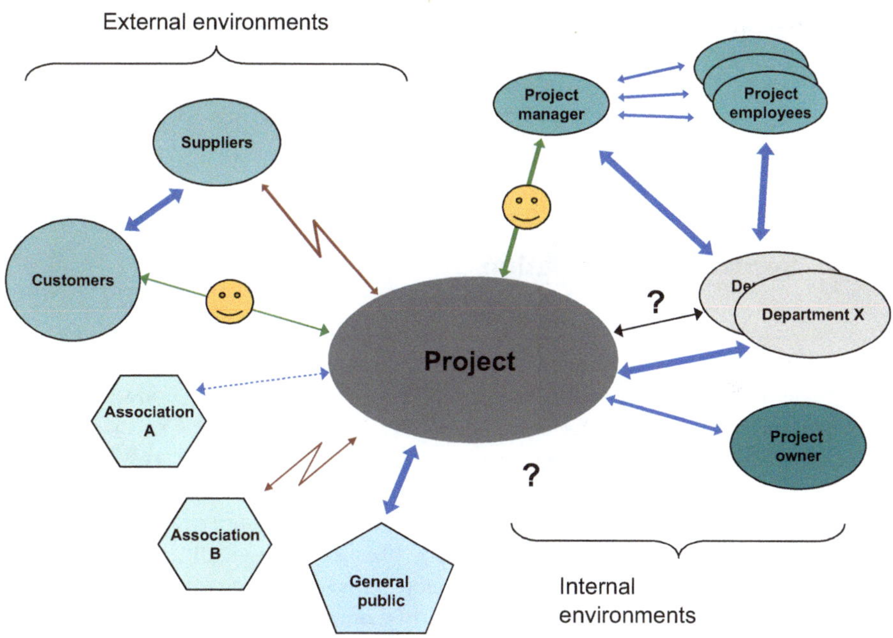

Fig. 18.3 Project environment analysis

A project environment analysis can be drawn up as follows:

- Determine internal and external stakeholder groups of relevance to the project, e.g. project owner, project manager, team members, specific departments, workers' council, customers, suppliers, associations, general public, etc.
- Illustrate the dependencies between the stakeholder groups and the project in a picture. Transverse relationships are important. The analysis should also show the quality of the relationships, e.g. the project owner and the development manager are at daggers drawn!

- Make assumptions about the possible interests, expectations or fears that the individual stakeholder groups might have, e.g. for whom is the project of fundamental importance?
- Who is authorised to influence the project? What are the feelings about it?
- Weigh up the risks and opportunities, e.g. who might offer the project conditional support only?
- Who has absolutely no interest in success or who will even be intentionally resistant?
- Who is officially or unofficially authorised to take decisions?

Measures for the project organisation, communication, risk analysis and for "relationship management", in particular, may be derived from this analysis, e.g.: Who should be informed and when? With whom does the project owner need to hold a one-off informal meeting?
And with respect to the "concept" and "implementation" phases might it be a good idea to suggest an information and communication plan as early as the preliminary study.

18.5 Verbal Communication

Verbal, i.e. face-to-face, communication is always the most effective method within any collaboration. It comprises words, images, non-verbal communication, immediate feedback and social integration. All other forms of communication, i.e. telephone, videoconference and e-mail, incorporate only some of these forms. At kick-off meetings, emotionally-charged situations, dispute resolutions or reflection sessions, these additional forms of communication are very important as they not only reduce misunderstandings, but are much more precise.
The disadvantage of verbal communication is that the parties all have to be in the same place at the same time. Standard projects with teams used to working together can thus put up with a greater proportion of communication by telephone or in writing than entirely new projects with intercultural teams or large geographical distances. Face-to-face communication can thus never be entirely replaced by electronic communication (Fig. 18.4).

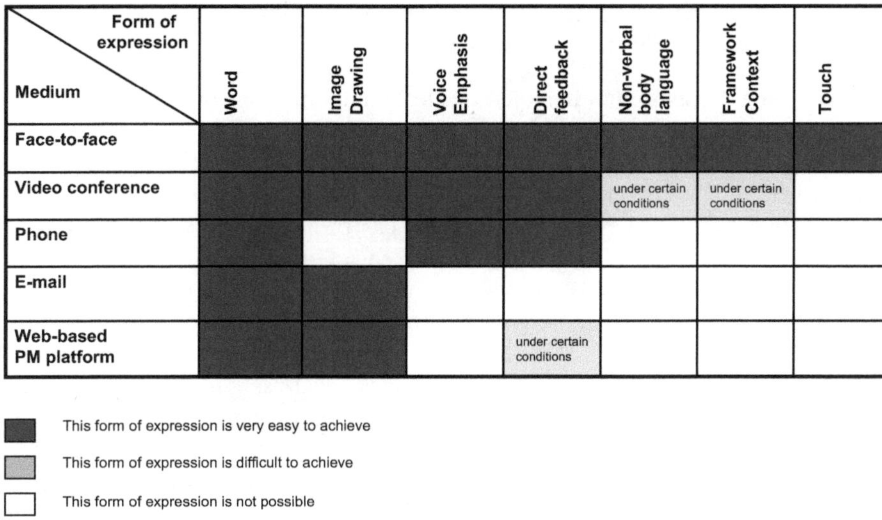

Fig. 18.4 Communication types and levels

In verbal communication we distinguish between formal and informal communication.

Formal Communication

- Workshops, i.e. any type of in-work workshops for the project kick-off, evaluating analyses, setting objectives, drawing up concepts, etc.
- Presentations
- Steering committee meetings
- Decision-making meetings
- Coordination meetings
- Reviews

Good preparation is essential for the success of all such meetings and workshops. Meetings and workshops are very time-consuming and expensive. They must therefore be set up and run effectively. Good preparation means, for example:

- Considering the goals for the event: what do we want to have achieved at the end of the meeting, workshop, etc.?
- Drafting a procedure and selecting methods. It is often worthwhile to consider and examine several alternatives.
- Selecting the suitable methods and structures
- Determining or suggesting a script and roles
- Making the infrastructure available (rooms, media, materials and catering)

Such a concept will lead to an invitation to the participants, possibly with preparatory tasks or advance information.

The following points should be noted when running meetings or workshops:

- Make sure that all the participants have agreed the goal, procedure, method and priorities
- Separate out from the event problems that can be clarified bilaterally
- Attempt to make specific decisions and agree actions
- The moderator should also pay attention to the process: collaboration, communication, addressing any unspoken conflicts

Post-meeting tasks must also not be neglected, such as sending out minutes, preparing workshop results, etc. For some meetings or workshops, it is a good idea to schedule these tasks at regular intervals. A weekly "project day" should be planned for project team meetings, for example, so that any problems or discrepancies will be quickly identified, allowing a fast response and coordination and maintaining commitment.

Informal Communication

This comprises any type of ad hoc meeting, whether between just two people, in groups, during a coffee break, etc. These meetings thus fulfil a very important function where informal relationships need to be established and maintained, where a problem needs to be quickly resolved or where information and feedback needs to be provided or obtained between formal meetings. The informal process of getting people on board and winning their support is also known as "lobbying", although it must not be confused with manipulation.
Informal communication is also possible between different teams, projects or project bodies. This free networking can even be encouraged as an essential element of an intentional communication culture.

18.6 Reporting

The written documentation of all project-related operations is a central issue in project information. "Reporting" attempts to document the available information as fully and comprehensively as possible, while the emphasis in verbal communication is placed on the current information.
Reporting serves as the basis for all assessment and control measures in the course of the project. A few basic rules need to be taken into account:

- Standardised structure for the various reports
- Adaptation of the information to suit the level at which it is disseminated
- No subjective wording

Reporting is often described in the project management guidelines, and appropriate forms and templates are made available.

Reports that are produced during the project process may include:

- Project request, project order, agreement
- Project process reports (goal-oriented reports)
- Minutes of meetings, memoranda
- Phase final report, project final report (result-oriented reports)
- Lists of problems and decisions

The progress report (status report) is used to provide a statement of the project's status and possible risks that is as current as possible, and to describe further developments. In standard projects it is worthwhile to make suitable templates available (electronic templates, checklists) (Fig. 18.5).

Recipients	The project owner The entire project organisation
Frequency	At specified times (e.g. monthly) At milestones, before reviews, etc.
Contents	Implementation progress, deadlines met or not met, cost situation Problems that have emerged Differences, unforeseen effects Need for a decision, support Important substantive results Further course of action, next steps Risk assessment for the planned process
Scope	Generally one or more pages

Fig. 18.5 "Progress report" template

The report may take a different format, depending on the purpose. The decision minutes from a meeting are one minimal format. Other options: Photo documentation (from workshops, for example), mind mapping, verbatim minutes (from political meetings), etc.
Here again standardisation is generally worthwhile in most projects as it allows the reader to quickly navigate the structure. A report should always contain the following elements:

- Attendance list, minute-taker
- Important discussion points, topics
- Resolutions and decisions

- Unresolved points
- What is to be done by whom and by when
- Date and time of next meeting

The minutes may always be taken by the same person or the task may be alternated. The project manager or moderator should not take the minutes as far as possible.

18.7 The Project Documentation

This is a project-specific repository for all documents produced during the project. It must be readily accessible to everyone involved in the project as a basic principle for the running of the project. Appropriate data should also remain available for future projects. Thus, the documentation makes it possible to compare projects, provides planning data and ensures that lessons are learned for further projects.
The project handbook or project folder contains the most important process rules, such as agreements, plans, structures, organisational charts and rules of the game. It should create transparency and be mandatory for everyone involved. The project handbook is kept right from the start, and the updated version is made accessible to the entire project organisation. The folder may be created on the intranet (project platform) or as a physical folder (Fig. 18.6).

Project handbook	
The project order and activity plan	Project order and project agreement Breakdown (project phases, project structure plan) Interfaces within the project Decision-making process
Project environment	Analysis of the environment and stakeholder groups Incorporation into the company's mission statement and corporate strategy
Project organisation	Organisational chart Description of the roles, tasks and responsibilities Contacts and addresses Ground rules for the collaboration
Project planning	Scheduling Resource planning Cost forecast
Controlling	Quality assurance Checking and control Action in response to deviations, change management
Information Communication	Information and communication concept Meeting planning Minutes of meetings, workshop reports Progress reports (may fall under controlling) Repository structures
Conclusion of the phase (of the project)	Close-off work Project evaluation Further course of action

Fig. 18.6 Possible content or breakdown of a project handbook

The handbook is broken down both into major topics (or work packages corresponding to the project structure plan) and by the type of document, e.g.:

- Process documents (minutes of meetings, memoranda, etc.)
- Procedural documents (project plan, etc.)
- Contracts
- Correspondence
- Results, "Lessons learned"
- Changes
- Other documents

The following recommendations will help to systematically and clearly compile a project's documentation:

- The individual documents should be arranged in chronological order
- Every document should be given a code that indicates, for example, the project, phase, structure plan item, work package, version and type of document
- It should be possible to clearly file every document within the documentation structure
- Every document should be recorded centrally
- Where possible and useful, documentation should be generated from standardised templates
- Change management procedures should be applied to revisions
- The documentation structure that is set up must not be changed without the project manager's agreement

For documents that are stored electronically, the access rights may be defined accordingly for the various project bodies.
The documentation should not be left until the last minute. It should be designed, set up and updated throughout the project. The project conclusion report is also needed at the end of the project. This sets out all the important experience gained from the project so that it can be analysed:

- Assessment as to whether goals were achieved, reasons for departures from goals, evaluation of the project's cost-effectiveness, final cost accounting, etc.
- Assessment of the collaboration, planning, choice of methods, etc.

The project logbook can be extremely helpful in projects where it is important that the detail be clear. The aim of this documentation is to clearly show, in chronological order, all the project activities, deliberations, decisions and agreements. It supplements the project documentation and is normally kept by the project manager in person, but should still be accessible to everyone. It may contain, for example, information sent and received (telephone calls, e-mails, contracts, etc.), ideas, drawings, possible solutions, decisions taken, agreements, etc.

18.8 Project Marketing

Project marketing comprises all the supporting activities that might positively affect acceptance and the course and progress of a project. If involves "selling" the project, which means:

- Communicating the meaning behind the project, endowing benefits, passing on own convictions
- Creating trust and acceptance

- Demonstrating fairness: also making possible disadvantages or problems transparent, taking fears and questions seriously and treating them accordingly
- Generating publicity, creating expectations that give the project "draw" or energy
- Exploiting resources and sales markets

Project marketing promotes the relationships between the project and its environment or stakeholder groups (Fig. 18.7).

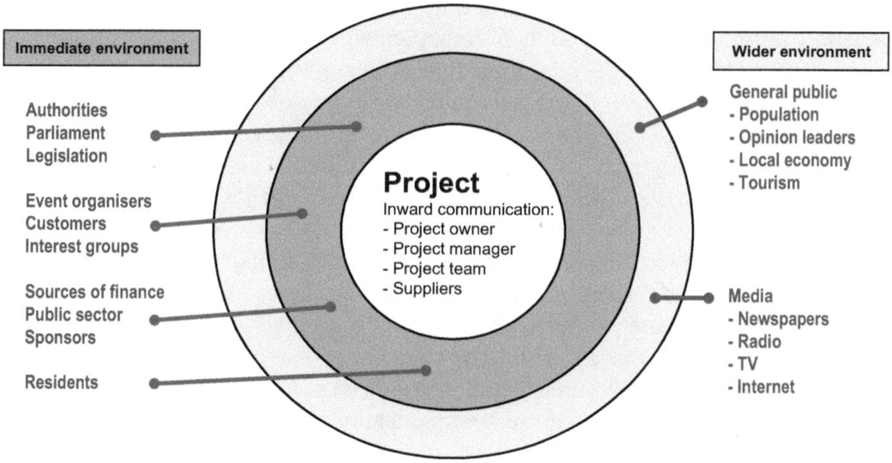

Fig. 18.7 Project marketing to stakeholder groups

Project marketing means shaping communication, and can be approached very creatively according to the situation and stakeholder groups, e.g.:

- Broad-brush information with the option of discussing the topic (e.g. open space event)
- Project magazine in which affected parties and users, i.e. not those involved in the project, can express their critical opinions
- Info mart: Information boards and stands with documents; people involved in the project justify themselves
- Creative workshop for interested parties. The results are recorded by the project team and worked on further.
- Visits, learning about the object of the project
- Using key people, promoters or decision-makers to pass on the information
- Speaking the language of the stakeholder groups, sensing what interests them

Project marketing often is not simply limited to "selling" the project. It sometimes also involves passing on management messages and values at the same time, e.g. to show that a new culture of collaboration is being initiated along the project.

The basic attitude when marketing is important: glossing over, deceiving, conning, fake partnerships will only be effective in the short term – if at all. It takes honesty, transparency and appreciation to create real trust.

18.9 Electronic Communication

The instruments and thus the options for electronic communication have developed enormously in recent years. The normal format for exchanging information, generally e-mail, has been gradually supplemented with new applications and instruments. These include platforms for computer-supported cooperative work (CSCW). But this generally unstructured information traffic often tends to become inflated. Not every piece of information is of relevance to everyone.
Managers are often covertly involved in e-mail traffic. It is a commonly held fallacy that a team can be "managed" via electronic media alone (known as "e-management"). Various tricks (such as "cc:" to a decision-maker) exert additional pressure on the recipients, or the sender is expecting the copied-in manager to respond. These detached actions promote unclear situations and represent a considerable potential for conflict.
To develop disruption-free and efficient communication within the project team it is very helpful to agree a few fundamental ground rules. An e-mail protocol is highly recommended since the majority of communication today uses this method:

- Meaningful subject lines (Information, Decision, Suggestion, etc.)
- Clear, informal salutation and closing
- Use correct spelling. All lower case is easy to type, but puts the recipient to unnecessary effort.
- Classify the recipients correctly (to:, cc:, bcc:)
- Keep text short, but make sure it clearly states what I as the recipient must do (supply information, process, take decision, about my information)
- Only use e-mails if it is efficient and effective to do so. In larger projects it should not replace direct contact (by phone or meetings)
- Be careful about using "emotional" wording (the keyboard does not pass on these emotions)

"Web 2.0" instruments such as blogs and wikis will certainly be used increasingly in project management. They support and encourage networking, i.e. dialog and exchange of experiences in order to jointly generate knowledge between the project participants. There are easy-to-use tools, similar to the existing social media such as Facebook. These are particularly suitable for change projects or ICT integration projects where interactive communication with the developers and users is particularly important. Blogs do not write themselves, however. They have to be designed, installed, communicated, maintained and controlled.

18.10 Project Management Portal

The idea of the project management portal (virtual project space) is not new, but the technical solutions have changed fundamentally and in quick succession over recent years. A comprehensive study on this topic was authored by Frederik Ahlemann.

Definition

Today, we generally understand this term to mean a browser-based application that allows all the different internal and external stakeholder groups of all the projects of a company, if possible, to have different levels of access to certain data, information, instruments and applications.
Introducing a project management portal into the company requires an intensive examination of the internal project processes, and also involves a non-negligible effort to develop, enter data and maintain the system. Such an instrument cannot wholly replace communication within the team. Quite the opposite, it demands a clear agreement on the rules for communication. This has the following significant advantages:

- All authorised stakeholders are directly included
- Reduced administration by the project manager
- Fewer potential errors due to faulty media
- Automation of standard procedures and processes
- All project data can be edited from any location
- Data and information about current projects are always up-to-date
- Standardised structure for storage, archiving and naming conventions
- Many different analysis methods to optimise the project processes

How It Works

A project management portal can be sketched out as follows (Fig. 18.8):

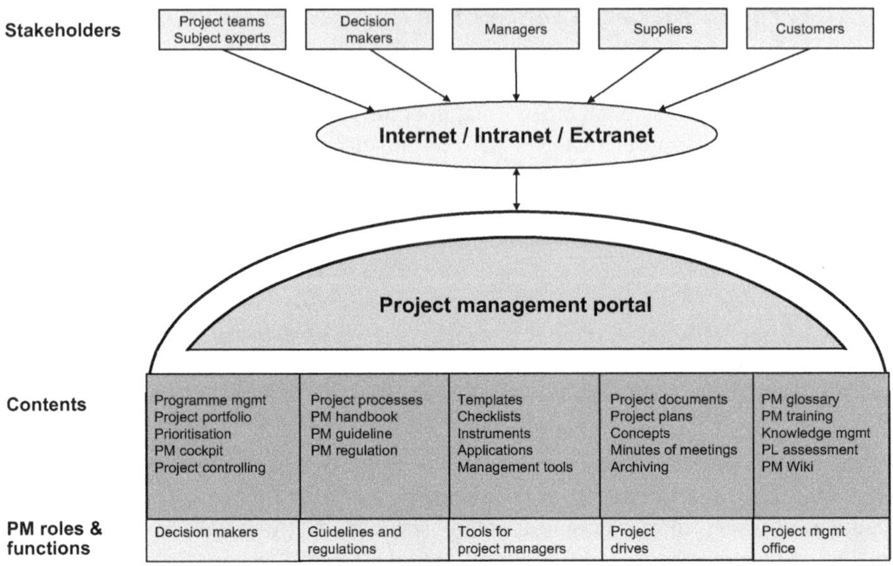

Fig. 18.8 Structure of a project management portal

Today the technical aspects of creating a project management portal can be purchased from a range of different products on the market. These products differ primarily in terms of their functionality and the interfaces they provide to other applications. A project management portal essential acts as a platform for:

- Collaboration within the project team (documents, workflows, PM processes)
- Representing the project status (results achieved, financial situation, identified risks)
- Exchanging experience between people involved in the project outside the project boundaries
- Communication with the company's internal and external world
- Creating and presenting the basic principles for management decisions (project cockpit)
- Setting up knowledge management for project management

18.11 Interpersonal Communication

What Is Communication?

By communication we mean the very complex mutual exchange of information between living things (people and social systems). If there is reciprocation, then we also call it "interaction". The important features of interpersonal communication are:

- Communication takes place via different "channels": language, images, body language, gestures, etc.
- Communication is always interpretation. What we perceive as relevant information, what we believe to be "true", is the result of a complicated selection from various possibilities (selective perception). The participants thus have different "truths".
- In communication, multiple messages are always sent or perceived at the same time. Some are open, while the others are indirect or hidden. The hidden message is often more important than the open, however.
- Communication is an interaction. A circle without start or end. Every reaction is also an action, i.e. a stimulus for a further reaction. There are always two sides involved.
- Communication does not only take place between the people who are present. Many absent parties are involved in the background in the "game of communication" as they affect the way those present think and respond (What would the steering committee say about that?).

Our Perception

What we perceive (or accept as true) is not objective. It is the result of highly complex processing carried out by a person or organisation. Every perception is selective and subjective. A few important phenomena in the perception process are:

- Perception selection as a protective function, e.g. error avoidance strategy out of fear that mistakes will be punished
- Tendency to create integrated wholes, order and sense: providing subjective additional information and interpretations
- Influencing perceptions by feelings, internal images, experiences: conflicts, e.g. narrow perception
- Personal (organisational) history: modes of behaviour or conditioning learned by experience
- Current situation

Models of Interpersonal Communication

Messages sent out by a person, team or organisation must express something using specific words: the message must be coded. This requires appropriate decoding by the recipients. This decoding essentially consists of four steps:

- Perception: what am I hearing and seeing?
- Interpretation: what does that mean?
- Feeling: what feeling does that trigger in me?
- Action: what does what I have heard prompt me to do?

The result of the decoding process then is what occurs in the recipient; it is the important information of relevance to him that he absorbs. Different people, groups, advisory boards and organisations listen and understand differently, give the same words different meanings and allow different feelings to develop. Misunderstandings are thus entirely predictable and normal (Fig. 18.9).

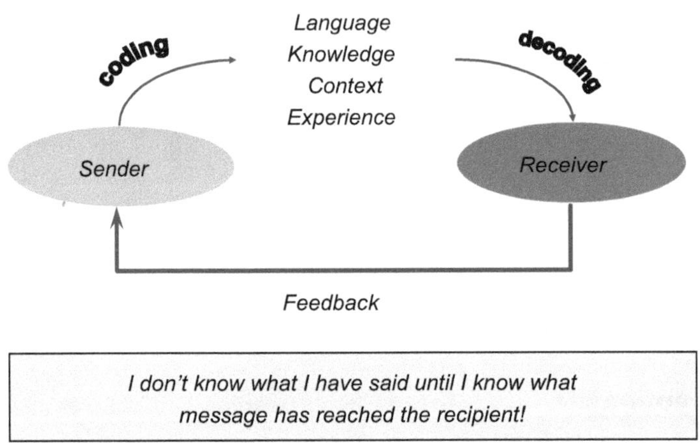

Fig. 18.9 Schematic representation of the communication process

A Communication Model

The known model "Anatomy of a message" by F. Schulz von Thun assumes that every message contains various sides or aspects (Fig. 18.10).

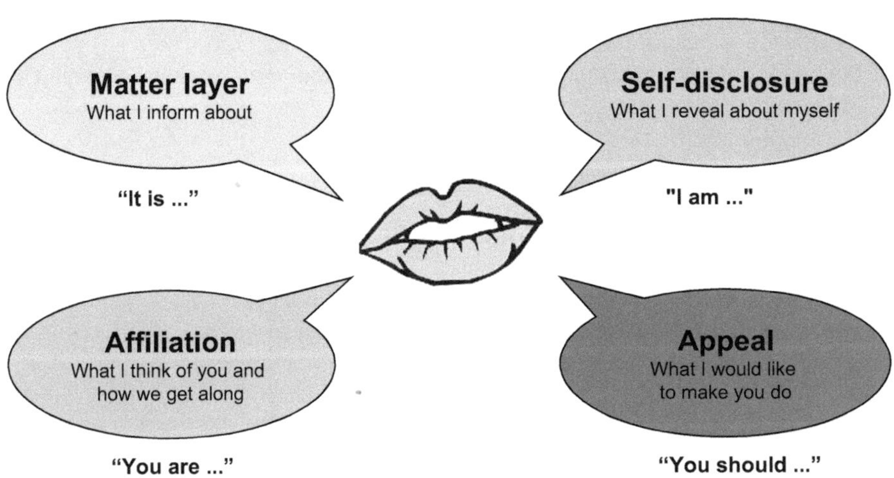

Fig. 18.10 The four sides of a message (Schulz von Thun, 2000)

The aspects of self-disclosure, affiliation and appeal are generally expressed in an encoded or non-verbal manner, rather than in words. Nevertheless they are still contained to a greater or lesser extent in every message. In Schulz von Thun's model, the recipient has four "ears" corresponding to the four-sided message. He always decides for himself to which message he is responding and how.
Problems can occur, however, if people are used to receiving one-sided messages. The desirable approach is a balanced "four-ears model", deciding according to the situation on which side(s) a particular response is required (Fig. 18.11).

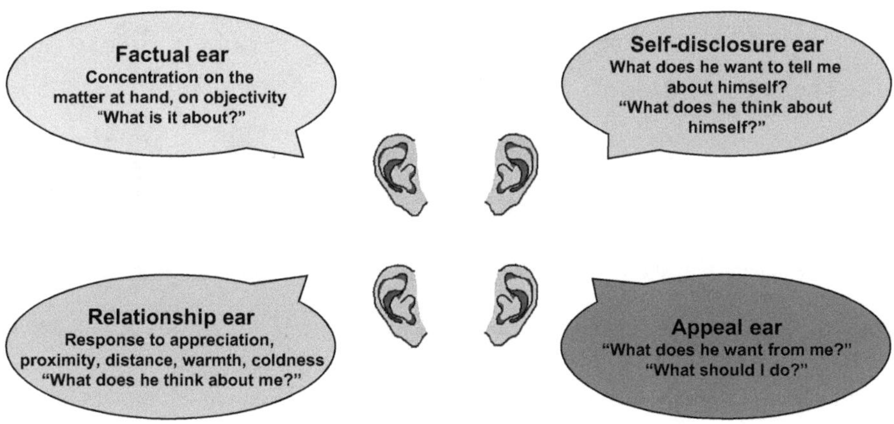

Fig. 18.11 The four-ears principle (Schulz von Thun, 2000)

The Communication Circle

The process of communication normally consists of the back and forth of sending and receiving. It is a shared game between two sides, between action and reaction. A comprehensive model must therefore be illustrated as a circle. Looking at this from a theoretical system viewpoint, this means that it is not just the behaviour of one side, but the rules for the interaction, the behaviour on both sides and their mutual dependency that should be considered.
An example: The project manager continuously grumbles at the project assistant. The assistant holds back and stops talking. The project manager says about this situation: "I have to criticise him because he always holds back". The assistant, on the other hand, says: "I hold myself back because he is forever grumbling at me". The two are thus involved in a "game" in which each side supports the other in his behaviour; the behaviour of one is dependent on that of the other. The game keeps being played out on its own and can only be stopped if the two get together to discuss the rules, e.g. what is each side doing to keep the game going?

This communication about communication is known as meta-communication: "Making the way in which we communicate with one another into the subject" (Fig. 18.12).

Fig. 18.12 Meta-communication

Conduct of Negotiations

One very common application for interpersonal communication is the negotiation. The actual content of the communication is not simply an issue in the narrowest sense of the word. It is also an interpersonal and emotional matter. Such meetings are held between two or more people. Negotiations can be held from different positions and with different attitudes. However, the aim of a negotiation should be to achieve a solution that is as forward-looking as possible for all parties in order to facilitate further collaboration. This is often a demanding task in projects, as they are naturally comprised of many conflicting goals and different interests.
The Harvard model offers a very commonly accepted structure for negotiations. It assumes that the people involved intend to work in partnership and build upon this accordingly.

Feedback

Johari Window

If we want to extend our social skills, we are forced to meet new people and experience new situations. We are also forced to repeatedly experience the effect our own behaviour has on other people. At the same time, we also check what the actions of other people trigger in us.
Indeed, there are some things that we know about ourselves, and others that we do not recognise. There are also some things that we make freely accessible to other

people, and those that we keep secret. The Johari window helps us to better perceive our interpersonal relationships. Area 3 (blind spot) is of particular relevance for feedback (Fig. 18.13).

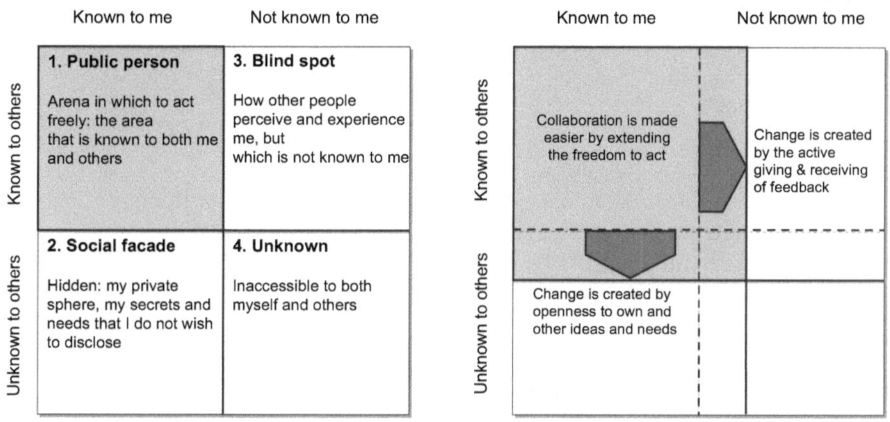

Fig. 18.13 Johari window

The greater the pressure to perform and career pressure are on a person, the more rigid his environment becomes, and the more "real life" is forced back into the hidden area. This means that a vast amount of energy needs to expend in maintaining the social facade. For the area of the public person (e.g. daily professional life, project work), this means a loss of action and motivation or energy block, and thus a restriction of the freedom to act.

Within the team, the project manager should therefore encourage the willingness and capability of all team players, open up himself to a certain degree and also be open to other people. On the other hand, he has to intentionally give feedback.

Of all the many definitions and descriptions for the term "feedback" in the narrower sense, the following wordings place particular emphasis on the interpersonal level:

- "Feedback is a message to a person that informs him how his conduct is perceived, understood and experienced by other people."
  (Klaus Antons, 2000).
- "I have no idea what I said if I do not listen to the response."

This claim points us towards the importance that feedback has for the interaction between people. Feedback is an opportunity to experience something of our effect on other people, to check and, if necessary, to modify our own conduct. Feedback can also clarify relationships. The response may be verbal, but it can also be a non-verbal response in the other person's behaviour to something we have said. In this respect we are constantly giving and receiving feedback, often without a single word being spoken.

Feedback Rules

- "I" messages, rather than "You" statements
- Specific rather than general
- Distinguish between perception and interpretation
- Accept feedback without being defensive or justificatory

Giving Feedback

- I formulate specific statements
- I give descriptions that are as accurate as possible and do not interpret
- I remain moderate and do not adapt my feedback to the situation
- I express everything with "I" messages

Receiving Feedback

- I express the information that I want to receive
- I do not argue, justify or defend myself
- I check the importance of the statements for myself personally
- I pass on my reactions

The Effect of Feedback

- Feedback reinforces positive conduct
- Feedback serves to clarify relationships between people
- Feedback corrects behaviour that will not help the group
- Feedback helps to understand the behaviour of other people better

Questioning Techniques

Open Questions, Questions That Stimulate

- Are intended to inspire the person questioned to express his thoughts
- Are intended to encourage new discoveries or to bypass mental barriers
- Are intended to deliver as much information as possible, and should have a clarifying effect

Example: "W" questions (who, what, when, how, where, why). What do you as the expert say about that? What does your experience tell you? In which situations have you observed that?

Information Questions

- Are generally short and precise
- Are good at obtaining missing, relevant information

Example: How late is it? What does he say?

Alternative Questions

- Allow the person questioned to choose between two or more options in his answer
- Force him to make a decision

Example: Do you want to... immediately or later?

Confirmation Questions

- The questioner wants to have his opinion, understanding or the result of a meeting confirmed
- Are useful for summarising or repeating what has been said so far

Example: By that, do you mean that...? Do I understand you correctly...?

Rhetorical Questions

- Do not allow any answer at all because the questioner provides it himself
- Can draw the hearer's attention; prompt him to reconsider; stimulating; common in speeches and monologues

Example: Do you know what it means to be a moderator? It means...

Suggestive Questions

- Intend to influence the person questioned by putting the answers in his mouth
- Contain one of the characteristic words: so, plus, certainly, indeed, etc.

Example: So you want to be sure...? So are you of the opinion that...?

Consequences from Considering the Model

Factors That Promote Good Interpersonal Communication

A number of behaviours are cited below that help people to understand one another in their communication:

- At the start of a meeting, clearly stating the aim and object of the meeting. Keeping to the matter at hand, not being diverted
- Not packaging opinions in questions (suggestive questions)
- Giving "I" (or "we" as a team) messages
- Ensuring that the other party is paying attention
- Having all "ears" open, listening with all the senses
- Actively listening: checking that you have been understood correctly; testing your own interpretation (repeating in your own words)
- Asking for clarification if anything is unclear
- Giving feedback and asking for feedback
- Expressing your own feelings, articulating suspected feelings
- Identifying and expressing your own opinion as such
- Making wishes, claims and consequences transparent and clear
- Placing hidden issues on the table and clarifying them
- Metacommunication (about the type, rules, how we communicate)
- Giving time, taking breaks

Factors That Obstruct

A discussion can only be constructive and open up new perspectives if both sides are equal in the discussion and are equally entitled to represent their point of view. Any attitudes and statements intended to emphasise one's own superiority or the inferiority of the other party will stand in the way of a fruitful discussion. This includes, for example: commands, downplaying, making accusations, talking about oneself, cajoling, comforting, volunteering pearls of wisdom, moralising, etc.

Communication at the Project Level

Of course, there are always people who communicate within organisations. Organisations or social systems, however, have their own experiences and histories and have their own encoding and decoding systems to such an extent that we can talk of communication between social systems. Even in projects, we have to deal with social systems, as well as just individual people (Fig. 18.14).

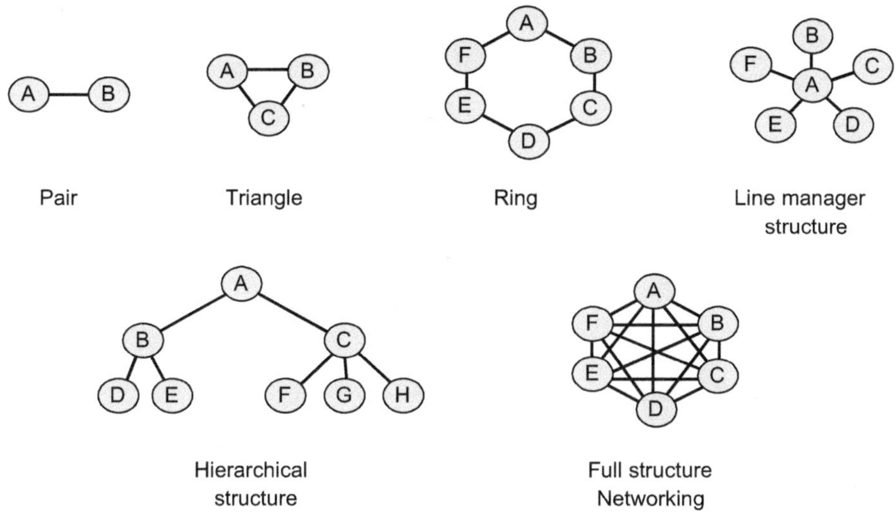

Fig. 18.14 Reporting channel and communication structures

Communication between systems is a dynamic that can inspire, arouse interest, and set in motion, depending on the structure. These dynamics can be intentionally shaped. For example, the rules for reporting channels between the project and originating organisation, how the communication culture should differ from the normal culture. In practice this can be set up by agreeing specific ground rules for communication and framework parameters, such as: Is there an obligation to provide or take? Who has which authority in the communication? What happens if there are differences of opinion? How are feedback circuits set up? How are decisions taken?

More complex structures with equal rights, such as networks, are capable of better communication than more simple and hierarchical structures. Just like any person, every social system (e.g. project, stakeholder groups, etc.) has its own reality and its own encoding and decoding systems. An interest in this can help to shape communication better. There are helpful methods for this, such as project environment or stakeholder analyses.

Communication also means establishing trust, and this is only possible if there is consensus between what is said and what is done, if problems and disadvantages are discussed as important topics, rather than being "glossed over". Within an organisation it is possible to work at creating a culture of transparency and trust, or not to do this at all. There are also communication specialists who can be employed to prepare and design a company's internal website with opportunity for feedback, or to create a concept for an information exchange or a large group event at which stakeholder groups can discuss aims and possible solutions.

Leadership and Teamwork 19

19.1 Leadership: What Is That?

The term "leadership" is used here in two ways: For the leadership of an organisation or company and for the leadership of employees. There are many different definitions of leadership, as shown by the following examples:

- Leadership means convincing people of an idea and empowering them to transform this conviction into action.
- Leadership always plays a part in relationships. The leading and led conduct themselves according to their own, subjective concepts, which are also group-specific and dependent on the situation.
- Leadership is the goal-oriented, social exertion of influence in order to fulfil shared tasks in or with a structured working situation.
- Leadership means ongoing problem solving in social systems.
- Leadership means identifying, shaping and controlling interpersonal processes, and can be divided into two parts: leadership activities and leadership behaviour.

Running and leading projects encompasses the leading and running both of a task and of employees. The project manager in the role of the "manager" leads the project in the business sense. He has to pursue the project goal and monitor the optimum use of resources and the time and financial means made available to him. The project manager also assumes the operational management role and the leadership of his project team. Many functions and roles are the same in normal line management and in project management. There are a couple of important differences for the project manager, however:

- He only accepts the leadership role "for a time"
- He generally leads within the matrix form of organisation with several established line managers
- He has little or no formal power (authority to issue instructions)

J. Kuster et al., *Project Management Handbook*, Management for Professionals,
DOI 10.1007/978-3-662-45373-5_19

- He has no or imprecisely defined authority for resources
- He is formally incorporated into the project organisation hierarchy for a temporary period only

The most important tools are:

- Clearly-defined project goals
- Only the authority that also has access to resources or can organise the necessary resources can guarantee commitment.
- Careful regulation of tasks, competencies and responsibility
- Support promised by the project owner
- A sophisticated communication concept
- Ground rules agreed with the project workers

This also includes clear transparency of the agreed measures and possible, disclosed consequences if the agreement is not kept. This must apply at every level of the project hierarchy.
There needs to be agreement between stakeholder groups and the project concerning what is to be achieved or not achieved (understanding of the goals), which measures can be used to achieve the goal and who has to complete which task and by when. Openness, a willingness to learn, persuasiveness, assertiveness, constructive problem solving and conflict resolution skills are needed in order to discuss and clarify contradictory opinions or viewpoints.

(a) Leadership = Management

In the overall business sense, this term means managing and leading a company, organisation or project, etc.
The company management concerns itself with securing the company's livelihood under changing framework conditions. It assesses factors that influence the environment and uses the knowledge thus gained to adapt the company to the new events. Inside, it takes account of the available strengths, skills and resources. The company's strategies, goals and policy are defined from the combination of external and internal factors of influence.

(b) Leadership = Employee Leadership

In the narrower sense, we talk of leadership in terms of intentionally influencing employees with a view to achieving goals. Employee leadership is characterised by the changes in the environment. Intelligent company leadership can rapidly adapt to new situations. It needs lots of competent people who know what they are doing and are prepared to think for themselves, act and assume responsibility. This requires a high level of transparency in terms of information and communication, coordinated interaction between all the participants and trust between employees and managers. Leading means influencing the behaviour of employees so that the goals can be achieved together with them. If it also involves issuing, asking for and correcting goal-oriented arrangements, then the project manager is performing the task of employee leadership (like a manager).

(c) Leadership = Coaching

The project manager also has the task of supporting and encouraging his staff and of developing their personalities with a view to making the most of the resources available to him. This is known as "coaching". It is an element of leadership based on partnership and a useful supplement to the other personnel development tools. In the role of coach, he is tutor and trainer, and pursues the aim of developing the people working on his project so that they can act independently and under their own responsibility. The coaching quality is characterised by realistic implementation, concrete results and sustainability.

Clearly the word "leadership" can be experienced in entirely different ways by the people involved. Some think about responsibility and challenge. Others think more in terms of domination and power to command, and thus of suppression and ways to assert personal demands. Actually, the word "leadership" suggests power and influence over other people. Anyone who has authority to lead will exert power over people or groups of people. However, it is the way in which this power is exercised that will determine the quality of the leadership and whether it will be successful.

It is increasingly expected of leaders that they will develop and deploy their coaching skills in addition to their management abilities. Top management should monitor this point very carefully, particularly when considering who to appoint as project manager. It is also true to say that the more mature the employee, the greater the manager's coaching skills will need to be. When work is organised in this way, and this organisation is encouraged, it will constantly grow and evolve to suit the current issues as a "learning organisation".

19.2 Leadership Organisation

The leadership activities essentially consist of three elements:

- Planning
- Steering and delegation
- Control

These activities take place at all leadership levels. They are needed in order to shape and direct pending tasks in a goal-oriented and coordinated manner. Various processes have to be planned, controlled and checked for effectiveness. Three management levels are essentially intertwined:

Strategic Management
(Sense level) commits to the general goals of the company, principles, standards and values that are directed towards ensuring that the company is able to survive and develop.

Dispositive Management
(Effectiveness level) advocates building up, maintaining and utilising the potential for success of the available resources.

Operational Management
(Efficiency level) converts the normative, strategic and performance concepts into efficient routine work.

Most companies and organisations have set up recurring routine processes and expedient organisational structures. This challenges them to deal with current changes on an ongoing basis. To secure their future, they continuously need new or updated procedures and sequences, new materials and products and new strategies.

Project Management Leadership Concept

Previously mentioned organisational structures are insufficient for project work. The project management leadership concept was developed as a result. This is a leadership tool that aims to resolve temporary, interdisciplinary tasks with a high proportion of innovation and complexity.
A project organisation is characterised especially by the fact that it is networked with different systems of authority within the producing company or the organisation providing the service, directed both inwards and outwards (Fig. 19.1).

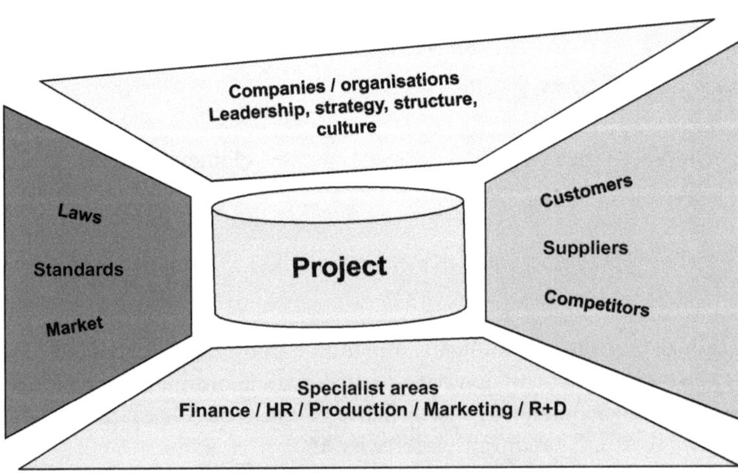

Fig. 19.1 Networking of the internal and external worlds of a project

19.3 Project Management Is a Form of Relationship Management

A project does not only incorporate a problem-solving process. It is also networked with the outside world. As a rule, we rarely acknowledge this, and run projects in relative isolation. A too introverted perspective, however, can have significant drawbacks:

- The users, affected persons, interested parties, etc. dismiss the project ideas and solutions because they feel them to be foreign and having nothing to do with reality. The realities of the project and the environment are not matched; the examination and the learning process do not start until the project has ended.
- There is little or no usable synergy, either with projects running in parallel or with other parts of the company.
- Stimuli coming from the project cannot be exploited effectively by the company. For example, the "alternative" project culture cannot have an inspirational effect on the overall organisation. Quite the opposite: the contrasts can block progress and create conflict.

Thus project work also means working on relationships with the environment. Within the company, this is comparable to the following, undisputed insight: Anyone who does not regularly undertake relationship-building work with customers, suppliers and employees will find it difficult to survive. Thus even a project has its own customers, suppliers and employees. Here, again, success is essentially dependent on the relationship-building work.

Analysing Project Networks

The project environment analysis is very good for illustrating network links with the environment. It is then a good idea to evaluate the relationships both qualitatively and quantitatively: positive, problematic, expectations, dependencies, directional impact, conflicts, logjams, fears, intensive, extensive, etc. Even the project manager and project team members as people are social environments, with quite specific interests, expectations, motives, etc. in association with the project. This analysis should therefore ideally be carried out and discussed with the whole project team right at the start of the project work.
The project environment analysis gives a good impression of the area of tension in which a project is moving. It promotes outward orientation, strategic thinking, marketing awareness, etc. and allows target group-specific measures to be taken. The following figure contains an example of this, with the promoting (+) and obstructing (−) links marked accordingly (Fig. 19.2).

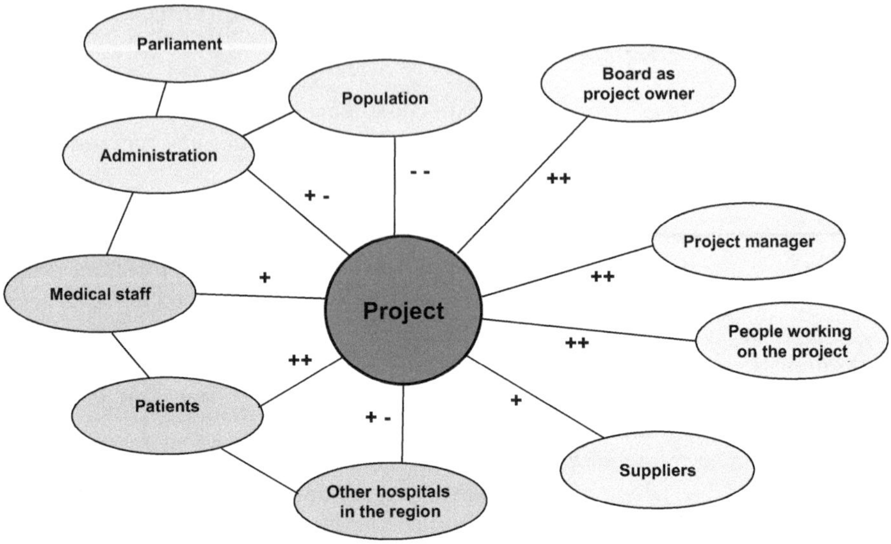

Fig. 19.2 Example: Network links for a project in a hospital

There are many ways to work on and shape relationships. At the project launch, the initial relationship-building work between project manager and team members is a fundamental requirement for effective project work. This relationship-building work can be extended to the wider environment.

Managing Stakeholder Groups

Stakeholder groups or stakeholders are people or organisations who have relevant relationships with the project. Such relationships include: affectedness, specific interests, legitimation or even experiences. Stakeholder groups outside the company are sometimes described as the invisible project team since they can significantly influence, support or even derail the project. Thus, different stakeholder groups often also place contradictory demands on a project.

Influential stakeholders attempt, for example, to put their stamp on the aims and project planning and to demand certain interim results earlier than planned. Or investors have an interest in minimising the project risks.

Figuratively speaking, the project is incorporated into a social network or force field that should not only be taken into consideration, but which may potentially be useful for the success of the project. The stakeholder groups can be handled in different ways. Communicative relationships can be established, but they can also be incorporated into the project organisation. Where there are very disparate interests, it is often advantageous to create opportunities for representatives of stakeholder groups to discuss their mutual interests and differences of opinion directly. This is a way to ensure that a broadly supported and thus robust solution can be found.

A stakeholder analysis is ideally drawn up with the project team at the start of the project. This will also create a shared “project reality” and a sense of how the project is networked. Stakeholder management consists of four steps:

1. Identify the stakeholders

- Who provides particularly important resources and who does not?
- Who can exert an influence over the project?
- Who is particularly affected by the project?
- Who can promote or obstruct the project success?
- Who needs the project results?
- Who must not be overlooked?

2. Analyse the stakeholders

- Image of the stakeholder structure: group the stakeholders around the project
- State and visualise the intensity and quality of the influence and interests of the individual stakeholders, e.g.: illustrate proximity to the project by positioning close to the project and show the quality of relationships with symbols for positive, neutral or negative

3. Evaluate the stakeholders

This task is illustrated most simply in the form of a table (Fig. 19.3).

Stakeholder	Interests Affectedness	Level	Power, influence, legitimation	Level	Expected reactions to the project	Evaluation	Actions
Electors	Open and multifunctional town hall is wanted, but should not be too expensive	3	Legitimated, to be decided for/against by public vote	3	Agreement appears likely	+	Professional project marketing, visits, communicate costs / benefits well
Customers, visitors	Customer-friendly service	1	Complaints, letters to the editor	1		+	As for electors
Local government employees	Don't want open-plan offices! Town hall to be easily accessible (distance from work)	3	Indirect via motivation, work detail	2	Different reactions	?	Treat with respect, include in the office planning
Clubs	Will finally have infrastructure for activities!	2	Influence the vote	1	Very positive	+	Awaken interest, include in the planning
Residents	Shadows cast by building are a nuisance	3	Legitimated to offer objections	2	Objections are likely (Will delay the project)	-	Early information, show what effects a delay will have
Regional historic buildings society	Must not impair the silhouette of the old town	2	Possibility of objections	3	Public debate, objections, influencing the vote	-	Incorporate into the project team
Environmental protection association	Reduced number of parking places -a nd must not be free	2	Association's right of appeal	3	Will probably appeal if number of parking places exceeds 20	-	Bilateral preliminary talks
Bidders: construction companies, tradespeople	Very interested in contracts	3		1		+	Hold out prospect of transparent submission process

Level of relationship with the project: 3 = high, 2 = average, 1 = low

Type of relationship: 1: + = positive, - = negative, ? = unknown, neutral

Fig. 19.3 A typical stakeholder analysis illustrated in table form (construction project)

The following thoughts may be helpful for working through these questions:

- What do the stakeholders think about the project? The problem should be considered from the perspectives of the various stakeholder groups.
- How are the relationships between the stakeholders? What interests and goals are the individual stakeholders pursuing? How are they interrelated? Where is there potential for conflict?
- What reactions and conduct from the stakeholders must the project manager expect?

A stakeholder map can be derived from the initial analysis (Fig. 19.4).

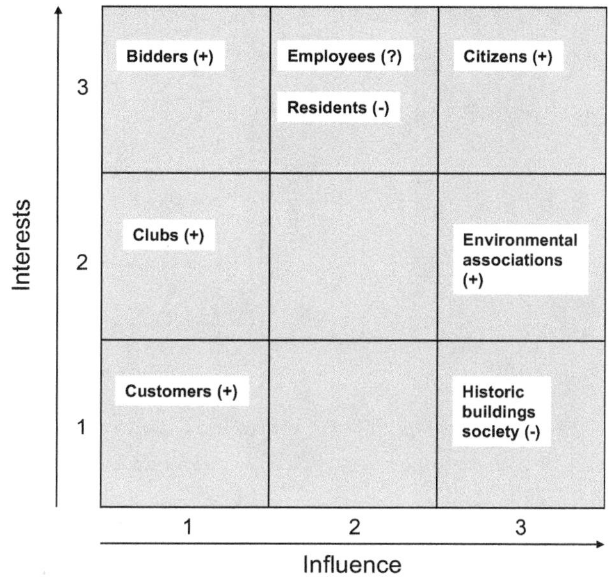

Fig. 19.4 Stakeholder analysis in matrix format

4. Control the stakeholders

Actions are formulated and implemented with reference to the analysis and evaluation of the stakeholders. Possible approaches for handling stakeholders include:

- Incorporating stakeholders into the project, i.e. creating a suitable project committee and defining tasks and roles for the stakeholder groups within the project
- Clarifying interests and goals, mediating (e.g. for opposing project goals or interests)
- Keeping stakeholders appropriately informed and maintaining the dialogue
- Creating additional communication platforms, e.g. workshop, large group session, meeting

The stakeholder map directly shows with whom the project manager has to deal and how (Fig. 19.5).

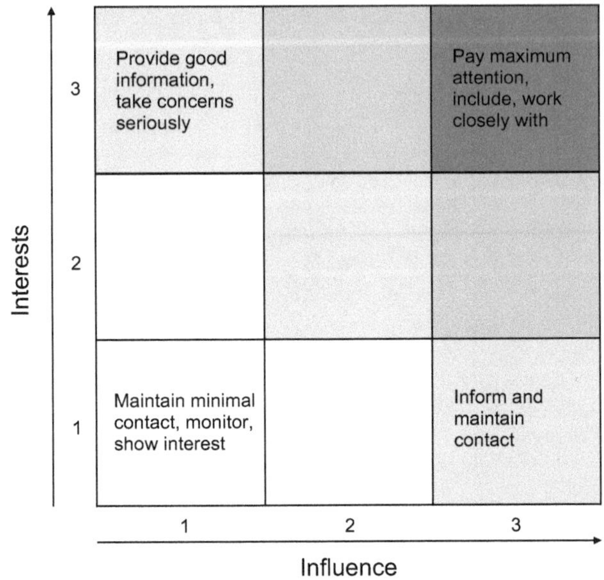

Fig. 19.5 Strategies for dealing with stakeholder groups

- The power and influence of stakeholders are difficult to control. What is controllable, however, is the relationship between the stakeholders and the project by relationship management
- Identifying changes and developments in relation to the stakeholders
- Maintaining regular communication and discussions with the stakeholders
- Periodically checking and adapting the actions taken

Incorporating the Project Owner into the Project

The aim is to ensure that the project owner side identifies fully with and values the project. This can be achieved by the following measures:

- Drawing up the project order together with the project owner (or interactively)
- Periodically informing the project owner (or his representative, e.g. committee) about the project progress and discussing pending questions or risk situations
- Involving the project owner in important project events, e.g. at the project launch, when important interim results are obtained, at the end of the project. He should be included as the "patron" and not simply in the role of decision-maker.

Of course, the intensity of the relationships between project owner and project manager always depend on the specific conditions. Under certain circumstances a department manager is closer to the project than a governing council, and can thus be better involved. In the case of the governing council, suitable "pivot points" need to be created.

Relationships Between Affected Parties and the Project Team

It is particularly useful to maintain these relationships since they will play a decisive role in success or failure in the introduction phase. The aim is to establish trust and offer opportunities to put expectations, fears and interests into words. The following options could be applied:

- Including important advocates in the project team
- Inclusion in a temporary working party (extended project team); this could also include groups to draw up goals, debate conflicts of interest, look for solutions, etc.
- Establish and promote reciprocal communication
- Info mart where future users or affected parties can find out and obtain first-hand information that is useful for them

It is also important to argue out differing goals, conflicts of interest, etc. in these environments. Under certain circumstances, special events or workshops must be organised for this purpose.

Agreements Concerning Work Within the Team

The greater the social complexity or the more open the goals of the project, the more individuals will become involved in the overall problem solving process. It is essential, therefore, for the common task of the group and its members to be recognised and felt to be important, i.e. the objective and subjective goals must be the same. As well as achieving the set project goals, the individual also needs to feel that he can reach personal goals (e.g. to satisfy personal needs).

For a project team, communication is the number one tool. Factual and relationship problems are discussed and analysed in joint meetings (interaction). Decisions are taken and individual interests are coordinated with respect to the whole within the team. This mutual discussion then allows the group to identify and resolve internal topics and disagreements.

If such all-round communication is prevented by bureaucratic or authoritarian leadership (e.g. everything in writing) or by a rigid, formalistic organisational structure, the group will find it difficult to develop into a powerful and willing team because there is no mutual exchange of information.

19.4 Leadership Work Within Project Management

There are other aspects involved in running a project, in addition to leading an organisational unit. In projects, this generally means orienting specialist staff that are working across departments towards a common goal and encouraging them to achieve this goal or provide a result with financial benefits for the company (Fig. 19.6).

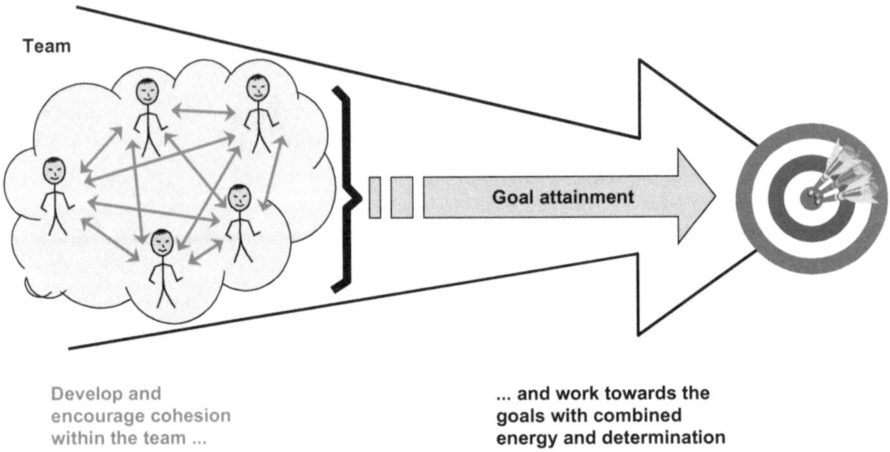

Fig. 19.6 Leadership process within the project team

Project workers are partners with equal rights who have committed themselves to solving the problem, so it is sensible to select a leader or project manager with excellent coaching skills. They must be capable of developing, coordinating and keeping abreast of the necessary work processes in the project team:

- The team is led with team-building processes in order to resolve the problem or achieve the goal as effectively and efficiently as possible.
- The collaboration is based on binding agreements (agreement culture). The team members are partners with equal rights, even if they come from different levels of the company hierarchy.
- The project manager primarily leads using information, communication and coordinating meetings. He ultimately achieves the greatest effect with a goal-oriented and appreciative leadership style.
- As part of his leadership responsibility, however, he must also be able to take clear decisions in deadlock situations.

The Different Tasks of Project Management

For a project team to work effectively, the project manager needs to perform a number of different leadership tasks and services. The services to be provided on an ongoing basis, which are not necessarily the responsibility of the project manager alone, comprise the following aspects:

Planning, Organisation and Checking

- Selection and use of problem solving and procedural systems. The team must be clear about where it is positioned within the problem solving process and who is working on which partial task (breakdown of tasks and roles)
- Scheduling, effort planning and adaptation on a rolling basis
- Define a time structure and make sure it is used
- Provide an appropriate infrastructure: rooms, money, equipment
- Set up a checking system that monitors when a goal is reached, progress (content, deadlines), effort (hours, costs) and signals discrepancies
- Determine a suitable instrument to provide an ongoing overview of all decisions taken and work packages assigned
- Indicate the consequences of changed goals or deviations from the work schedule
- Ensure documentation is created (reports, project plans, descriptions, programmes, etc.)
- Indicate the consequences of changing goals
- Establish risk management

Facilitation

- Structure and lead meetings
- Include all members and involve them in the problem solving process
- Avoid "treading water", provoke decisions and drive the group forward
- Ensure the right mix of methods (e.g. individual, small group and plenary work; brainstorming, metaplan, etc.) and use the right tools (pin wall, flipchart, overhead projector, etc.)
- Visualise and document the work process on an ongoing basis
- Initiate and lead evaluation sessions: "How did we do?"

Team Development

- Look after the relational structure and atmosphere within the group, i.e. make sure that the atmosphere allows the members to work and that it remains that way.
- Create opportunities for members to meet, bringing them closer together
- Help to break down tensions and resolve conflicts, i.e. assist people to work through subliminal themes
- Make sure that no member's dignity is hurt and that nobody is treated contemptuously

Information and Communication

- The team must be supplied with the necessary information on an ongoing basis
- The flow of information and communication must be defined and ensured within the team

- The environment, i.e. all stakeholder groups must be provided with timely information about the project
- Ongoing contact must be maintained with the other project managers and managers of sub-projects, with members of affected organisational units, with senior management and with the project owner

Controlling
- Controlling is an ongoing process, for which the whole team shares responsibility
- Departures from goals are quickly detected and action initiated
- Allows a response to breaches of the ground rules, non-compliance with framework conditions, etc.
- The project manager supervises and monitors the work process of the project team from a higher level position ("helicopter view")

Crisis Management
- Ongoing monitoring for the possible signs of crises and conflicts
- Clearly address suspected or obvious crises as such
- Tackle and resolve crises within the team

Promote and Maintain Performance

Most projects today are under pressure. Where resources are tight or there is little time available to achieve the goal, this leads to tensions that put pressure on the affected people. Thus, pressure of work and the associated stress are important variables that have to be managed to avoid significant drops in performance, whether in individual people or the team as a whole.
People who are very willing to put in extraordinary effort are also prone to stress. These people "give their all", so there is considerable risk that they will suffer from chronic stress over time. Acute, natural stress consists of three phases (preparation, service provision, recovery). With chronic stress, however, there is no recovery phase, and the person finds that the stress is continuous. This is very hard to cope with, and there is considerable danger of burning out. Burnout can be described as work-related psychological and physical exhaustion. The sufferer exhibits some or all of the following features:

- Decreasing efficiency
- Cynicism
- Persistent tiredness
- No sense of own personality
- Apathy, no energy left for constructive criticism
- Uncontrolled willingness to work (not simply to resolve this problem)

When the affected person loses access to himself, he slips ever deeper into burnout. One of the project manager's tasks is to keep an eye on the condition of his team and step in where intervention is needed. Even if he is not actually the line manager, the project manager can, together with the responsible HR manager, ensure that the essential requirements for good working conditions exist as far as possible. He can also act preventively by, for example:

- Giving more room to manoeuvre
- Offering social support
- Avoiding contradictory tasks or at least making the tasks transparent
- Allocating tasks in their entirety
- Facilitating participation
- Reducing work intensification/time pressure
- Targeted coaching in time management, personal boundaries/assertiveness (ability to say Yes/No), reducing unrealistic expectations (in self and other people)

A Project Manager Wears a Number of Different Hats

The various leadership functions can be imagined as hats that the project manager wears at different times. He generally has to wear two or more hats at once, and so risks being constantly overloaded. This is particularly true if he is also wants or has to be intensively involved with the substantive problem solving process (Fig. 19.7).

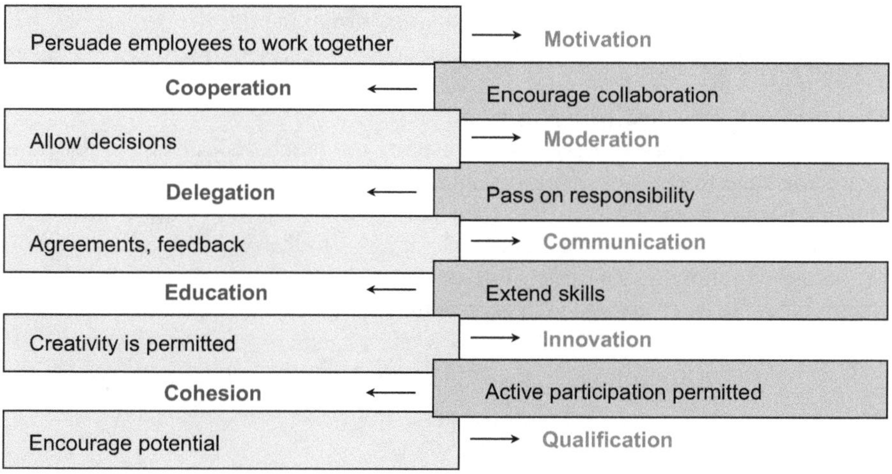

Fig. 19.7 Elements of leadership within the project

Rather than carrying out all the leadership work himself, it is a good idea for the project manager to involve team members in this. For a division of labour to work, it must be expressly discussed and agreed between the affected parties, at least within a new group. Without this discussion, such a procedure very often leads to misunderstanding, rivalry, logjam or even total paralysis.
The more the project manager is able to share the hats between different team members, the more he can step back from the leadership function and apply himself to the problem solving process. Of course, not every hat can be easily delegated and some will not fit all team members equally well. There are some tasks that "people" simply expect the project manager to do. For example, he cannot delegate the overall responsibility, as then he would no longer be doing the job he had accepted from the project owner.

- The success or failure of a project is highly dependent on the skills and personality of the project manager.
- Viewed realistically, it is difficult to find a project manager who is able to wear all the hats. He will therefore rely on the full support and cooperation of his team to ensure successful project work.
- For the project manager it is important that he is always aware of and states the role in which he is currently acting, e.g. "My opinion as a specialist is. . ." or "My opinion as the project manager is. . .".

Leadership Style in Projects

Why should a project manager who has intentionally accepted the leadership of a project share the leadership tasks with the other group members?
One advantage has already been mentioned: the project manager can then devote more time to holistic problem solving. On the other hand, the leadership task is always associated with power and prestige.
If the project manager emphasises these aspects too much, he can trigger the same conduct in the project team members that we see in very hierarchically structured line organisations: People will jostle for personal prestige and status, rather than trying to find goal-oriented and beneficial solutions to the problem. Self-interest is put before the common interest, while cover-up strategies block open and honest communication about errors, defects or failures (Fig. 19.8).

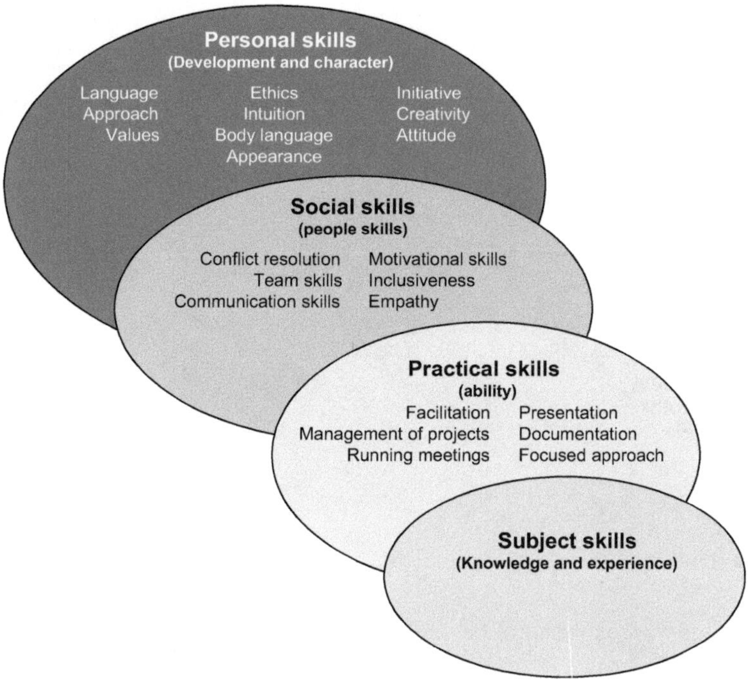

Fig. 19.8 Interpersonal skills

The project's success is primarily measured in terms of the benefits that it achieved for the company. The extent to which the project manager performed his leadership tasks on his own is for now of secondary importance.

Project owners should, however, look very carefully at the mood within the project team. Enthusiastic project workers express themselves positively about the teamwork with their project manager. They feel they are appreciated and recognised, and they are already looking forward to working on the next project with "their" project manager. As the saying goes, "Never change a winning team".

From Self-organisation to Greater Flexibility

A team that works well together is to a large extent capable of self-organisation. The consequences are that it can quickly adapt to changes, is highly committed and consistently identifies greatly with the project and is highly motivated. This leads to correspondingly high capability. In such a situation, the project manager only has to intervene with corrective measures in exceptional circumstances (Fig. 19.9).

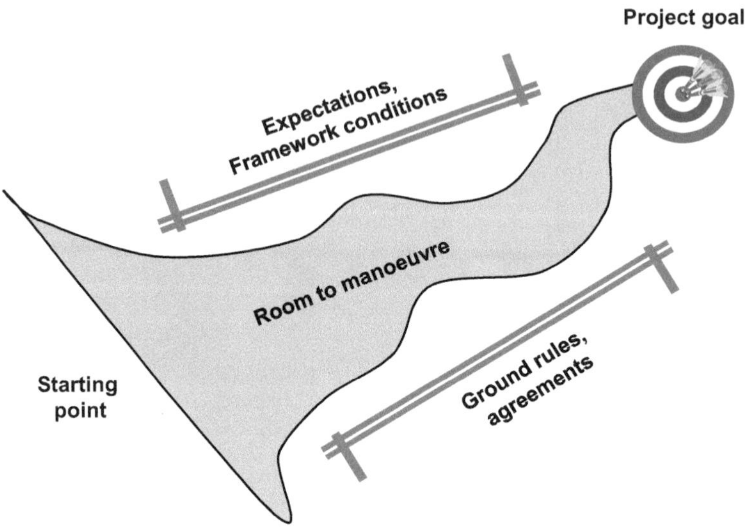

Fig. 19.9 Creating room for manoeuvre

This form of work is found to be very satisfactory by many people. This requires:

- The project owner to be fully behind the project
- The technical skills of the people involved to be appropriate for the task at hand
- The team members to be motivated and identify with the project
- There to be no or little rivalry between the members
- The roles and ground rules to have been clarified
- The project manager to give enough room for manoeuvre
- The necessary resources to be available

Under certain circumstances, it takes a great deal of energy and time to fulfil these requirements. The initiative generally comes from the project manager. He can encourage and allow team-oriented work. But he can also prevent it. It is also important to see, however, that the greater the time pressure or pressure to perform, the more leadership and coordination is needed from the project manager.

Clear Framework Conditions Create Room to Manoeuvre

Leadership and shaping of the organisation generally have the aim of controlling work processes with a view to achieving goals. Generally binding rules and regulations reduce the complexity and increase efficiency. In projects, goals can sometimes by definition only be described inexactly, e.g. "Develop an option for diversification" or "Discover a new possible solution".

It is difficult to structure the work process in advance from this starting position. Nevertheless, to achieve usable results, clearly described framework conditions (limits) are needed within which the project can move.

Anything that applies to the project in general, also applies to the project team. If the project manager and team members come to clear agreements (ground rules) about the applicable framework conditions, then the team members will have considerable room to manoeuvre within which they can act independently. It is then the employees' task to look for and select the optimum way. The project manager can be relieved of this task without any fear of losing control.

Leading by Delegation

For the project manager, successful delegation requires first and foremost the ability to meaningfully isolate individual tasks from one another, to communicate them in a comprehensible way, to coordinate the different task areas and to integrate the results obtained. He must also be able to allow the workers to take their own route to the goal, even if he would not have chosen it himself. In other words, the ability to delegate needs trust, openness, tolerance and certainly the necessary sangfroid to avoid becoming enmeshed in details.

There are a number of arguments in favour of delegation, particularly in project management:

- It eases the burden on the project manager, who then has more time for the strategic, planning and performance decisions.
- The team members are generally subject experts who are more skilled in their specialist area than the project manager, who is more of a "generalist".
- Decisions can be prepared and taken by those specialists who will be directly affected by the consequences.
- Delegation often opens up space in the project work needed to identify creative solutions.
- Delegation also means including employees. They are challenged and thus encouraged. By accepting responsibility, employees also enhance their qualifications.

Qualified employees generally want to be able to use and demonstrate their potential. With the delegation principle, the project manager emphasises his trust to the employee, at the same time reinforcing the individual's autonomy and ability to act. The delegation principle also increasingly corresponds to today's image of the mature and capable human being.

These arguments are all in favour of the delegation principle, but it is worthwhile to mention the risks as well. In this context, the project manager needs to ensure that:

- The "right" people are used, i.e. the risk of overloading must be clarified in advance.
- Employees are comprehensively informed, i.e. they have understood the context and relationships, in addition to the goal and purpose.
- Implementation and results are regularly and systematically monitored (status reports, milestones).

When tasks are assigned, the corresponding authorities should also be assigned and responsibility transferred. The following questions should be clarified in order to delegate tasks:

- What should be done?
- Who should perform the task?
- Why should the employee perform the task?
- How should the employee perform the task? Clarify any framework conditions that will have a detrimental effect on the method of implementation.
- When should the task be completed?

The picture below shows in a simplified form the process for deciding on how a task is to be performed. On the one hand there is the decision whether to carry out the work oneself or even omit it altogether and on the other hand there is the decision to delegate (Fig. 19.10).

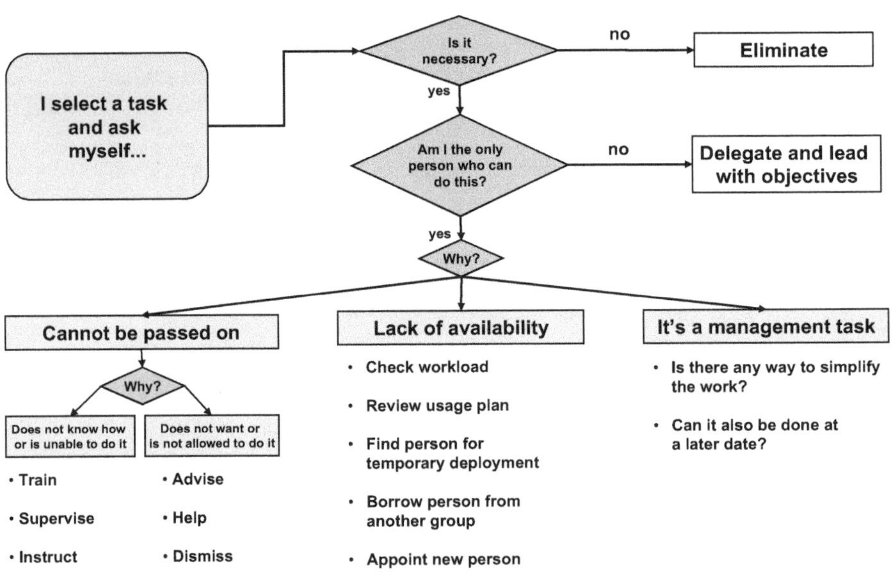

Fig. 19.10 Decision-making process for delegation

Delegation Is a Matter of Congruence

It is not unusual for the wrong type of delegation to be used in practice. For example, tasks are delegated, but authority and responsibility are either not discussed or are not discussed in sufficient detail, and there is also silence concerning the time resources that will be needed. It should be ensured here that the specific task always incorporates a responsibility that is geared towards this task (Fig. 19.11).

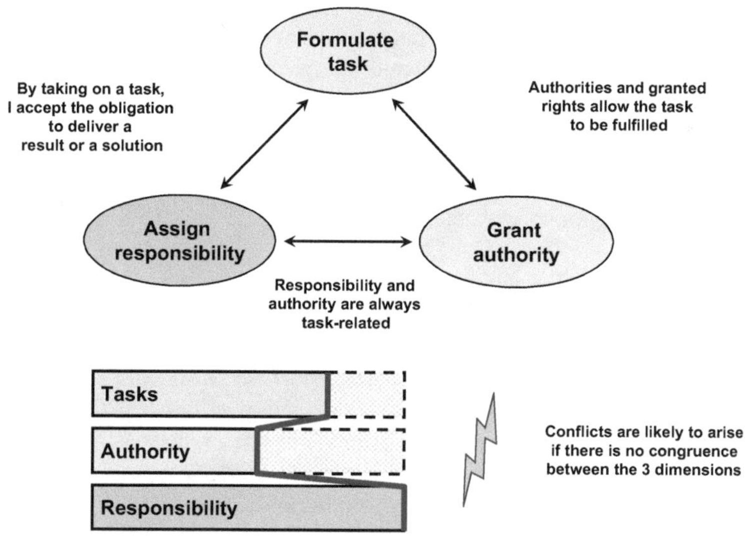

Fig. 19.11 Interdependency between tasks, authorities and responsibility

At the same time, the employee must be given certain authorities that will allow him to resolve problems autonomously, i.e. the task, authority and responsibility (T-A-R) must be substantively matched to one another. This idea is thus often described as the "congruence principle".

19.5 From the Start of the Project to Its Completion

The project start phase is a decisive and highly defining aspect of the subsequent project work. On the one hand, the project definition is a clear project order for the decision-making bodies. On the other, it involves understanding and successfully implementing together with the project team the various tasks resulting from the order. The initialisation of the team process is thus of particular importance. It involves shaping the processes at the psychosocial level such that they allow performance and solution-oriented work at the technical level.

Developing the Group's Capacity for Work at the Project Start

Any group that is newly formed or has changed its composition undergoes a development process.
The sooner the team succeeds in "getting it together", the faster it will achieve its full potential to perform. Dependency on the manager is still great at the start, so the team leader is responsible for ensuring that the work in this phase is good and done with care. A good start requires the processes of importance for the project's success to be initiated at three levels: relationship, content and organisation.
Throughout the course of the project, it is important to keep the processes alive, to check them frequently and to take alternative action if required. The aforementioned processes are completed when the project ends. This includes reflecting on and analysing the project's progress whether it has succeeded or failed. It is important to identify the reasons for and causes of the result and to learn lessons from them.

A positive working atmosphere can be promoted by:

- Opportunities to meet at which the team members can get to know one another better, prejudices can be broken down and shared positive experiences can be gained.
- Discussions about the shared understanding of the work and tasks, about the sense and benefits and the question of appreciation of the project.
- Improvement of the framework conditions and open meetings in which people can talk about their feelings (doubts, anxieties, fears, worries or hurts) and the group works together to improve the collaboration.
- Frequently updated agreements on the collaboration at the content and relationship levels

Topics at the Kick-Off Meeting

The kick-off (launch event or kick-on) sets the process of "turning up" in motion. It can be speeded up by an intensive get-together, e.g. in the form of a two-day out-of-town launch event with overnight accommodation. The project owner very often appears at such meetings to emphasise the importance of the project for the company and to underline his personal support for it (Fig. 19.12).

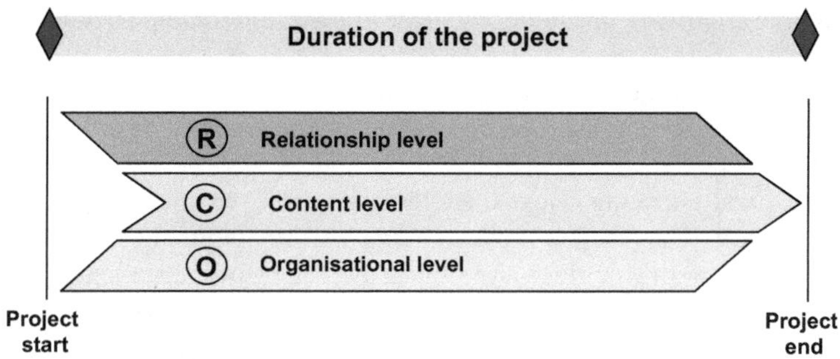

Fig. 19.12 From the start to the end of the project

Relationship Level

The way in which questions of content and organisation are resolved is defined at the relationship level:

- Can people speak out and are they listened to attentively? Or is the person who speaks the loudest and takes the floor first the one who gets his way.
- How are differences of opinion and opposing interests handled? Is it simply a matter of "Might is right" or is there a discussion and a joint effort to find consensus?
- Are people interested in the concerns and feelings of other team members or are they treated as mere functionaries who are there to fulfil a project order?
- How should we deal with the different hierarchical levels?

Content Level

- Be familiar with the project order
- Put the expectations on both sides into words
- Achieve a common understanding of the problem and tasks
- Reduce the different perceptions of the goals to a single definition
- Discuss the different courses of action
- Indicate possible approaches to finding solutions and opt for one
- Ask individual members about their prior knowledge and skills

Organisation Level

- Clarify the form of collaboration and specify the ground rules
- Establish responsibility and areas of authority
- Define and allocate roles and functions
- Discuss effort required and deadlines
- Define rules for and means of information and communication
- Agree on mandatory checking mechanisms: checking of effort and costs
- Describe the content and format of reports and project documentation

Answering the following questions can lead to better mutual understanding:

- How did I come to be in this project team?
- What are my expectations, feelings and attitude towards it?

By observing and checking out one another, we discover at an early stage in the collaboration how this group is working, who will be important in which respect, and what are the applicable rules and standards. The informal standards are agreed in this phase. The generally applicable system of standards within the company organisation will play an important role in this. This basic system will, however, have a different emphasis according to the composition of the group.

Overcoming Prejudices

Even when a new project team is brought together, the members are not as a rule entirely unknown. People have already heard of one person or another, or they have an opinion of the department from which the person comes. This knowledge is often rather superficial or even simply prejudice derived from circumstantial evidence or personal experience. Opportunities to meet in person in an informal setting will help group members to get to know one another better and make new experiences together.

Of course, sympathy and antipathy play significant roles in the whole process of establishing relationships. This phenomenon can be explained as follows:

- We all carry positive or negative memories with us from previous meetings with other people. If we now meet someone who, in one way or another, reminds us of these earlier encounters, then they awaken positive or negative feelings accordingly which may be transferred to that person.
- Such memories can also be triggered by conduct, facial expressions, gestures, vocal pitch or even by smells.
- Rather than people, situations (e.g. testing, experience from earlier project teams, dependency relationships) can also trigger such feelings. These are then transferred to the people involved in the new situation.
- Antipathy can thus be explained by the discovery in the person who is felt to be unsympathetic of conduct or personality features that one would reject in oneself or not permit to continue.

The processes described above are often entirely subconscious, so they are difficult for the affected parties to identify and understand. The following conduct can help in overcoming prejudices (Fig. 19.13):

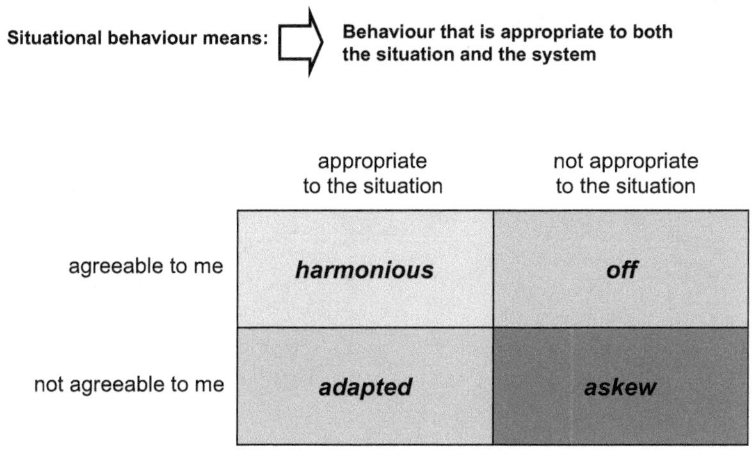

Fig. 19.13 The concept of agreement (Schulz von Thun, 2000)

- By actively listening, we concentrate fully on what the other person is saying and attempt to understand their viewpoint
- This will not work if we are already putting together a response in our head while the other person is talking
- Of course, this also requires us to allow other people to finish speaking and not constantly interrupt
- It is particularly important to create informal opportunities for all the participants to meet before the start

Checking Processes

The team leader must be aware that every team is always active at both the content level and the relationship level. Unfortunately there are still many people who attend solely to the content level and are then astonished when, despite good work at the content level, the team consistently hampers itself and makes life difficult with conflicts, indifference and running around in circles.

Process checks can prevent this happening. This means specifically and intentionally examining what is going on at the relationship level in the team. Process checks mean knowing what is going on in the team. They create awareness of the hidden behaviours of team members or project environments, and thus make the handling of problems and conflicts transparent. It is thus a promising way to detect and thus positively handle conflicts at an early stage.

The effectiveness of the process check is also highly dependent on the project manager's perceptiveness and communication skills. The technical issues may be in the foreground, but ultimately the problem has more to do with the sensitivity of the participants. Successful process checks not only improve the relationships and working atmosphere between team members, but they also create the basis for effective project work (Fig. 19.14).

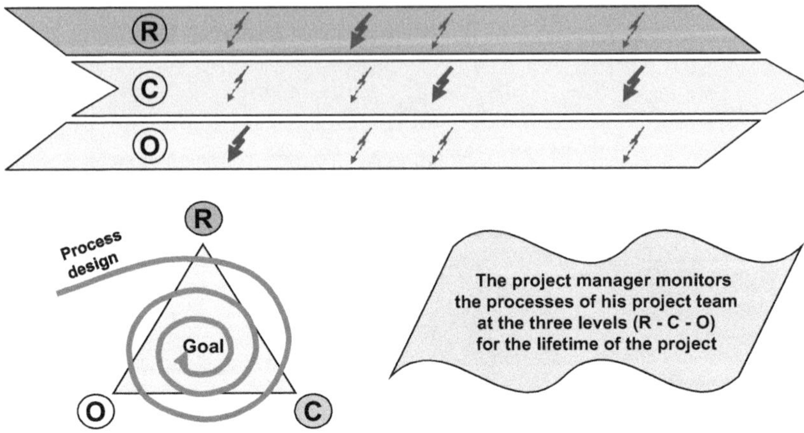

Fig. 19.14 The principle of ongoing process checks

Interventions

By "intervention", we mean the intentional interruption of a work process (e.g. a meeting) with the aim of changing something in a goal-oriented manner (structure, process, etc.). According to Paul Watzlawick:

I cannot not communicate
I cannot not behave.

This can also be applied to interventions: "I cannot not intervene as the team leader".

Interventions are:

- Subject-specific
- Intended to control a process
- Intended to create a structure
- Confrontational questions
- Crisis and conflict interventions

Interventions are not:

- Purely intended to pass on information
- Active participation by the manager in the work on the topic

Interventions are sensible if:

- Problems are identified in the interaction
- The group is no longer working towards the goal

Before I intervene, I consider:

- What I am perceiving
- The feelings, images, imaginings, ideas I have
- The possible consequences of the intervention
- What I would like to change or allow as a result
- How I intend to proceed
- What the intervention could trigger

I ask myself how the intervention works:

- For me?
- For the affected person?
- For the current group process?

Critical Topics at the End of a Project

- Technical and organisational problems because the successor organisation (training, service, etc.) was forgotten
- The project manager and team feel as though the end of the project is always being pushed back because the solution can be made even more perfect. This is a point at which the necessary resource check is omitted.
- Projects are suddenly no longer current and simply fizzle out, rather than being cleanly concluded
- Failed projects are "swept under the carpet". There is no review. Bad feelings are not worked through (the project leaves "bodies" behind)
- Where projects do not have clearly defined goals, an unceasing stream of adaptations is demanded of the project team. The project completion date is thus continuously put back.

Aspects of Teams 20

20.1 Teamwork Within the Project

Interdisciplinary solutions often have to be found for the pending problems within projects. This cross-discipline collaboration can only be done by project teams. It takes their energy to work in a goal-oriented manner within the team. At the start of the project work, the newly established team is generally not yet focused. It is worthwhile building a picture of the affected system within the project that is as complete as possible, even if it means that the project gets off to a slower start. If all the affected parties are relatively integrated within the project, this will generally create a heterogeneous project team composition. The various systems with their different cultures and realities are then exposed to one another in this microcosm. A basic understanding of different personality types will support and encourage a constructive examination. The team energy can develop.

Project Groups Have Valuable Advantages in Terms of Output and Better Achieve Complex Goals

Complex goals incorporate a high proportion of unknowns. Many different subject experts and representatives of different company functions have to work together to devise the solution. This task is ideally passed on to project teams.

With their wide variety of knowledge and scope of information and experience, interdisciplinary and cross-organisation project teams develop a powerful advantage in terms of performance that would otherwise not exist anywhere within the organisation. Obstacles to communication within the hierarchy are broken down by direct discussions in the team. The collaboration of all the participants supports the problem solving process. With the right capabilities and level of commitment, the team quickly comes up with solutions that are based on good technical skills and are broadly accepted.

J. Kuster et al., *Project Management Handbook*, Management for Professionals,
DOI 10.1007/978-3-662-45373-5_20

Ability to Achieve Consensus and Acceptance Are Greater with Team Solutions

Projects often take place in an environment of entirely conflicting interests and demands. To develop common impetus and to implement the devised solutions, it is essential that the people representing the interests of the various organisational units identify with the project's results.
The shared development process increases the learning experience and knowledge of the limits of organisational, financial or personal feasibility among the team members and the various offices within the organisation. Willingness to seek consensus and accept the solutions devised in the project team increase. This development only occurs outside the project team, however, if the project team members succeed at least partly in taking the solution process back to their originating organisation.

Together It Is Easier to Set Off for New Shores

Sailing in uncharted waters means leaving the familiar behind. And this always gives rise to uncertainty and fear. The participants derive strength from doing this together, and are braver and more willing to take risks. This can lead to a departure from the known paths, and an ability to think and try out truly innovative and unusual possible solutions.
The enthusiasm of a team can encourage individuals to overstep their previous boundaries and motivate them to achieve outstanding performance. This reciprocal incitement and challenging, combined with encouragement and support, makes creativity, originality and innovation possible to an incredible extent within a team that works well together.

This feeling of "United we stand" does have its risks, however:

- For a start, studies have shown that a strong group spirit in the individual leads to suggestions, ideas and assumptions that come from the group no longer being critically checked or questioned with the necessary care. Everyone assumes that the others know what they are talking about, and they are fortified by the mutual feeling of grandiosity. This leads to decisions that may have catastrophic consequences for the project and organisation under certain circumstances. This effect is further intensified if the team includes people who are high up in the organisation's hierarchy. A systematic problem solving method, sophisticated decision-making mechanisms and the inclusion of critical thinkers in the project team reliably help to avoid serious errors or ensure that they are discovered at an early stage.
- Second, as the sense of belonging increases within groups, a tendency develops for the members to match their performance levels to one another. As a consequence, outstanding members can only give their best performance if

they are integrated into a group of similar top stars. It makes little sense to include a top performer in a project team that otherwise consists of average players.

Working in the Project Team Can Be Fun

Man is a social animal. Working in a team responds to the innate basic need for contact and involvement with other people. Belonging to a group in which great work is achieved together can trigger a powerful sense of satisfaction and even happiness. The members immediately see that their contribution is appreciated by other members and is helping the team to move forward together. Such experiences give meaning to what people are doing. These are all known to be important motivating factors that can lead to great commitment and dedication.

Project Teamwork Is Time-Consuming and Cost-Intensive

A considerable amount of effort is required to ensure reciprocal agreement, information and coordination within project teams. The effort involved must not be underestimated, even if the organisation is optimal and suitable tools are used. Project team meetings are generally expensive and time-consuming. This fact may discourage many managers from delegating their best employees to projects. The managers thus ensure their own success and the success of their department in the short term, but they also take from the company the opportunity to develop in its own right in the medium and long term. However, if a project team is used appropriately to resolve complex problems, the cost/benefit ratio will immediately be more positive.

Project Teamwork Can Be Very Frustrating

When things are not going as they should, project teamwork can also trigger a great deal of disappointment. Even if the members are highly motivated and identify with the work, it can still be very difficult to drop ideas members have developed and adapt to new turns of event. In these situations it makes little difference whether the stimulus comes from within the team's own ranks or from outside.

Be Careful What You Wish for...

Successful project work demands considerable freedom to act and good faith. Within the context of the project order, the members must be able to develop and try out their own ideas and determine their own ways and methods of working. In the course of the project, the members are intensively exposed to the strengths and weaknesses of existing products, sequences, organisational structures, etc.

These experiences often turn project members into critical and independent-minded employees. And this is something that every organisation should want, but it does require managers to be able to correctly handle these employees, who are not always easy to lead.
Conversely, the project worker learns exactly what he is capable of doing in the freedom of the project work. There are times when he will find it difficult to respect the stricter standards of the line organisation. Strengths that may help the line organisation to develop further can be derived from this area of tension. Viewed in this way, project work can be an opportunity for the company to develop in the widest sense of the word.

20.2 Criteria for Putting Together the Project Team

The following key qualifications should be ensured when selecting the project team members:

- Team leader and project manager
- Team members and subject experts
- Ambassadors

Team Leaders, Project Managers

The team leader is there to maintain collaboration within the team. He ensures that all members of the project team have functions and roles suitable for them. In the broadest sense, the project manager is the most important ambassador in the whole project team. He must be particularly good at this relationship-building work. He needs to be able to lobby people, look for support, bring decision-makers into contact with one another and encourage good discussions between them. This criterion relates more to the personalities of future project team members, but there are other aspects that need to be taken into account when putting the team together.

Team Members and Subject Experts

All the important areas of knowledge needed to solve the problem must be covered by the team members. The members should be established specialists in their area and be fully aware of operational reality. Although they are insiders, they must not have lost their feel for the operational environment. They should be informed about the latest developments in their subject areas and know where to obtain the appropriate expertise and experience, even if they do not have it themselves.

Interdisciplinary collaboration requires openness towards and respect for other disciplines. A changed perspective creates new "truths". A team member should be a good communicator, i.e. able both to listen and to inform other people.
Project team members should be able to represent with dignity or bring their originating organisation or interest group (trade unions, users, women, men, general population, etc.) into the project team.
Their role will expand as the project progresses. Now they can no longer simply bring their own causes to the project. They increasingly have to contribute to ensuring that the devised solutions are understood and accepted in the project surroundings and in their originating organisations.
Teamwork can sometimes be charged with tension and very contentious. This requires a healthy dose of self-awareness and steadfastness, combined with flexibility and adaptability. Often people have to lower their sights in favour of achieving a common whole. This can be a painful and difficult process for some people.

Ambassadors

Ambassadors can also have their own interests or be the mouthpiece for a cause. They often have a good network of contacts. The causes of affected parties and users must be properly represented in the project team by a delegated member. An ambassador increases the chance that the team will take full account of the operational reality. It should be ensured, particularly with problems which will need broad acceptance in order to be resolved, that the forces opposing the project are identified at an early stage and are brought into the project.

Who Puts the Project Team Together?

It is officially deployed and then relieved of its responsibility on completion of the project by the project owner. The project manager must have a decisive influence on the choice of members, however. In all organisations, he is reliant on the cooperation of senior managers. To allow them to delegate the right people to the project teams, the project owner and project manager will need to draw up a requirement profile for the future project team members. The criteria described below can help with the selection:

The Goal for the Project Phase Determines the Team Size

The composition of the project team can change from one phase to another. A smaller "pioneering" group is beneficial for feasibility studies and procedure planning. On the other hand, if the aim is to achieve the widest possible acceptance, then a larger, representative team will be better.

This can create a dilemma for the project manager: If all affected areas need to be represented or the project owner puts together a massive team, the group will be much too large and cumbersome. It has proved to be useful in such cases to create a project management team (core team) that does the essential preliminary work, ensures coordination with all sides and then submits the preliminary work to the "extended project team" for its decision.

Different Skills Are Needed at Different Times

Various skills are needed in the different project phases. For example, a lot of creative thinking and "tinkering" are required during the concept phase, while the implementation phase demands organisational talent and marketing expertise. These skills cannot always be found in the same people, so it can be sensible to change the project team members in the course of the various project phases.

The following critical points should be noted when changing project team members:

- There must be careful and detailed handover work carried out and recorded between the team members
- The handover must determine for which problems the previous team members are "liable", and the extent to which they will have to be available to their successors to help resolve them.

People with No Time Are Best Left Alone

Team members who have too little time for the project or do not wish to spend too much time on it (say $\leq$20 %) will hamper the collaboration and progress. It is thus important to clarify at the start:

- How much effort the members are likely to have to put in
- Whether their managers are willing to release them to the necessary extent

Experienced Realism Versus Youthful Enthusiasm

Experience helps the company to avoid repeating mistakes. Inexperience, in contrast, encourages more critical questioning, braver implementation and more unconventional thinking. Every effort should therefore be made to create a balanced team. Another possibility would be to split this task between the project management committee and the project team.

Not Too Much Harmony in the Team

Every project manager would wish for the "chemistry" to be right in his team, allowing people to work together harmoniously. This wish is understandable, but not always sensible. Tensions that arise from the problem or the opposing interests of different organisational units are not necessarily a bad thing. Indeed, they can be used within the project team as an opportunity for clarification.
Problems that are avoided or shelved while working on the problem have a habit of dramatically resurfacing later. Dramatically because the changes needed to costs and schedules increase disproportionately as the project progresses.
The recommendation for the project manager could thus be expressed as "Save your need for harmony for out of hours and bring critical and unorthodox thinkers into your team. Ask for fresh impetus and realistic thought in the project work." Care is needed not to ask too much of oneself and the other project team members. If the right balance is not struck, the work will be hindered too much, resulting in losers rather than winners.

20.3 Team Leadership in the Project: A Service

Project teams consist of people who are working together for a common goal. The project manager needs to succeed in creating a climate of open and honest discussion. Good information about the closeness to reaching goals and the time frame, communication, clear division of labour and the creation of a culture of collaboration are the critical success factors. They will determine whether the goals can be achieved.
Leading a project team is exciting because it is not generally a disciplinary position, so there is no responsibility for personnel. This means that the team leader needs to have and use special capabilities. Leading a project team differs from a line management position in various respects:

- The project members do not generally report directly to the project manager to the full extent, i.e. he does not have comparable authority to give instructions (formal power)
- The project team members are subject experts and thus often have superior technical knowledge to the project manager
- The team may include members who are above the project manager within the organisation's hierarchy
- The project team members represent the ideas of their originating organisations, interest or function groups, and identify initially with their role as representatives
- The project team only exists for a limited time, i.e. the team members consider the extent to which they want to get involved with and commit to the project

For the project manager to lead the team successfully, he must understand the processes that occur in working groups. He should know which leadership behaviour triggers which reactions. The following figure provides an overview of the various aspects of team leadership. These are often models and summary versions of highly complex topics, i.e. reduction has resulted in a degree of simplification and a few important aspects have been omitted (Fig. 20.1).

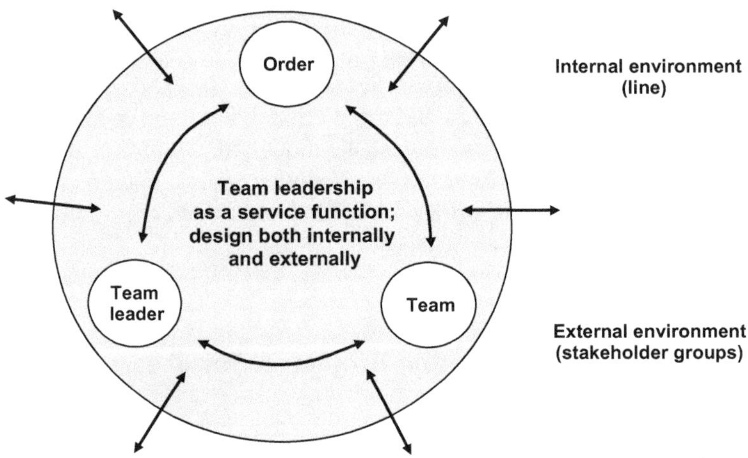

Fig. 20.1 Aspects of team leadership

Internal Environment

The project manager has accepted the role of achieving the agreed goals together with the project team. This comprises a wide range of tasks such as planning, organisation, facilitation, team development, coaching, information and communication, controlling and resource management with the line. It is important to optimise the working conditions, forms of collaboration and flows of communication so that the project team can work effectively and successfully.

Processes that can be influenced and controlled are in play within a team. Knowledge of group dynamics helps us to better understand the mechanisms of these group processes and to learn what can be used to promote positive collaboration.

A team consists of individual members who all bring their personal interests, fears and hopes to the project. Examining the topic of motivation will show how commitment and enjoyment can be awakened and maintained.

Opposing interests, differences of opinion, incompatible personalities and numerous other reasons can cause tensions and disputes to arise time and time again in project teams. The project manager's ability to deal with conflict situations within his team is thus particularly important.

External Environment

Experience shows that projects can often trigger uncertainty, resistance or even conflicts within their environment on account of their function (change or renewal). Whether sufficient attention was paid to the conditions in the environment will come to light no later than the implementation process in which the delicate project solution seedlings leave the warm greenhouse of the project team and are planted out in the harsh reality of everyday life. Here it is the further growth will ruthlessly determine whether the project succeeds or fails.

20.4 Roles and Functions in the Project Team

The term "role" hides a whole wealth of expectations of a position or function (=role expectation).
Roles are defined (in a job description or specification) by describing the task, authority and responsibility (T-A-R) and expectations in terms of conduct (observance of standards, values, etc.).
It is relatively unimportant for collaboration and integration whether this task distribution is specified from outside or whether the group structures itself. What are important, however, are the following points:

- The individual group member must be accepted in his role and functions
- The role holder must himself accept and understand his role and functions
- Unclear allocation of tasks and an inadequate understanding of the role will prevent the group from achieving its full capacity to perform
- If a single person is performing multiple roles at the same time, this must be transparent.

Note

With inadequate controlling or a poor choice of resources there is often the risk that individual people will be overloaded or role conflicts (concerning goals or interests) will arise:

- The expectations of individual project roles should be considered jointly by the project team at the earliest possible time. This will involve clarifying and making transparent the different expectations of individual roles (calibrated perspective) and the communication process within the team.
- It is important to differentiate between roles that are performed by individuals and those carried out by the project team or parts of it (sub-project teams).
- Role definitions are always project-specific, even if certain individuals always carry out the same project roles.

Role definitions (job description, specifications) are always aimed at the specific role holder and his deputy. There are things that a role holder (e.g. project manager):

- MUST do (he must manage the project and lead the team)
- SHOULD do (treat the team members equally)
- CAN do (organise a shared event outside the project work)

20.5 Promoting Collaboration Within the Team

First bring the best people together to form a team. It is not yet ready to work, however. Teams pass through various stages before they are fully capable of performing. Managers have the task of gradually encouraging collaboration. Autonomy, trust, shared goals, an effective flow of information and accepted standards and values must grow in order to enable cooperation. At the start, the main topic should therefore be the different perspectives and realities of the individual members. This discussion will also promote mutual understanding. The polygon of rapprochement shows how the project manager can develop the area of common understanding (Fig. 20.2).

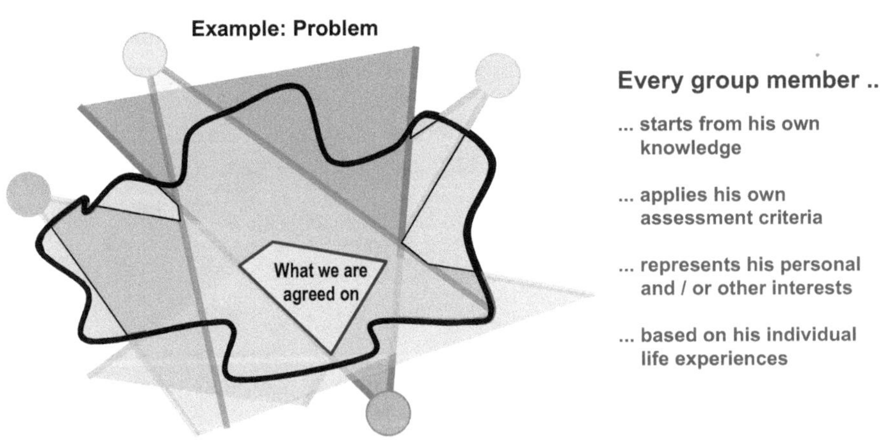

Fig. 20.2 Polygon of rapprochement

Conduct and Behaviour, Commitment

In addition to technical and methodological knowledge, a successful project manager needs to be able to implement this knowledge to suit the situation. The more personable and authentic his conduct, the more sustainable the effect of his leadership will be. In this context, the IPMA International Competence Baseline (ICB) speaks of a project manager's behavioural competence.

The way in which a project manager acts as a leader depends both on the other person's cognitive ability and on the conduct of the project manager or the way in which he acts as a manager. The behaviour depends on the personal inner attitude and mindset he has towards other people (Fig. 20.3).

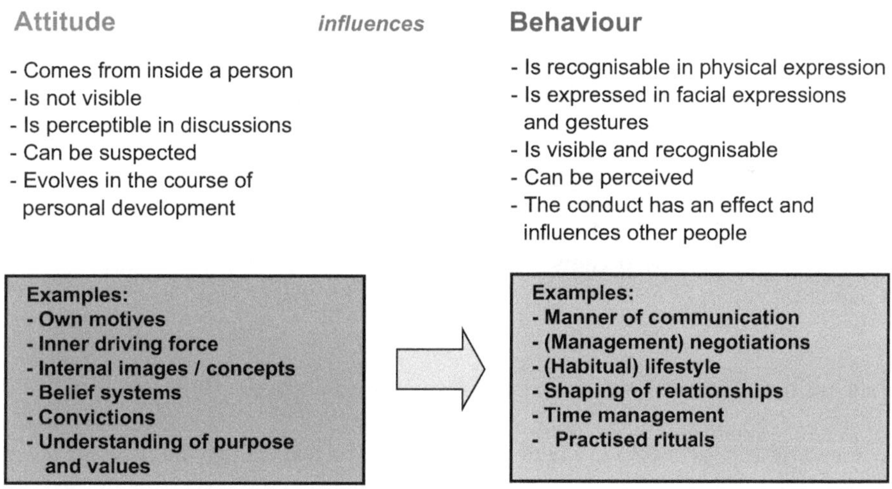

Fig. 20.3 Attitude and behaviour

Examples

Someone who is open, interested and approaches other people cordially will generally have a larger network of contacts than someone who is asocial, distrustful, reserved and reticent.

A project manager who ostensibly appears cooperative and open, but dismisses new ideas as unrealistic (because he is distrustful at heart and regards other people as incompetent), is acting contradictorily. The project workers perceive the incongruent leadership behaviour of the project manager as inauthentic. This makes him implausible and his influence as a leader diminishes.

The project manager's ability to lead and motivate is part of social competence and essentially depends on the image of humanity, inner mindset and attitude. A person's image of humanity develops along with his personality. It is not visible, but it is undergone and experienced over the course of working together.

Mindset and attitude shape the way a person behaves. A project manager should be credible, exude self-confidence and self-assurance and should be perceived by his project workers as a natural authority with appreciation. He will then generally have considerable influence over his project workers' readiness for action.

Commitment

If a project manager is respected as a person of natural authority, the project team members will normally endeavour to adopt this behaviour as well. In this sense, the way in which a project manager behaves has a big effect, acts as a role model and encourages courtesy. The project workers become even more committed and thus carry out their work with a greater sense of personal responsibility, and are more likely to finish on time and within budget.
Personality and team development require an atmosphere in which an open and honest exchange of familiar and different viewpoints can take place. A recently created project team is not yet generally able to discuss this and similar interpersonal topics directly and openly.
If the project manager has sufficient social competence, he will be able to develop the necessary atmosphere in the team himself. However, if he has a blind spot with respect to his incongruent leadership behaviour, it may only be possible to draw this to his attention through feedback. This can generally be provided by an uninvolved person (e.g. a coach or organisation developer) provided that the affected parties agree to this being done.

Teamwork Takes Place at the Content and Relationship Levels

When a group of people works together, this affects both the content level (technical and organisational level) and the relationship level.
The content part of the examination takes place **at the content level**. Project goals are defined, analysis results are discussed, approaches to solutions are sought, the project is planned and organised and meeting dates are agreed. Intelligence plays a dominant role in the work at this level.
At the relationship level, feelings, needs, sympathy, antipathy, values and standards play the more important role. This level is thus also known as the psychosocial level.
In a situation in which several people meet for a discussion, they define the relationship that they assume with respect to one another by the way in which they deal with one another. Who likes whom how much or how little, how much is which member appreciated, who has the say here? Informal standards also determine how the group members behave. We all know them, without ever having to discuss them. Who needs to be called Sir or Madam? Will people be allowed to get away with turning up to meetings late? Should we all go for coffee together? We shall discuss later how these standards come to exist. Such topics can also be discussed and agreements made. They will then become the formal standards or ground rules of the group. Things that must not be discussed, e.g. who is in a relationship with whom, who earns how much, who messed up which project and when, are designated as taboo subjects.

The content and relationship levels are always linked to one another, and the latter is the dominant level, so severe interpersonal disruptions will always have a negative effect on the content-related work. Such disruptions (differences of opinion, resentment, violent discussions, affectedness, bad blood) may be minor and negligible. They are essentially a normal occurrence when groups work together, and are not an indication that the team leader or team members are inadequate.
It is not until the disruptions increase and the group is unable to resolve the difficulties or until the same problems keep on recurring that the project manager needs to ask himself what is behind them.

Disruptions at the content level are often caused by problems at the relationship level. Possible indications of relationship problems are:

- Mutual recriminations or attempts to justify behaviour
- It is impossible to agree dates; attendance is very tentative
- Discussions about differences of opinion are mixed with personal attacks: "How can anyone talk such rubbish? Why does your department always receive special treatment? What you are telling us is old news? Where would we be if we all wanted that?"
- The team becomes bogged down with irrelevances. Decisions made on one day are questioned the next. The teamwork feels like wading through treacle. Two steps forwards and one step back

The job of the team leader includes making sure that problems at the relationship level are identified and discussed. The following resources are available:

- Both the content level AND the relationship level are included in any evaluation of the work process. Everyone gets say what they feel about the collaboration
- The team agrees ground rules that allow every member to discuss disruptions
- When tensions increase too far, time is taken to clarify the disruptions. The process analysis may be used as a tool

Some people may initially be unaccustomed to this type of discussion about the sensitivities of collaboration and find it rather peculiar. But if such conversations are held regularly within a group, the participants discover how helpful such an exchange can be in clarifying issues. After a certain time, problems at the emotional level are discussed at an early stage, making them easier to resolve since positions have not yet become entrenched and a whole series of misunderstandings can be avoided (Fig. 20.4).

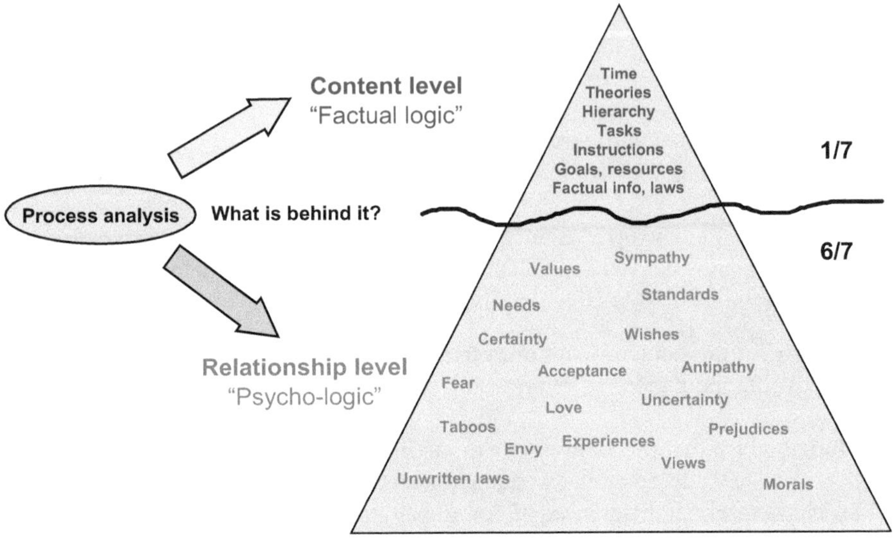

Fig. 20.4 Relationship diagnostics (iceberg model)

Theme-Centred Interaction (TCI)

Teams that define their working agreements (ground rules) according to this model described by Ruth Cohn create for themselves an instrument that will both facilitate and enliven interactions, even if it is first necessary to learn and practise using them. Three factors are treated equally in groups that work using TCI:

- The person or individual subject (the "I")
- The group as a whole (the "We")
- The theme as a task of the group (the "It")

As long as there is always a dynamical balance between these three factors, the conditions will be perfect for the members as people, for the interaction of the group and for the completion of the task to be done. Personal fulfilment, cooperation and task resolution go hand in hand.
The triangle in this model is surrounded by a transparent, multi-layered globe that symbolises the surroundings. The surroundings in terms of time, space, nature, people and everything that is, was or will be (Fig. 20.5).

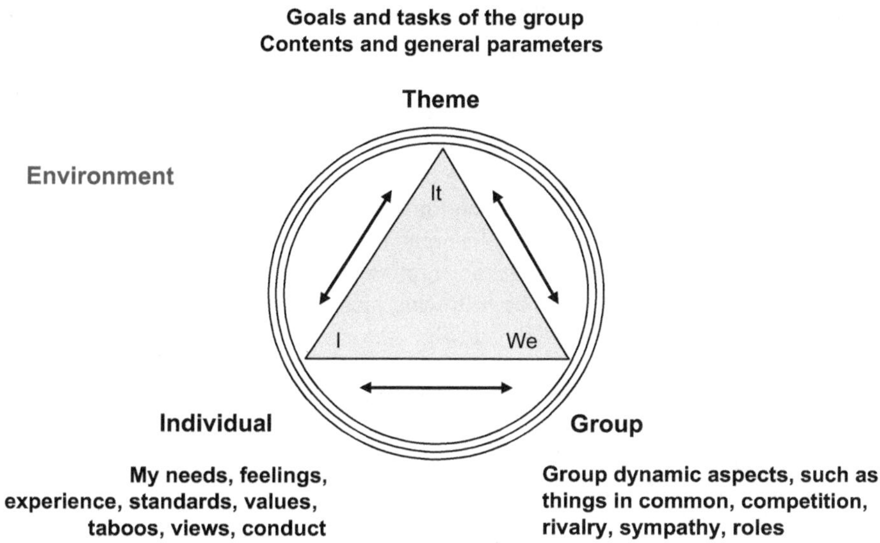

Fig. 20.5 The TCI triangle (Ruth Cohn, 2011)

One of the tasks of the successful project manager is to balance within the project work the needs and demands of the three factors (I, We and It) with due regard to the relevant surroundings. This is not always possible. There will always be phases that are marked by one of the three factors. Under normal circumstances within the company (including project work), very many people tend, for example, to concentrate as much as possible on the content level or the theme, and thus neglect the relationship level (I and We factors).

The work and development process of a group will be more successful the more balanced (=treated equally) the three factors are. To allow this to happen, we apply the following rules to our common agreement concerning work within the team:

- **Disruptions have priority**: They are symptoms of unspoken interests, tensions and conflicts. They hinder productive collaboration and priority must therefore be given to eliminating them.
- **Side discussions take precedence**: They are annoying and usually important.
- **Be your own chairman**: Be aware of your inner realities and your surroundings.
- **Represent yourself in your statements**: Assume personal responsibility for whatever you say.
- **Be authentic and selective**: Be aware of what you think and feel.
- **Avoid interpretations**: Speak instead about your personal feeling or response.
- **Make your questions as genuine as possible**: and give a little background about your question. Information questions are important for understanding something. On the other hand, questions that do not really ask for information are dishonest.

20.6 Motivation in the Project Team

Projects often take the participants to the very limits of their performance. The "carrot and stick" approach has unwanted side effects. This is why it is a good idea to make people aware of and balance expectations and visions.

By a motivated project team member, we mean a person who is distinguished by a behaviour that looks on the tasks at hand positively and is active and committed. The motives that lead to such behaviour may differ greatly. A motivation theory-based approach assumes that the employees are looking to satisfy personal wishes as a result of their work. The following picture shows five motivation fields to which such wishes may be assigned (Fig. 20.6).

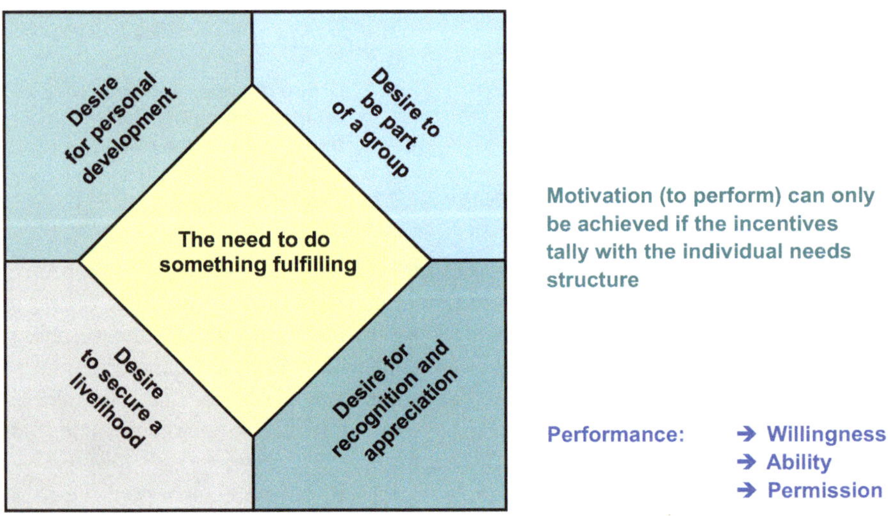

Fig. 20.6 Principles of motivation

When a decision is made to commit to a project, needs from all five fields of motivation may apply. It can be assumed from this that people with many different motivation structures come together in a project team. The team leader will find it advantageous to understand the various fields of motivation and be able to create the appropriate framework conditions that allow them to be satisfied.

Let us now consider the individual fields of motivation in greater detail and discuss the question of the extent to which the project work recognises and permits the satisfaction of the relevant needs.

The Need to Do Something Fulfilling

There is a direct connection between the meaning and the personal commitment that accrues for someone from their activity. No one can champion something for a

long time if he is not convinced about it or if he is unable to see why his contribution is important and sensible for the matter to succeed.
The strength of the project team members' commitment thus depends on whether and to what extent the project goal, the procedure, the solutions, etc. are identifiable and sensible to them. If these elements are not perceived sufficiently or if they are contrary to their own interests (e.g. job cuts, merger negotiations, organisational changes), then it is very important for a discussion about them to take place between the project owner and the project team members, clarifying whether the right people have actually been delegated to the project team. If it has not been possible to establish the necessary motivation right from the start, particularly for demanding projects, the chances of success are correspondingly low.
If a project team member feels that his contribution is not really necessary or sufficiently qualified (e.g. his commitment makes no sense), this person will not be able to commit himself properly. Thus the expectations on both sides must be clarified when the collaboration starts, and it must be made clear to the potential project team member what is expected of his work.
General statements of disinclination or despondency may be signs that the team is starting to wonder how important the project actually is or why this or that step needs to be done. A joint review, checking goals and procedures, can help to get back on track or restore the association with operations from the project environment.
If this is worked through with care when the project starts, i.e. if a situation analysis and goal definition are carried out, then it should be easy for most people convey the meaning of project initiatives and thus to have a positive influence on the motivation of team members.

The Need for Development Opportunities

There are several reasons why the opportunity for personal development can have a powerful motivational effect:

- The significance of ongoing further development increases in line with changes in the professional environment. At this time, the world of work is changing so fast and fundamentally that the development aspect is becoming even more of a question of survival.
- The image of man applied in humanistic psychology assumes that the life of a person is a continuous process of development and learning. Spiritual, mental and physical disruptions will occur if this process stops. Work assumes a central role in the human existence. For this reason, some work psychologists claim that work should be made sufficiently demanding and challenging to suit employees' skills and abilities. If this is done, work will be fun, found to be interesting and will elicit corresponding commitment.

Projects are generally initiatives, a large proportion of which is new and unknown. They also offer equally extensive learning opportunities to the participants.
If the project manager succeeds in making team members aware of this opportunity and supports them with suitable actions, then he will have a "motivational trump card" that makes him the envy of many managers. Support measures as part of a project can include:

- Delegating interesting jobs with correspondingly high responsibility and authority
- Visiting events on technical and methodological topics
- Visiting other companies and organisations, trade fairs, organising technical lectures, book studies
- Collaborating with colleges, research organisations, consultants, supervision
- Supporting and maintaining intentional and visible opportunities for on-the-job learning

The costs of such measures should be taken into account when drawing up the budget. In many companies, project work is often viewed as a performance test or opportunity to shine for potential managers. But this is contradicted by the fact that there is rarely a suitable feedback system that allows the performance of the "test subjects" to be reported systematically. It is important for motivating the affected team members that the project manager both discusses their performance with them directly and ensures that this information flows back to the appropriate offices within the organisation.

The Need to Belong

We have already discussed that man, as a social being, is reliant on being in regular contact with other people. Of course, there are large differences between individuals, but studies show time and time again that the membership of a work community is of extraordinarily high importance in our society. For example, very many people find their life partner at work, and people who suddenly become unemployed often lose their entire network of social contacts, not just their livelihood.
There is often a constant feedback process between the members of a team: ideas, individuals' suggestions are accepted, and may be further developed or rejected in the group. The individual is compared with the other members and in this way can establish a realistic picture of their own strengths and weaknesses. This examination contributes to the further development of every individual, and of the group as a whole.
If the project manager succeeds in turning the encounters between team members into a positive group event, this can create high and permanent commitment in the individual member. Everyone will take care not to disappoint their colleagues, with whom they enjoy working and are important as persons of reference, with a poor contribution to the work.

The Need for Recognition and Appreciation

It is common knowledge that motivation can be encouraged with praise and recognition. What is more surprising is that often very little is done to allow praise and recognition to be expressed, such as:

- Assigning clear and verifiable jobs. This starts with the project manager who, when given instructions such as "Just sort that out" or "Just draft something", has little prospect of being successful. He has no concrete instructions against which his result will be assessed.
- Even the most positive recognition will not have a motivating effect without a thorough critical examination of the results with the person, as it cannot really be taken seriously. However, involvement with an issue will also bring forth aspects on which people can have different opinions or have to be criticised. Nuanced criticism will not harm the motivational effect since the intensive examination of the project shows the affected employees that their work is important to and appreciated by their managers.

Recognition is often expressed after presentations to the management committee or in meetings with the project owner. The team is generally represented by the project manager at this point, so it is extraordinarily important for promoting motivation and team spirit for the project manager to pass on such messages to the team and share the success with his colleagues. He also needs to ensure that he does not pass off the result to third parties as his own personal work, and consciously refers to the team contribution.
In long-running projects, there are often phases without obvious progress or successes, or the work may be full of difficulties and setbacks. To improve motivation and support the will to persevere, it is helpful if the project manager also highlights smaller "stage wins" and mentions progress already made from time to time. He can also encourage the team to give each other positive and encouraging feedback. It is in these phases, in particular, that all participants are grateful for small congenial get-togethers that contrast sharply with the energy-sapping everyday course of business.

The Wish to Secure a Livelihood

If a person wants to make a lot of money quickly, there are more lucrative activities in most companies than project work. Nevertheless, the project workers must not get a raw deal when the "cake" is shared out. Particularly with large, long-term project initiatives there is a risk that the performance of project workers will not come to the attention of those senior managers who are responsible for salaries, bonuses and promotion.
Again it is the project manager who needs to draw attention to this by sending suitable feedback to his project workers' line managers. He will naturally find this task easier if he has access to an extensive and systemic feedback system.

Insights Derived from Considering the Models

The interfaces between the leadership work of senior management and that of the project manager can be clearly seen from the description of the motivating framework conditions. If these interfaces are not clearly defined, there is a risk that:

- The project manager and senior managers will get in one another's way (because both are doing the same thing, quite separately from one another)
- No one feels responsible and it is the employees who suffer

Despite the recommendation to the project manager to allow team members to participate in the management task, there are still tasks that he cannot delegate. For example, a team member will find it very difficult to suggest a colleague to senior management for promotion. Motivation as a process can be represented as follows (Fig. 20.7).

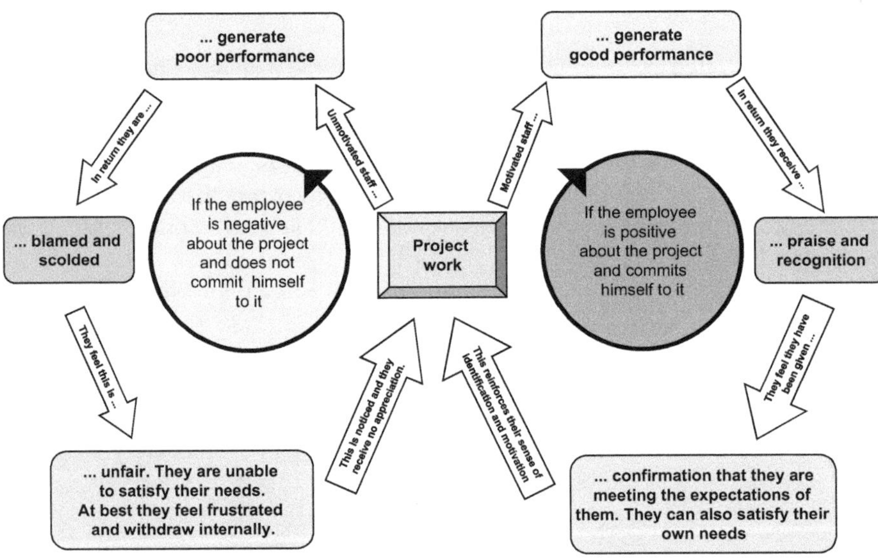

Fig. 20.7 Motivation process

Individual team members have different needs and expectations of their collaboration on the project. Greatest identification and commitment are achieved if everyone finds satisfaction in the project work according to his own personal needs structure. This requires framework conditions that cover all five fields of motivation and give the project manager room to manoeuvre and the team members options.

20.7 Group Development Phases

Just as each person has developed over the course of their own journey, groups also develop their own "life story". The theme, situation, context, surroundings and, above all, the personality of the group members and manager ensure that no group process is ever repeated identically.
A group of people never starts out as a "team" in the sense of a real "We" to which the members perceive an emotional belonging. It is not until this "We feeling" is achieved that the participants are able not only to listen to the others' concerns, but to consider them as well, identify and accept common features and differences and decide how the group wishes to work. This emphasises the fact that it is not possible to work on the topic immediately from the formation of a group. The manager must first help the group to develop such a "We feeling".
Every new interaction as a group means that the group is passing through the development process again. Productivity can be achieved faster depending on how the previous collaborations were perceived and completed or how the new process is managed by the project manager.

The Way to the Team

People need agreed goals and the knowledge that they can be achieved better together in order to become a working and effective team. These are the three levels at which the group constantly has to work as it develops into a team if it wants to complete its job successfully (Fig. 20.8).

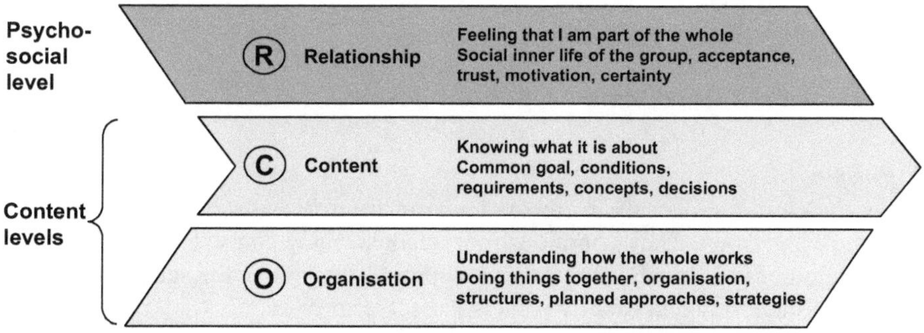

Fig. 20.8 The three levels of team development

All three levels interact with one another. The needs of the members are different. Some members want to work on the themes at the content level immediately as this is often the level at which they feel safe. Other members cannot yet devote themselves to this level at all because things are not yet right at the relationship level. Then there are the uncertainties and logjams, the necessary information is missing or the perspectives needed to grasp the whole are not there. There is still insufficient trust, and people do not yet feel part of the whole.

These are, in the main, the operations at the relationship level that decide whether the individual is ready to actively collaborate, make his contribution and trust other people. The healthier this level proves to be and as the individual perceives himself to be accepted and understood, the more he will feel secure and the more he personally and the team as a whole will be able to do, particularly in terms of quality.

Five Phases of Group Development

Tuckman (1977) developed and subsequently expanded the following model, which breaks the development process of a group down into five phases (Fig. 20.9).

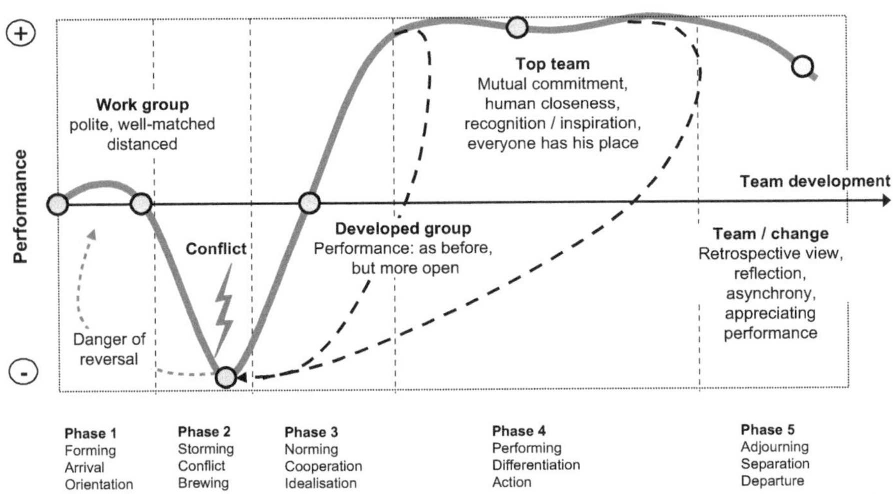

Fig. 20.9 Phases of team development (B. W. Tuckman et al.)

1. Forming

When they join a group, the members harbour different expectations of what will happen in the group. They consciously or unconsciously concern themselves with the question of how much contact and acceptance they have, the status they might achieve in the group and exactly what is expected of them.

Phase 1 is characterised by polite and reserved behaviour by the members.

2. Storming

Once the social standards within the group have been defined, the members start to open up and increasingly demonstrate their "normal" behaviour. Many participants now want to exert influence. They react critically to one another, and personal and technical areas of authority become important. The team leader's style is often criticised and his failings are discussed. The participants are looking to find their place in the social structure of the team and their specific role or function in accomplishing the tasks.

In **phase 2** the participants see one another more clearly, they come to trust one another and are prepared to reveal more of themselves. The new situation has somehow become "more normal", and so has the individual's behaviour. Everyone attempts to assert their position and interests. Roles and status are shared out. Aggression is more tolerated, particularly towards the manager.

3. Norming

After reverting to aggressive types of feeling as in the storming phase, the participants are now able to express appreciation and acceptance for one another. Belonging to the team starts to become fun. This relaxation makes communication more open and the individual participants appear more genuine and enhanced by the room to manoeuvre they now have. Cooperation and consensus are the dominant features of the group's work.

Phase 3 is characterised by a large degree of cohesion. The team has become attractive for its members and new group standards promote more open and personal behaviour.

4. Performing

The participants dedicate their activities more and more to the task. The group energy thus benefits the work almost exclusively. The question of how the team members' needs can be appropriately met has become secondary. Through the division of labour, delegation of responsibility, an effective communication system and a reliable feedback procedure, the group of individuals has become a dynamically working team. Although it can always drop back to phase 2, it now knows the ways to overcome its conflicts.

Phase 4 clearly shows that the development of a group is a cyclical and open process that raises the important themes and problems of interaction time and time again.

5. Adjourning

In the final phase of team development, the emphasis is on the breakup of the team. This phase is characterised by a highly emotionally charged asynchrony of the participants. While some are already devoting their energies to the next project, others are still indulging themselves with the memories of their previous intensive collaboration. In this phase, the project manager provides a comprehensive review and a careful evaluation of the team's performance. He acknowledges this performance and provides structured feedback for all participants.

Phase 5 consists of important steps, such as ensuring that the team members reintegrate back into the line and arranging a clear wind-up at the relationship level. These phases do not necessarily take place in that order, and some phases may be repeated on account of changes and disruptions. Such disruptions occur when the goals of the task change or when the structures or procedures are modified, group members change, i.e. people leave, are absent temporarily or new people arrive, or if the work is unsuccessful, team members become demotivated and unenthusiastic or find themselves under pressure.

The following Fig. 20.10 is intended to help identify the phase in which the group finds itself and to find the right themes and development measures.

Phase	Features	Needs of the participants	Possible themes	Important points for team leaders
Forming, arrival, orientation	Waiting, "sniffing" one another, testing, appreciating the task	Wish for orientation, what applies here, dos and don'ts, who are the others?	Information, orientation, gentle easing into content topics, getting to know one another.	Relatively highly structured, visible to the participants.
Storming, conflict, brewing	Finding roles, "battles", rivalries, attempts to change the task. Criticism about the procedure, concept.	Self-portrayal, self-assertion, getting own opinion across.	Procedural matters, dealing with rivalries and power, creating transparency.	The leader is the guardian of the task. Allowing space for psychosocial issues. Working situationally, i.e. in a less structured manner.
Norming, clarification, cooperation	Good and open communication. "We" feeling, proximity and identification with the team and task, flexibility of roles.	Mutual give and take, the team wants to work autonomously (less centred on the leader).	Both content and relationship themes.	Clearly define the problem, but leave the "How" open. Restrained management. Unmanaged teams increasingly possible.
Performing, enthusiasm for work, productivity	Concentration on results, less structure needed.	The incentives for collaboration are progress, results and successes.	Keep competition within limits. Give support in procedural matters.	Question "harmonious" decisions taken too quickly.
Adjourning, conclusion, departure	"Asynchrony" of the participants. Considerable closeness. Depression on ending.	Frequent need to stay together for longer.	Give help on transferring, conclude the work. What happens next? What is still unresolved?	Manager is guardian of a "clear conclusion, taking account of the content and relationship levels. Learning groups as follow-up!

Fig. 20.10 Group development process

Summary

Teams are social systems that are capable of learning, control and develop themselves and can thus be extraordinarily flexible and powerful.

Change Management and Dealing with Resistance 21

21.1 Change Management

These days, almost every project initiative leads to small or large changes in the core organisation. Innovation nearly always affects some of a company's business processes. It leads to new operational structures, to different organisational units, to previous work becoming unnecessary, to services being defined differently. The changes always lead to the people involved having to change their activities, their behaviour or even their attitude. And in order to take people along on that journey of change, a project management approach that concentrates purely on objective goals will not suffice. More focused project activities and project leadership are needed in order to help people engage with the changes that impact on behaviour. And this becomes all the more important given that expectations around the time taken to implement change are getting ever shorter. The ongoing coordinated measures taken around these changes can be referred to as change management (Fig. 21.1).

J. Kuster et al., *Project Management Handbook*, Management for Professionals,
DOI 10.1007/978-3-662-45373-5_21

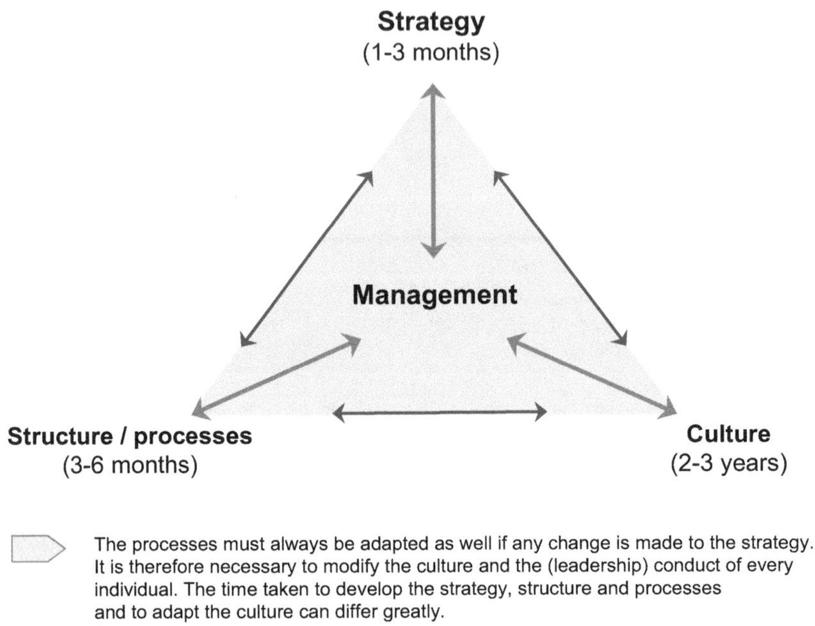

Fig. 21.1 Changes to strategy, structure and culture

The general business model shown above illustrates the necessary changes of strategy, the corrections and adaptations to the structures and processes, and the small or large cultural changes that this always brings with it. All three steps can be part of one project, or they can be parts of different projects. Whereas the classic, fact-based instruments and measures may suffice when pursuing strategic and structural objectives, changes in the cultural dimensions will increasingly demand psychological instruments and measures. This means that as well as classic project management skills, the project manager needs the skills and expertise to shape and manage these processes. The behaviours of each individual and of the group as a whole are what makes up the culture. The extent to which such further development succeeds will depend largely on the extent to which the familiar, the known and trusted, can be adapted to new circumstances, or even built up from scratch again (Fig. 21.2).

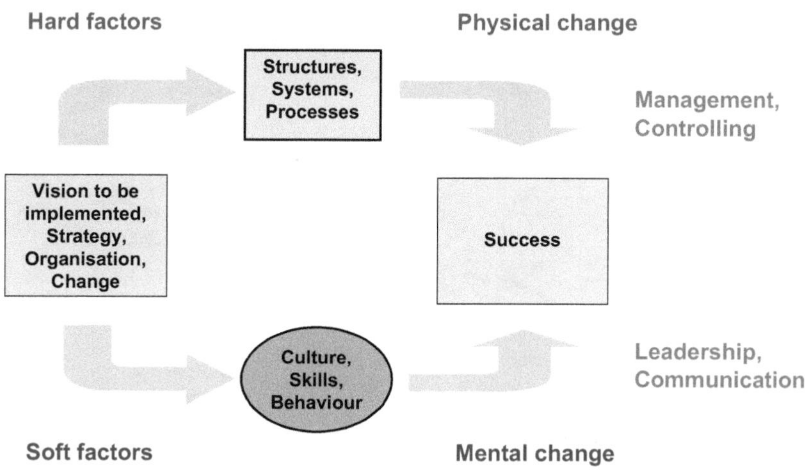

Fig. 21.2 Two levels of change

In order for the changes to really work, to become established, it is crucial that senior management identify fully with the changes, and give their full commitment. A lot of change initiatives fail or are ineffective precisely because this commitment from the top is missing. In turn, that means that the crucial message about the necessity of the changes and the reasoning behind the changes is not being constantly communicated and confirmed. The biggest impact is made when management actively participate in the change process, which means they are affected by the changes to the same extent as every other staff member in the organisation. The changes that the organisation is seeking to implement will only find wider acceptance if management set the right example.
An important leadership task is to recognise developments at an early stage, in order to act by introducing appropriate and proportionate changes. Changes take place all the time. The question is, how actively are they being managed? If we do not deliberately engage with it, change will simply pass us by. Changes in organisations are actually brought about by the people that work there and by their readiness to change things about themselves and to engage with new developments and learning processes. Significant innovations are both a challenge and an opportunity. They represent a big challenge for everyone involved, and they also need time. When the impact of change processes is underestimated, there is a very real risk that only the negative aspects will be felt.

Change Process Model

Using Virginia Satir's change process model, the development of individuals, groups or organisations has five phases. Each phase involves different degrees of readiness to engage, and has different time frames (Fig. 21.3).

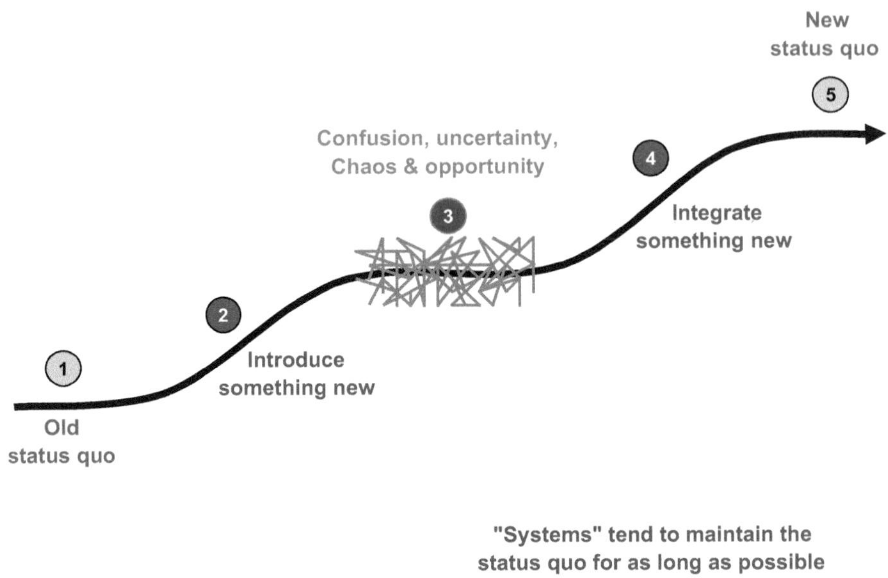

Fig. 21.3 Stages of change processes (Virginia Satir)

These changes normally begin at the an organisation's highest hierarchical level, and are then implemented right the way down to the lowest level.

Phase 1: Status Quo or Equilibrium

This phase (which is usually where things are at the start of the process) features equilibrium and continuity, and gives a feeling of safety. The familiarity of the routine gives people a feeling of being efficient. In practice, we frequently filter out a need for change in order not to endanger this stability by not acknowledging or refusing to acknowledge reality. The longer this phase lasts, the harder it will be for the people involved to engage with change.

Phase 2: Upheaval, Something New

In this phase, there is a growing awareness that changes are necessary. There is usually just a small circle (e.g. management) that recognises this need for change, then plans and initiates something new. It is during this "thawing out" phase that the willingness to embrace change is developed.

Phase 3: Confusion, Uncertainty, Chaos and Opportunity

When something new is introduced, there is always confusion to begin with. This is often the first stage in which an organisation's staff become involved in the change process. They have to jump on board what is effectively a moving train, and then (with varying degrees of willingness) try and adapt to the changes and to new things. This is usually done using old methods and behaviours, which are often not

really suitable for the new situation. They will encounter situations where nothing works any more: the old, familiar ways no longer work, and the new ways are unfamiliar, and maybe even unknown.
Which gives rise to chaos, both for the people and for the whole organisation. The only way to real change is by positively living through this phase. However, it is important that this phase should not last too long, as it involves anxiety, uncertainty and a lot of expensive frictional attrition because it is often not clear whether the old still applies and whether the new applies yet.

Phase 4: Integration
People's ability to approach phases of confusion and crisis positively, despite all the inherent difficulties, is a key factor in reaching phase 4. Once the first successes arising from the changes begin to become visible, the new will become more attractive. Then comes an integration phase, when the new reality is tried out and when new behaviours are tested out, until a feeling of safety gradually sets in again. The new status quo then becomes "frozen in place". Mistakes are also part of the learning process. It is important to recognise them and reflect on them, but they do not necessarily need to lead to an immediate change of course. If the management team cannot offer this tolerant approach to mistakes, then instead of progress things will fall back to stage 3.

Phase 5: Stability and Status Quo
After the integration phase there is a new equilibrium phase. This consolidation phase is important for deepening of the experiences and so that the new knowledge and skills become deeply rooted. But in today's world, it is safe to assume that the phases are becoming ever shorter as they get caught and overtaken by other changes.
Progress through the five phases will not be linear. Each time a major difficulty is encountered, some people will fall back to the chaos phase. Anxiety and resignation gain the upper hand again. But over time, the frequency and the duration of these relapses will decline. The change process begins to take hold.

Asynchrony
The change phases in the organisation will occur at different times for different groups. That means that not all of the people or hierarchical levels involved in the change process will be in the same phase at the same time. That can lead to additional dynamics in the project. This asynchrony comes about because the people who have been involved with the changes for a long time, and are therefore further along in the process, have knowledge that newer participants do not have. They therefore become impatient for quick results, which in turn increases the pressure on their colleagues (Fig. 21.4).

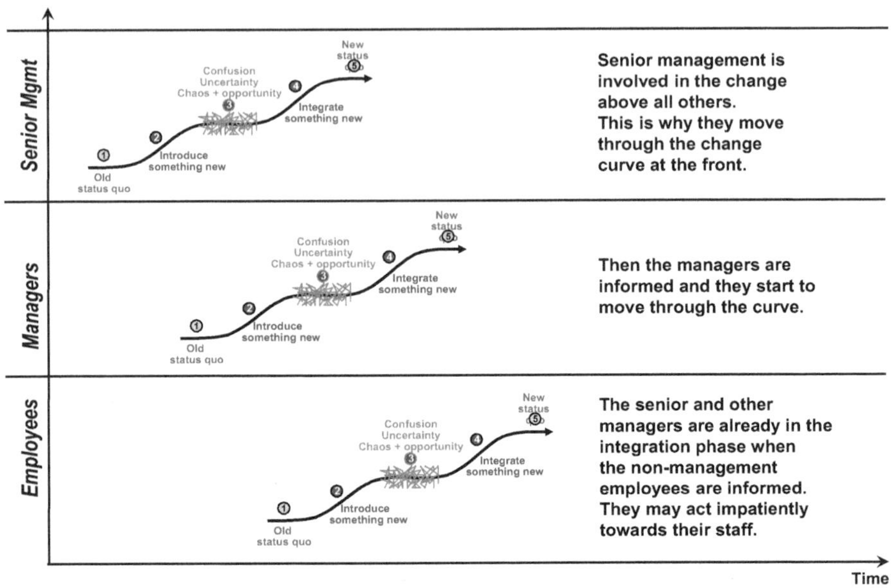

Fig. 21.4 Asynchrony of change phases

This asynchrony places high demands on the project manager, and leads to tensions created by the differences between senior management's expectations of fast implementation and the creation of powerful solutions on the one hand, and on the other hand the need on the part of the project staff for information, time to process the changes and to feel involved in shaping them.

The Effects of Change

Changes involve many small adaptations. These can lead to chaos, uncertainty and conflict during the transition phase:

- People have to take on board and become familiar with new unfamiliar tasks and behaviours. The old values no longer apply.
- People have to give up familiar, routine processes and the associated safety. Previously established boundaries change.
- Everything has to be examined, changed and if necessary, thrown overboard.
- Key people change (management, colleagues, team members).
- Established information flows and relationships no longer work.

Basic Principles of Change

- There is no such thing as change without resistance. The thing that should cause concern is resistance itself, but an absence of resistance. Without change, human development stagnates.

- Constant change without limits, without resistance leads to uncontrolled growth, chaos and breakup.
- Resistance is a (fair) offer to find the solution in a different way.
- Everyone has a limit to the amount of change they can process.
- Involve the people who are involved and affected, and build in feedback loops.
- Transparency and openness in your approach are essential.
- Make sure you also communicate about the drawbacks.
- Make the advantages and long-term benefits understandable.
- Carefully prepare staff for new situations and enable them to deal with the changes.

Logic and Psycho-logic

If the project is characterised by a lot of changes that affect people, it is useful to adapt the project to match the interventions that will be necessary as part of change management. These may initially appear to delay the project's progress, but they actually make up for that by the time the project is implemented or fully handed over to the line organisation because they lead to higher satisfaction levels and to better acceptance of new ideas. The following aspects therefore become more important:

- Developing commitment
- Dealing with resistance
- Communication
- Developing skills and attitudes

This is basically about ensuring that the areas that are of deep concern to someone when their person or position is changed are addressed, and that the naturally occurring resistance is accepted and processed. That involves widening our perspective, adding a psychological dimension alongside the logic dimension in the running of the project (Fig. 21.5).

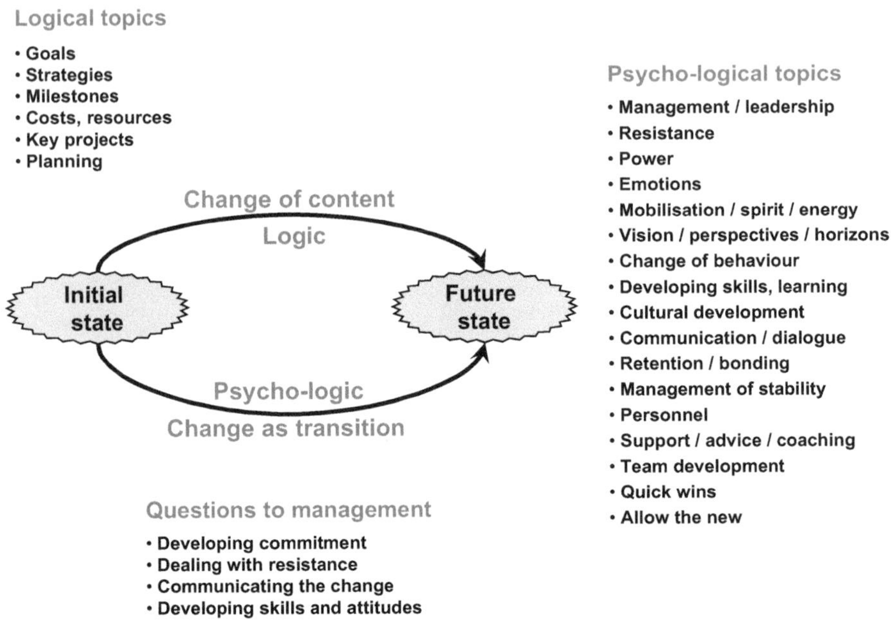

Fig. 21.5 Overview of change management

There is no such thing as change without resistance. An important objective is to help the affected people to move from resistance to a productive attitude. This usually involves more than a one-off action. Depending on what happens in the project, it may need to be repeated several times.

Ideally, we would like everyone to become fellow campaigners. That won't actually happen. But by using targeted interventions, we can try to convert the energy of resistance or of uncommitted people into active cooperation, and we can increasingly involve and confront people who are passive or who are simply following (Fig. 21.6).

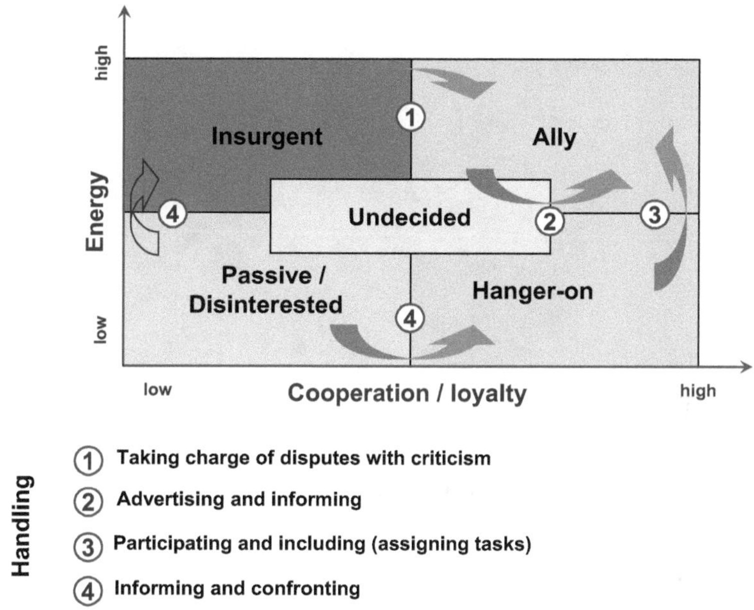

Fig. 21.6 Attitudes in change processes

21.2 Resistance

Social systems tend to maintain the status quo or equilibrium for as long as possible. What this means for the staff is that the changes are imposed on them, and they are forced to deal with them. However, it can be clearly seen that such conflict does not take place without leaving marks. In particular, resistance will increase when staff feel that their livelihood is endangered.
Change means upheaval and opportunities for self-development, or to let oneself go, to bid farewell to old familiar ways. Failing to recognise resistance will lead to blockades. Social pressure leads to counter pressure. When resistance (see conflict) is carefully handled, it can lead to the emergence of new resources and skills.

Forms of Resistance

Resistance always contains a coded message. The underlying reasons are often in the emotional area. Resistance from someone who feels threatened by change can manifest itself in a number of different ways. Some sort of existing order, structure, relationship or similar is questioned. This threatens our inner balance, which can lead to defensiveness. But whatever form the resistance takes, it is rooted in individual or social needs. Resistance generally contains potential energy that can be used constructively (Fig. 21.7).

Fig. 21.7 Forms of resistance

How can project managers deal with resistance, how can they handle it? The project team or other affected people will be more willing to change if:

- The change is perceived as an opportunity, rather than a threat
- There is a good basis of trust, and open communication
- They feel safe, i.e. if they don't fear sanctions and misuses of power if they uncover or address any wrongdoing or malpractice
- The changes are meaningful for them, and something positive can come out of them
- Their contributions are recognised
- They are involved at an early stage in the change process
- They are kept well informed and receive the necessary support
- Their resistance is met with engagement. The "suppressed emotional energy" cannot be channelled constructively until it has been taken seriously.

They will be less inclined to play an active role in the changes if:

- The working atmosphere is overshadowed by fear and uncertainty
- Patterns of behaviour that have been used in the past have become automated
- Ignorance or contempt prevail
- There is a lack of awareness of the problem, lack of understanding, isolation
- There is no communication in the team
- There is no clear reason

So successful change can only occur in a climate of trust. This calls for a much greater than normal commitment by senior management to communication, information, guidance and educational work. This works when management recognise the importance of this work, and take the time for it.

When dealing with resistance, the change process can be managed more consciously. The Swedish social psychologist Claes Janssen uses his "Four rooms model" to show how people experience a change process (Fig. 21.8). It always follows the same sequence from contentment, through denial and confusion and on to renewal.

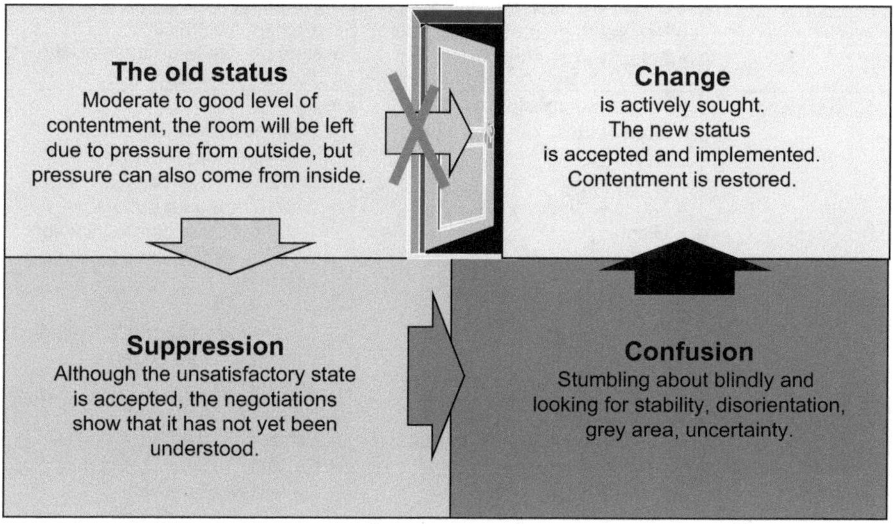

Fig. 21.8 The four rooms of change (Claes Janssen)

When supporting people who are affected, this change process can be sped up by project managers recognising the typical reactions and behaviours, enabling people to pass through the rooms more quickly. Supporting measures and interventions are very helpful in this context (Fig. 21.9).

	Indicators	Responses
Defensiveness	• "Everything's fine as it is" • It doesn't affect me • "Let's have a look...." • "Things are not as bad as they seem" • Irritability, confusion • Devaluing the information • Refusal to hear new information • Selective hearing	• Make clear that changes will take place • Confront with information and with reality • Explain what is expected • Allow people time to process the information and then arrange a discussion • Help people to understand the situation, and offer support
Resistance	• Anger, aggression • Fear of loss, fear about the future • Injured pride • Narrow-mindedness • Blaming • Complaining • Sick leave • Self-doubt • Role of a victim • Passivity • Delay	• Listen • Acknowledge feelings • Show compassion and understanding • Create pressure outlets • Facilitate open communication • Show understanding for their concerns • Not helpful: • Talking down people's feelings • Putting forward counter-arguments
Exploration	• Recognition of possibilities • Decision-making difficulties • Energy is released • Lack of focus, chaos • Goal setting • Awareness of resources • Looking for and trying out alternatives • Learning new skills • Being open • Developing new sets of perceptions and attitudes	• Help with clarification • Establish priorities • Provide / facilitating training • Develop alternatives • Hold vision meetings and target setting meetings, brainstorming sessions, planning meetings • Get projects underway • Create short-term targets and successes
Commitment	• Focus • Working together • Balance • Stability • Vision • Goal-oriented • Activity-oriented • Success stories • Reliability • Optimism and realism	• Agree long-term goals • Team-building measures • Produce a mission statement • Agree new ground rules • Fine tune, stabilise the new • Look to the future • Recognise the work involved in change: Give praise • Institutionalise the change

Fig. 21.9 Characteristics and leadership behaviour is the change process (taken from Potocnik, Neulinger and partners)

Dealing with Resistance

Change can also trigger conflict. When people develop, it gives rise to friction, anxieties and new expectations. These conflicts must be taken seriously. They must not be put down or glossed over because of operational constraints or a lack of time.

In discussions to clarify resistance, particular attention should be paid to ensuring that the negotiation is not about positions, but that the focus is on addressing the real interests behind the resistance. These are not generally disclosed openly, and must be drawn out gradually. The following careful three-stage approach is recommended for this (Fig. 21.10).

Order of approach ↓

Cause of resistance	Dealing with resistance
Step 1: Concerns about content	- Provide information - Put forward arguments - Provide clarification
Step 2: Fears and misgivings	- Understand, rather than explain - Listen, rather than argue - Enquire and reflect
Step 3: Self-interest	- Adress clearly and compromise - Confrontation and backing down - Clarification about the hierarchy

Fig. 21.10 Dealing with resistance and its causes

Stage 1

As a first step, assume that the resistance is due to objective reasons. It is good to ask what the person understood, and then to inform, to discuss with them, and to clarify any factual misunderstandings.

If it is really down to objective reasons, this will enable the content issues to be clarified and the arguments to be considered. If the other person just puts one argument after another forward, and if there are other reasons behind the resistance, then the discussion will go round in circles. They'll tend to jump between topics and always put forward different arguments. In that case, proceed to the next stage.

Stage 2

At this point, assume that the person has fears or anxieties. Nowadays, we assume that well over half of all resistance is due to fear. It is important to change the discussion strategy at this point. Instead of putting forward arguments, listen. The key issue now is understanding, rather than explaining. It is important to ask follow-up questions and to make sure that you have correctly understood the other person by repeating what they have said back to them in your own words. Only once the other person has the real impression that you have understood their fears or anxieties and that you are taking them seriously will it be possible to look for solutions. Here again, it is important to ask the other person what might help them, rather than offering solutions too quickly.

The most certain way of learning nothing at all about someone's fears and anxieties is try to calm them, to comfort them, to console them, or to factually state that "there is absolutely nothing to worry about".
If the resistance is really based on fears, the person will respond to a compassionate, sensitive approach and will work with you to find practical ways forward. If several conversations of this sort don't move things forward, then the cause of the resistance is probably not fears and anxieties. In that case, move on to stage 3.

Stage 3

Assume now that the cause of the resistance lies in the other person's self-interest. It is therefore important that you keep the other person's interests at the front of your mind: what advantages do they stand to lose, what hard-won rights or benefits, what incentives, what status or prestige? Or what little benefits and freedoms in terms of working time or car sharing? The way into the conversation is by a clear statement of these interests. So one way to start might be to ask about the benefits or privileges that the person might lose or have to forego because of the changes. But we cannot assume that the other person will be able to speak openly about it, or that they will be ready to negotiate a compromise. Sometimes an open confrontation is necessary so that the person can be shown their egoistic behaviour, and how it is preventing important changes occurring. And there are times when the only way to resolve the situation is to use the weight of hierarchical authority. Again, this needs careful handling. It is always helpful to give the person a chance to engage and to revert to cooperation before taking the hierarchical authority route. That becomes even more important if you still need or want to work with the person after the changes.
Following this series of stages is recommended because it avoids doing any damage. If you tell someone who has objective reasons that are experiencing fear or anxiety, they could feel that they are not being taken seriously. And assuming that someone experiencing fears has their own self-interest at heart could often lead to illness and make it impossible to talk about their fears.

Checklist: Interventions When Encountering Resistance

- Gather information
- Allow people to develop the best alternatives themselves
- Acknowledge the other party's best alternatives
- Develop options that offer mutual benefits
- Formulate objective criteria
- Evaluate possible negotiated solutions

Inform about: Actual situation: without judgment Planned changes and goals: in good time, and transparently Consequences for staff: fully and ongoing	**Listen to/explore**: What the staff say Unusual behaviour Rumours, etc. Any forms of resistance
Involve in: Discussion sessions Evaluating the current situation Decision-making processes Hearings and presentations	**Explore**: The needs of those affected Wishes and opinions Good ideas Cases of hardship
Negotiate about: Issues of contention Differences of opinion: in a cooperative and consensus-building manner	**Training**: As early as possible Tailored to suit needs, both for content and at relationship level

When team members signal resistance, they expect it to be taken seriously and to be examined. If project managers do not pay any attention to the resistance, team members may feel dismissed, offended or not taken seriously. Their disaffection and unhappiness will increase. The negative emotions will be projected on to other people. The resistance will become personalised. That can lead to a social conflict with its own dynamics, such as “It’s the project manager’s fault that. . .!”

Conflict

22

Wherever people work together, there is a potential for conflict. In projects, this can be at relationship level, content level or organisational level. If they are to develop productively, project teams need both phases of harmony and phases of conflict.

Many unnecessary conflicts occur because:

- Issues are not explained early enough
- There is a lack of transparency and openness
- Interfaces are not explained properly
- People do not communicate enough with each other
- Differing positions are not talked through together
- Other people's needs are not taken seriously or are ignored
- Resistance and conflict signals are not recognised

The term "conflict" is used in many different ways in everyday conversation, and it is helpful to define it and describe it more clearly in the context of teamwork.

22.1 Definition of Conflict and Inner Attitude

The word "conflict" comes from the Latin "confligere" (clash, collide) and means a meeting of opposing interests that cannot be resolved with equal satisfaction. What normally begins as a factual difference between two people often gradually escalates into a conflict. It is frequently not possible to look back and clearly establish the start of such social conflict. Using Friedrich Glasl's formula, conflict is defined as:

A social conflict between two people is when at least one of them feels affected in his own actions by the other person.

Whether what the person experienced is justified or just felt does not actually make any difference. The single important factor is the personal perception.

J. Kuster et al., *Project Management Handbook*, Management for Professionals,
DOI 10.1007/978-3-662-45373-5_22

22.2 Causes of Conflict and Types of Conflict

Organisational conflicts are much easier to resolve than social conflicts. If organisational conflicts are not examined and processed, then over time they develop into social conflicts.

Typical causes of conflicts in organisations might include:

Conflicting goals and motivational conflict	**Conflict of positions**
• Different project objectives and department objectives • Different motivation levels about achieving the goal	• Hierarchy, power • Status • Influence
Organisational conflict	**Role conflict**
• Unclear processes or instructions • Unclear authority and responsibilities • Inadequate flow of information	• Formal and informal roles • Different roles in different situations
Relational or interpersonal conflict	**Intrapersonal conflict**
• Standards and values • Sympathy/antipathy • Chemistry	• Envy • Uncertainty • Fear • Being over-challenged or under-challenged

A distinction can be made between three types of conflict:

Judgement conflict	**Value conflict**	**Distribution conflict**
Different evaluations about, for example, information, solutions, necessary resources	Different values with regard to consequences	Incorrect allocation of resources such as time, resources, capacity

In projects, the following dynamics also come into play:

- Projects always involve something new (products, procedures, organisational structures, etc.), and the existing things become old. The changes that are unleashed by introducing something new trigger emotions such as uncertainty, fear, sometimes rejection and resistance, sometimes euphoria and pressure. These tensions feed into the project work, and cause conflict (Fig. 22.1).

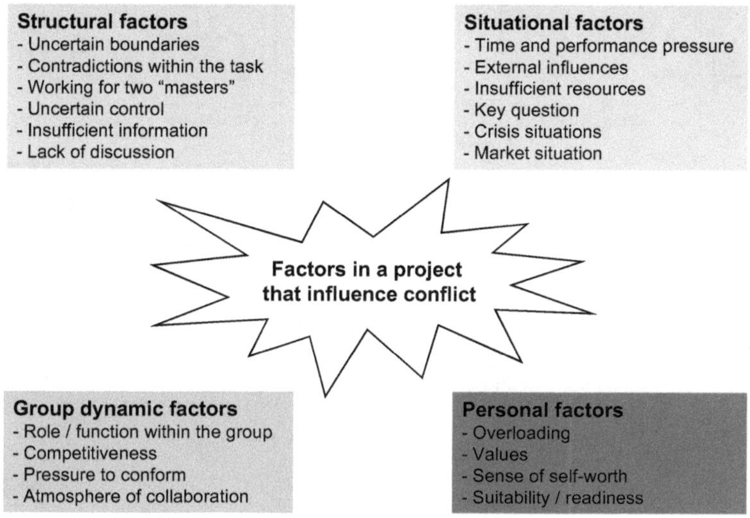

Fig. 22.1 Factors that influence conflict in a project

- The division of tasks within a company often means that there may be conflicts of interest between different departments, or that they may have different objectives. The project team members representing their core organisation in the project will feel these contradictions.
- Different corporate cultures and value sets may develop in different organisational units of the same company because people from the different departments meet in the project team. If such differences in working together are slyly "sent upstairs" (to Board meetings, for example), then they often collide spectacularly in projects. Such situations are known as "proxy wars" or deferred conflicts.
- The common good is also often raised. This is what it is all about at the end of the day, although individual interests and group interests are often felt more strongly in everyday activities. Even project teams are not immune to that.

22.3 Overview of Conflict Issues

Conflict signals are rarely clear and easy to detect. They can be observed and recognised when someone's behaviour changes, or when somebody notices it or perceives it as different to normal. In many cases, it is only possible to conclude that there is a conflict if several symptoms manifest themselves at the same time. This overview uses symbols to show the most important aspects of a conflict (Fig. 22.2).

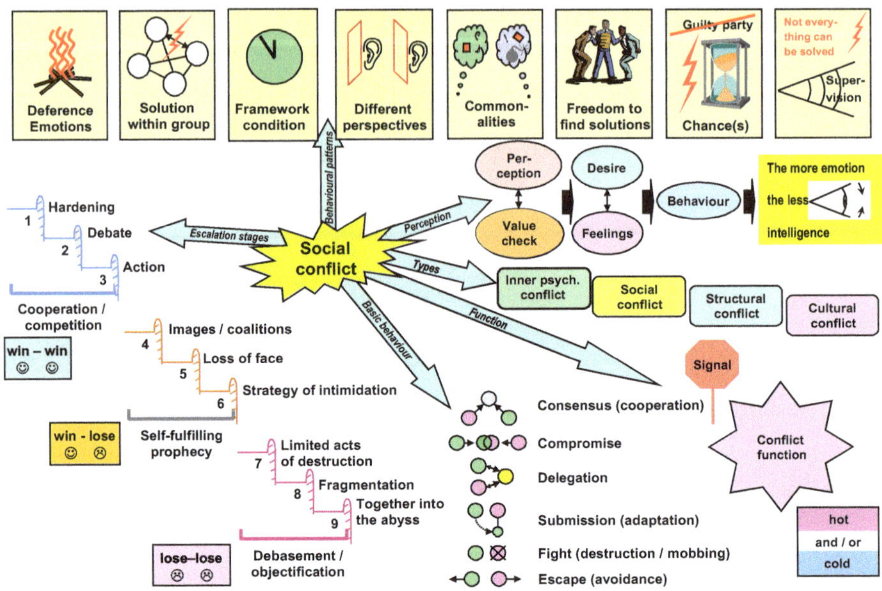

Fig. 22.2 Conflict overview

22.4 Perception

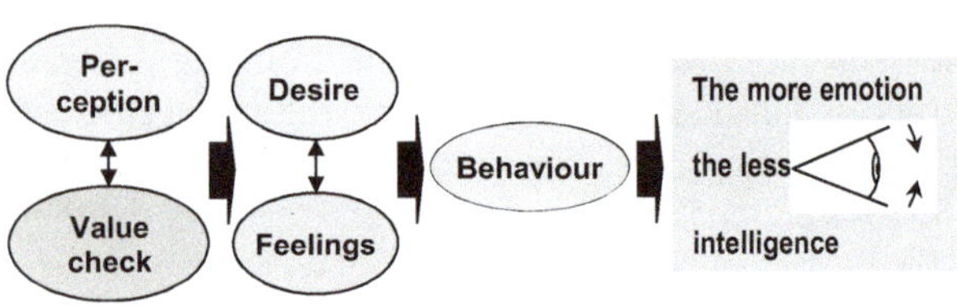

Perception relates to what someone sees or hears and then thinks to themselves, or imagines. A value check then enables us to check these perceptions against our own internal values, against our own previous experiences. The "perceived" thus becomes an "image" of what the person is used to seeing. The person generally responds to that at an automatic level with non-verbal body language. This behavioural response is directly influenced by the person's feelings and beliefs, and is rarely consciously controlled. If someone is relaxed, they can normally consciously control their behaviour. But if someone is tense, experiencing pressure or stress, or if they are overwhelmed by emotions, they will be less able to consciously control their behaviour. Their behaviour becomes more aggressive and destructive. Or to put it another way, through their words and actions, people can trigger effects that they generally (when they look back on them) did not intend. The effect of this exaggerated behaviour on the other person (or people) is that they then become more forceful, inflexible and inconsiderate in their approach to the conflict. Which, of course, simple annoys or pressurises the attacking person or party even more. The result is that both sides become embroiled in mutually escalating the conflict. It can ultimately become so intense that the parties feel completely overwhelmed by the conflict: Instead of the people having a conflict, the conflict "has" the people (Fig. 22.3).

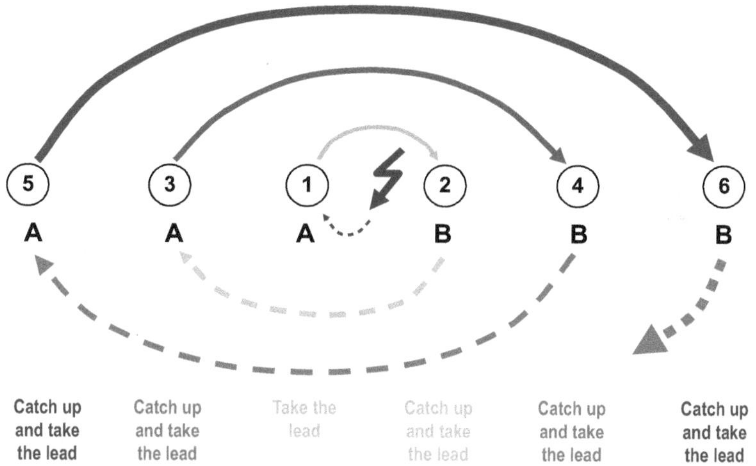

Fig. 22.3 Circular process of escalation (by F. Glasl)

But the origins of many conflicts lie within ourselves. This may be because I don't listen carefully enough while the other person is talking because I'm already formulating my answer or thinking about the next question, or that I only hear what I want to hear (Fig. 22.4).

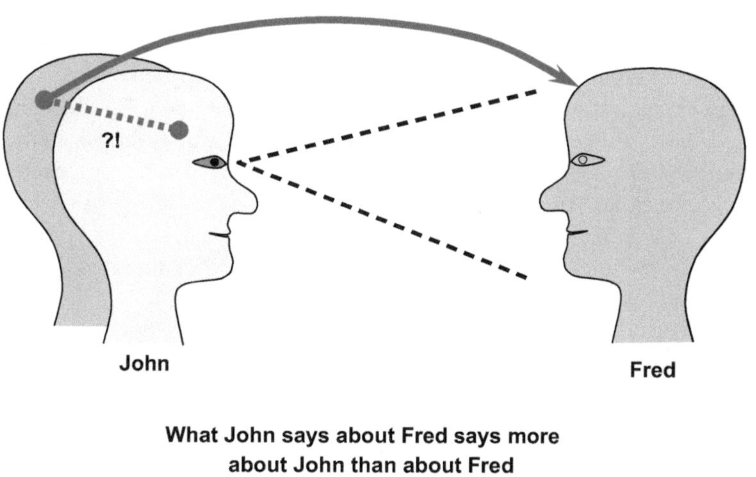

Fig. 22.4 Projection of one's own shadow onto the other party

Or it may be because I perceive the things that I see through my own filter and focus too much on my own internal image or what I'd like to see, rather than on the reality of what is actually there. In truth, the two parties are not struggling with one another so much as with the image that each person forms of the other person, and with that developing into an image of "the enemy".

22.5 Conflict Types

As with resistance, there are a lot of different symptoms that can point to conflicts. They may manifest themselves on their own, in combination, or even all at once. The important thing is the project manager's cognitive ability. He must recognise that a conflict is usually about a fundamental behavioural change, and that this only becomes visible gradually. Suddenly one realises that nothing is as it was before:

Rejection, resistance	Constant objections, "yes but" behaviour, grumpy reactions
Aggression, hostility	Harsh words, "angry" looks, disparaging comments, deliberate mistakes, "walls", sabotage
Stubbornness, intransigence	Dogmatic behaviour, following rules to the letter, "That's not how we do things here"
Flight	Avoiding contact, keeping out of the way, taciturn behaviour, disengagement, changing the topic of conversation
Over-compliance	Not contributing any ideas, avoiding criticism, simply agreeing with everyone else's opinions
Disinterest	Formal politeness, withdrawal, passive approach to work, ignorance
Formality	Doing the bare minimum, following instructions to the letter (but no more), keeping written records of every step and expecting everyone else to do the same

The potential for conflict should never be underestimated, especially in projects. Conflicts are absolutely normal. They are not a reflection on the quality of the project work. Quite the opposite! If there were no conflicts, then one would have to wonder if critical issues in the project work are being avoided. But difficulties and problems always show themselves in the end in such cases, at the very latest when the results are implemented. But by then, they are a lot harder to resolve and it takes a lot more money and time to resolve them.
The quality of the project manager and the project team can actually be seen in how they deal with conflicts, and in how they behave in conflict situations.

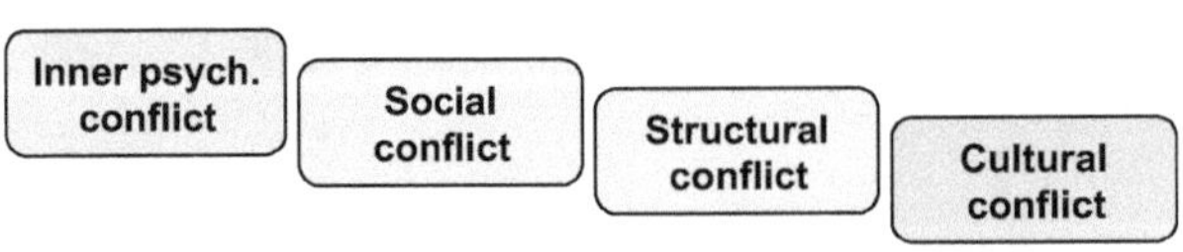

The first thing to establish is what type of conflict the current conflict is. It is normally best to do this on the same level as the conflict has occurred. This initial clarification will make it easier to select the right intervention.

22.6 Functions of Conflict

Whilst there are many possible causes to a conflict, it always has a signal effect, a specific function or a significance. Viewed in that way, conflicts are more of a constructive potential, based on a personal need, on settling something, on a provocation or a hidden intent (Fig. 22.5).

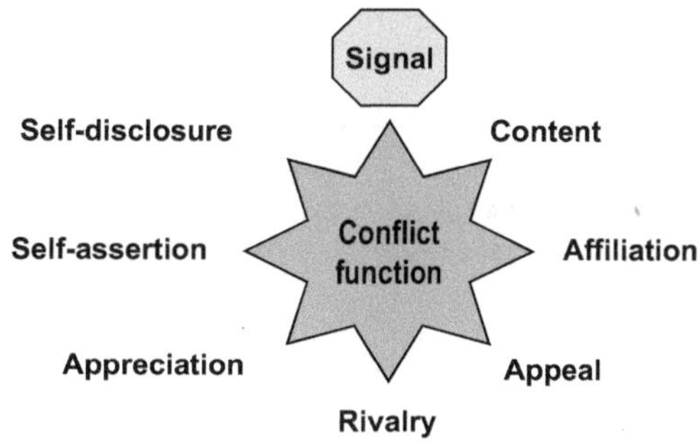

Fig. 22.5 Signal effects and functions of conflict

The intent may relate to:

- **Giving a signal**
  One side wants the other side to listen, to pay attention and states: "There is a problem!"
- **The content**
  There is a difference in objective views/expert opinions/perception
- **The relationship**
  Making it clear what I think of others/opponents, and how I see it
- **An appeal**
  Pent-up anger and disappointment must now be released before any more damage is done
- **Rivalry**
  Ultimately, asserting your own requirements and at the same time preventing a unilateral imposition of "conflicting interests"
- **Appreciation**
  Showing the other party that I want to be taken seriously and that I want to be compensated for the injustice I have suffered, want to see my dignity restored
- **Self-assertion**
  Standing up for your my rights and signalling that I will not give up without a fight
- **Self-perception**
  Explaining to the other party how I see myself and how I wish to be seen by them

22.7 Hot and Cold Conflicts

Friedrich Glasl's work on interactions between the parties in conflict can be used to identify whether it is a hot or cold conflict:

hot
and / or
cold

In hot conflicts, the interaction is very active and the parties tend to react oversensitively. Attack and defence are clearly perceptible and visible for all to see. The behaviours are impulsive and emotional. The energy of both parties is fuelled by the overly positive self-images. Interventions in hot conflicts are most successful when the perceptions, attitudes and behaviour of the individual parties can be explored and mutually recognised.

In cold conflicts, the interaction is paralysed. Disappointment, frustration and feelings of hatred are suppressed and have a destructive effect on the parties. There is no arguing, or only indirect arguing, since the parties keep out of each other's way and avoid direct contact. In the case of cold conflicts, it is helpful to use constructive interventions that reinforce the sense of self-worth of all parties, and to create framework conditions that enable the parties who are in conflict to re-engage in face-to-face interactions. Sometimes the parties that are in conflict may not wish to recognise the conflict, or they may dispute that the conflict exists. Where that happens, it is legitimate to start by defrosting the conflict by confronting the parties with the tense situation (Fig. 22.6).

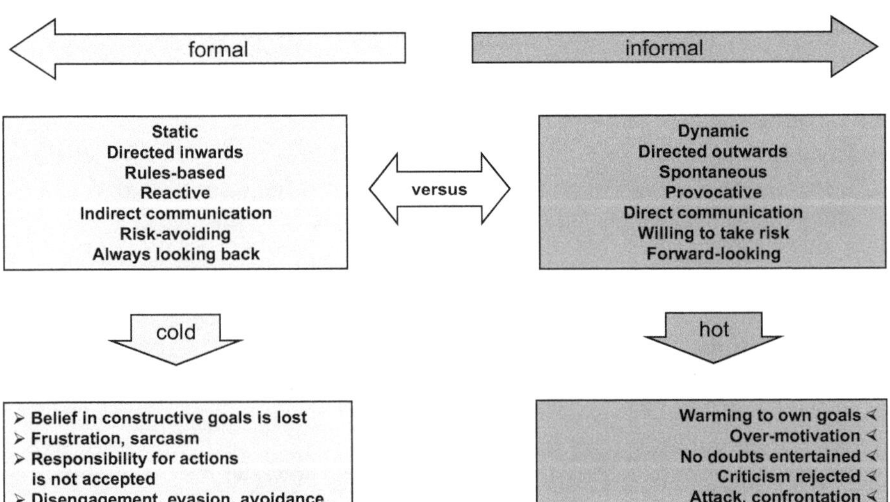

Fig. 22.6 Hot and cold conflicts (Gerhard Schwarz)

22.8 Basic Model of Conflict Behaviour

There are various possible ways of reacting to conflict situations. These reactions depend on the one hand on the situation. On the other hand, people tend to behave consistently, to repeat past behaviours. The following model by Gerhard Schwarz shows basic behaviours in conflict situations (Fig. 22.7).

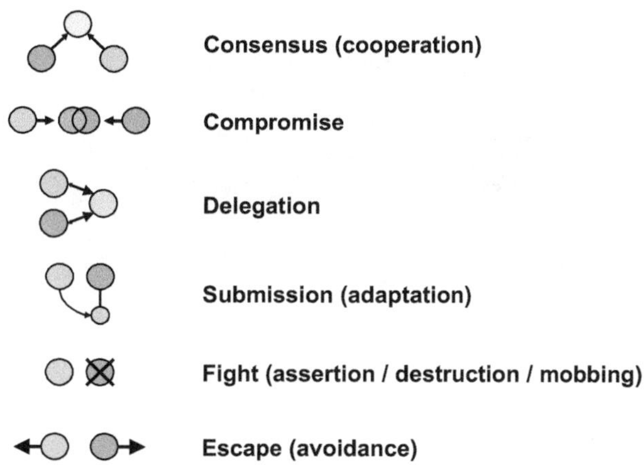

Fig. 22.7 Basic model of conflict behaviour (G. Schwarz)

Escape, destruction and submission are the three instinctive behavioural patterns. Delegation, compromise and consensus are learned behaviours in more highly socially developed people.

Escape

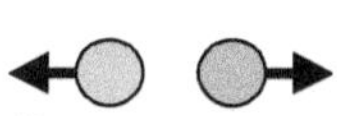

One or more parties escape from the conflict by running away. This includes not recognising the conflict or when nobody is willing to discuss the conflict. But escaping doesn't solve the conflict. There is a risk that it will spread further. On the other hand, escape does at least allow the situation to calm down, which is useful because when emotions run too high they get in the way of constructive conflict resolution. When that happens, it can be helpful to have some distance so that people can engage again at an emotional level. Another form of escape is to devalue or belittle problems.

If there are simply opposing, incompatible elements, ambivalent feelings or conflicting needs and inclinations, then there is no conflict. We only have a conflict once personal positions have been taken (Fig. 22.8).

If I devalue … *problems … feelings … needs … sensitivities …* *of my own … in other people … in a situation …*

If I deny … *the existence of a problem … the importance of a problem … the solvability of a problem … options for solving a problem …*

If I … do nothing …
adapt or over-adapt myself …
agitate … → *inwards (headaches)*
→ *outwards (grumbling, calling out)*
turn violent … → *inwards (stomach ulcer, heart attack)*
→ *outwards (explosion, hot conflict)*

… I can satisfy my emotions and "let off steam"!

→ But it will not solve the problem
→ No problem-solving process can be started

Fig. 22.8 Consequences of devaluing oneself

Fight/Destruction

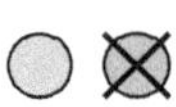

By destroying the conflict partner, the opportunity to engage with the different viewpoints is lost. This actually hinders our own development. Conflicts help to show up mistakes, and open the way for new behaviours. If conflicts are solved by destruction, the winner must always win in the future, because the slightest weakness could be exploited by a new opponent. That means using a lot of energy just to defend yourself. In our society, behaviour that is intended to mess a colleague around or to hurt them, torment them or ignore them is considered to be bullying.

Submission

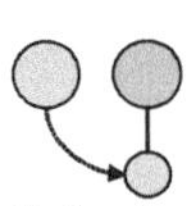

Submission is a recognised democratic principle: Minorities submit to majorities, people submit to entire groups by suppressing a part of their individuality. As a way of resolving conflict, constant submission is dangerous. It prevents development, and is an ideal breeding ground for the next conflict.

Delegation

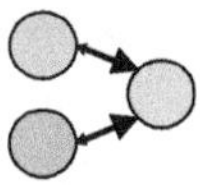

With delegation, the conflict is often denied, or people refuse to recognise it. A third instance (manager, judge, etc.) is therefore summoned to decide who is right. The durability of the solution depends on the degree of acceptance for this third person, and on whether the parties view their decision as just. If the third person does not have the authority to decide, or if they don't want to or if their decision is not accepted, then this works in the same way as escape. In effect, they're not really looking for a solution, or don't want to engage seriously with the issues. By delegating the conflict resolution decision to a third person, the two parties will never learn how to independently resolve their conflict for themselves. Instead, they become dependent on the decision-making body or person. The two parties miss the opportunity to become closer to each other, to deal with the conflict and to engage with each other.

Compromise

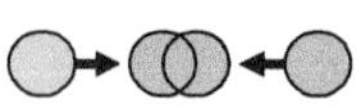

Compromise is a widely used form of solution. Both parties agree to give up part of what they are seeking in order to enable the conflict to be resolved. However, their own opinions and what they actually want remain unchanged in essence. So if both partners give up a large amount of what they want, this can lead to latent dissatisfaction. The important thing for the success of this solution is finding a balance that both sides feel is fair: There must be balanced give and take on both sides.

Consensus

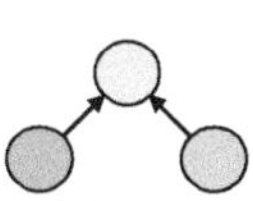

With the consensus approach, a new, third solution is developed and then fully accepted and implemented by both parties. This type of conflict resolution involves the participants being willing to let go of their old positions, putting themselves in the other party's position, and understanding their perspective. Only then will they be in a position to jointly develop a viable solution in the true spirit of the project. This sort of solution is very durable, since both parties will feel it is their own and they will work to defend it. This sort of conflict handling needs a long-term focus on the goals, understanding, patience, and a lot of time.

22.9 The Dynamics of Conflict Escalation

In his conflict escalation model, Friedrich Glasl clearly describes the downward spiral. An unresolved increase in conflict (conflict escalation) increasingly activates people's destructive forces. The behaviour of the parties who are in conflict becomes more and more emotional and unpredictable. And when that happens, the potential for managing the situation disappears. It becomes harder and harder for the parties who are in conflict to distinguish between the objective truth and their experienced perception, and harder for them to independently and neutrally resolve the conflict (Fig. 22.9).

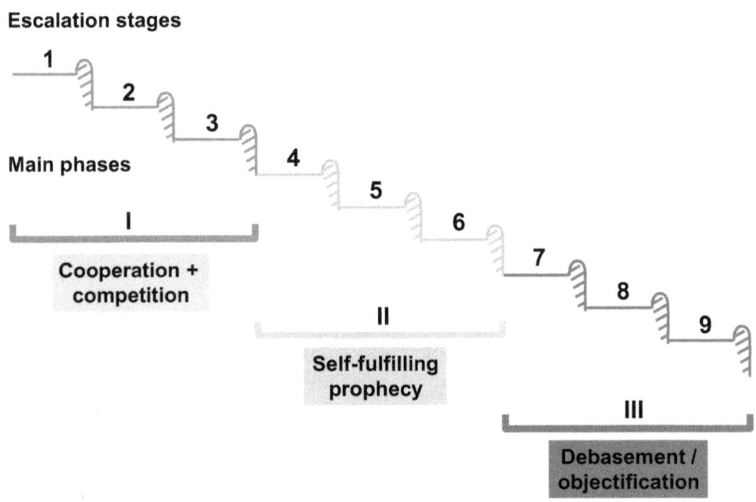

Fig. 22.9 Escalation stages, and their main phases (F. Glasl)

Phase I: Cooperation and competition → "win–win"		
Stage	How to recognise the escalation stage	Appropriate interventions
Stage 1 Hardening	• Positions become more rigid, harden and clash with each other • Awareness of the conflict leads to tension, but there is still a belief that the conflict can be resolved through discussion. • Discussions are broken off, then started again • No intransigent positions yet	• Ask what it is about • Support the discussions
Stage 2 Debate	• Polarisation in thinking, emotion and desire • Black and white thinking • Perspectives from positions of	• Make people aware of their actual behaviour • Listen and support • Make people aware of what is

(continued)

Phase I: Cooperation and competition → "win–win"		
Stage	How to recognise the escalation stage	Appropriate interventions
	perceived superiority/inferiority • Use of arguments that are intended to affect the other party at emotional level • "Intellectual tennis"	driving them • Transactional analysis explanatory models are very helpful for this stage
Stage 3 Actions, not words	• An increasing conviction that "talking doesn't help" • Strategy of confronting each other with "*faits accomplis*" • Loss of empathy with "them", increasing danger of misinterpretation because non-verbal behaviour no longer matches verbal behaviour	• Hold individual discussions to establish patterns • Discuss the non-verbal behaviour

Phase II: Self-fulfilling prophecy → "win–lose"		
Phase/stage	How to recognise the escalation stage	Appropriate interventions
Stage 4 Images and coalitions	• Lots of rumours • Seeing negative characteristics in the other party (projection) that people can't/won't recognise in themselves • Positive self-image, negative image of others • Parties manoeuvre each other into negative roles and fight these roles • People attempting to confirm their own prejudices • People trying to win supporters for their own side	• At this point, it can be useful to bring in a neutral facilitator • Correct distorted perceptions by working on self-image and image of others • Role negotiation
Stage 5 Loss of face	• Public and direct (forbidden) attacks intended to cause the other party to lose face • The other party is seen as the personification of evil • Unmasking of the opponent • Increasing self pity	• Explore the history of the polarisation • Analyse the critical incidents • Agree limits: What must not be allowed to happen at any cost? • Who is at which stage, and with whom?
Stage 6 Strategies of intimidation	• Spiral of threats and counter-threats • Ultimate acceleration of conflict escalation • Increase in pressure so that the stress increases • The participants no longer think about their actions	• At this stage, it is essential that a neutral facilitator is brought in (if not before). Someone with training and experience in conflict management

Phase III: Debasement/objectification → "lose–lose"		
Phase/stage	How to recognise the escalation stage	Appropriate interventions
Stage 7 Limited acts of destruction	• The opponent is no longer viewed as a person • Limited acts of destruction have become an "appropriate" response • Reversal of values: Inflicting a relatively small amount of damage starts to count as a victory • Taking increased pleasure from seeing the other party in difficulty • Lying becomes a virtue, provided it damages the other party	• From stage 7 onwards, things have gone beyond constructive intervention • The only way to resolve things is through the use of external authority decisions. This in turn confirms the sense of: "It might be the end of me, but at least it will be the end of you, too!"
Stage 8 Fragmentation	• The destruction and dissolution of the enemy system becomes the goal • Relationships are systematically destroyed • Destroying vital system functions to make the system unmanageable • Making it impossible to refuel or resupply, to regain strength	
Stage 9 Together into the abyss	• Total confrontation • No way back • Total confrontation, unlimited and destructive • The opponent must be destroyed at any price, even if the price is self-destruction	

When resolving conflicts, both sides always have something to gain or something to lose. So appropriate strategies must be used. Normally, people try to maximise their own gains (Fig. 22.10).

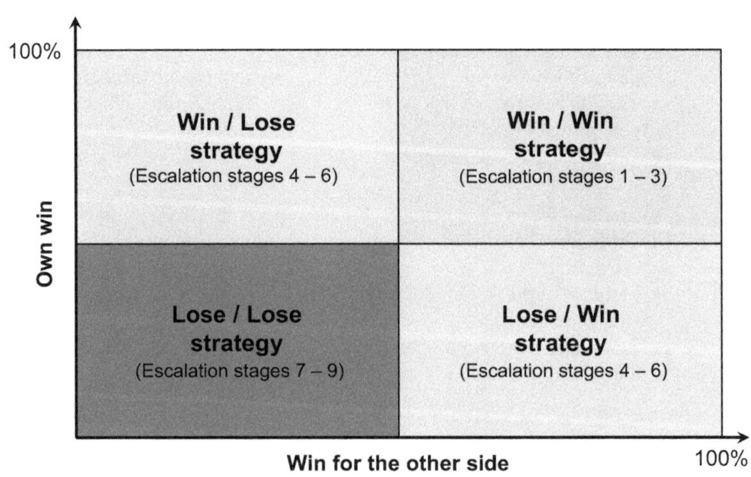

Fig. 22.10 Negotiating strategies (based on: Ponschab/Schweizer)

But at the end of the day, in terms of a viable long-term solution, this is unrealistic. Therefore, as a general rule of thumb:

- With support from the project manager, conflicts in a project that come under stages 1–3 can normally be successfully discussed and resolved to the satisfaction of both parties.
- In conflicts that are in stages 4–6, at least one of the two parties loses out in the search for a solution. The conflict resolution should be led by an experienced specialist.
- For conflicts that are in stages 7–9, both parties lose (to a greater or lesser extent). This conflict can only be resolved by a higher instance with considerable authority.

22.10 Resolving Conflict as a Project Leader and Team Leader

To start with, we will examine a couple of basic behavioural ideas that will help the project manager to avoid some common mistakes.

Behavioural Patterns for Discussions When Resolving Conflicts

Talking is the most important tool in resolving conflicts. Take a look at: "Successful negotiation using the Harvard approach..." These basic principles on how to behave in conflict situations/conflict discussions may also be helpful.

Do not Discuss Conflict Solutions When Emotions Are Running High

The higher the emotional involvement (red faces, strong words), the less we are able to perceive things clearly and to control our behaviour. Discussions that will lead to solutions call for a clear head. Which is why it can be useful to have a pause, a little breathing space, to allow things to calm down a bit. After that, the conflict should be resolved as soon as possible.

An Impartial Person (Mediator) Can Offer Help

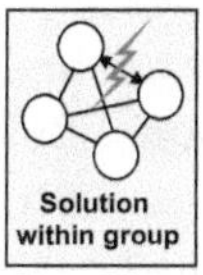

Conflict makes the people involved feel trapped. This then inhibits their perception and their ability to think. If the team leader doesn't feel confident enough about resolving the conflict, it can be helpful to bring in a neutral third party as a mediator. Seeking support from a third party is a reasonable thing to do, and is not a sign of failure. The neutral person's role is to accompany the parties who are in conflict as they journey towards a solution, to encourage them and to support them. Getting both parties round a table with a mediator helps reach a mutually agreed solution, rather than taking things to law.

The Earlier the Better

Conflicts first have to become noticeable. But once a conflict is present, it is advisable not to wait any longer, and to act as soon as possible because conflicts tend to spread.

Both Sides Put Their Points of View

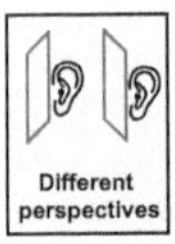

First have one party explain their point of view, then the other. They should each be allowed to speak without being interrupted. Mediators make sure that each party is allowed the same amount of time to speak. An egg timer can be used, for example. Each side should have an opportunity to talk about the situation from their perspective, what their feelings are, and about what has made them feel injured, and why. This stage is about understanding the subjective realities of each party. That includes the various backgrounds, ancillary aspects, fantasies and fears that are related to the parties' experiences of the conflict situation.
If they are allowed enough time and space when giving their points of view, one of the parties involved in the discussions might even say "And the really key thing is....". That could be the key to resolving the conflict. Gaining a better understanding of the background issues to the conflict is of decisive importance.

Emphasise Points in Common, not Differences

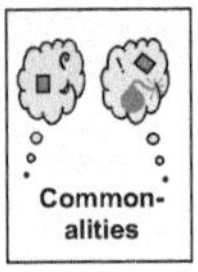

When discussing conflicts, the positive feelings for the other party often become lost along the way. The focus is on problems, anger, disappointment and injury. But there may also be positive points that the parties have in common. It is useful to jointly identify these, and to remind people of them. The key issue is to activate the energies that are needed for a sense of commitment towards one another and to enable the parties to move from their different perspectives towards finding a solution, or at least getting them to express a willingness to engage in constructive steps.

Identify Room to Manoeuvre

If a conflict situation can be discussed and explored enough, then it becomes about working together to seek out solutions. The issue of having room to manoeuvre can be useful in this respect. To what extent is one of the party's willing to move away from their position? But if you establish that there is no room to manoeuvre, then any amount of negotiation is pointless. At that point, the team leader should encourage both parties to reflect on their own positions. Further discussions will then be needed after that.

Look Forward, Don't Dwell on the Past

Dwelling on the past implies looking for "guilty parties" or someone to blame. That does not help resolve the conflict. In order to resolve a conflict, we simply need to understand its origins. Once the background has been carefully explored and is known, the next step is to work together to find ways of improving things. That presents an opportunity to move forward. If the people involved have not had a chance to have their say, or to express themselves fully enough, or if the conflict has not been taken seriously enough, then the conflict will keep reigniting.

Stay Focused on Goals

There is no point in holding conflict discussions unless the parties agree to reach a binding agreement at the end of the process. Putting an agreement in place should eliminate (or at least limit) the dysfunctional effects of the conflict. It is useful to end a discussion with some sort of "agreement". This should describe in as much detail as possible the behaviour that is expected on both sides, and should be signed by both parties. This sort of agreement or set of rules increases the sense of commitment.

Not All Conflicts Can Be Resolved

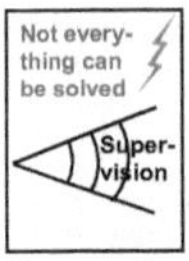

Sometimes human limitations (meaning "not able to"), organisational framework conditions, cultural hurdles or environmental considerations simply don't allow a conflict to be resolved. Highly complex projects with irreconcilable, contradictory requirements tend to be very prone to conflict. These tensions sometimes have such an impact on the project team that they suffer under it, or they even become an image of the outside world. In that sort of situation, if the company's senior management are not fully behind the project, the resulting conflicts cannot be resolved in a constructive manner. The pressure on individuals can be so strong that they are simply overwhelmed. If people under that sort of pressure begin to behave oddly, it is better to replace them.

22.11 Practical Recommendations for Resolving Conflicts

Before intervening in a conflict, it is worth finding out exactly what type of conflict it is. Different types of conflict need approaching in different ways (Fig. 22.11).

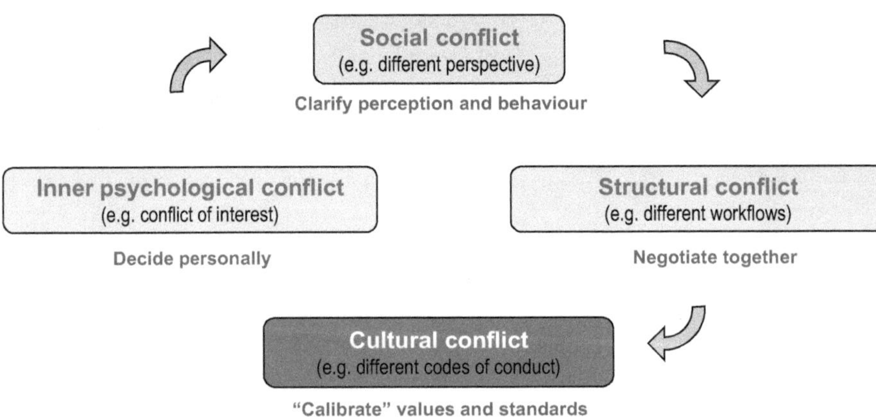

Fig. 22.11 Conflict types and appropriate measures

The chance of reaching a solution to the conflict is greatest if the reason for or function of a conflict is clearly understood, and if it is clear whether the parties are involved in a hot or cold conflict.

For practitioners, this means: In the first main phase (escalation stages 1–3), it should be possible to find a solution yourself. That in turn means that it is important to recognise and address conflicts as early as possible. The project manager should have a highly developed ability to pick up on these things. That will enable him to intervene effectively and at an early stage if he...

- really wants to resolve the conflict
- trusts in himself, takes decisions and takes time
- acts in an appropriate way for this specific type of conflict
- supports his team in discussing topics that are difficult to talk about
- structures his communication clearly and moderates in a disciplined manner

22.12 Conflict Management Strategies

A strategy is a concept that enables a defined goal to be achieved. In order to formulate a conflict resolution strategy, we must first know the goals that we wish to achieve. In contrast, tactics refers to the sequence of individual steps within a strategy. A person may not be fully aware of the strategy, but it can be reconstructed from their behaviour. There are two basic strategies that explain how people behave in conflict situations (Fig. 22.12):

	Problem-solving strategy	Poker strategy
Definition	I view the conflict as our shared problem	I work on the assumption that someone will come out of the conflict on top. → I want to be the winner
Intentions	• I know my wishes, interests and goals, and intend to openly and clearly state them • I am looking for a solution that satisfied both of us • I am working towards common goals	• I know my wishes, interests and goals, but I am going to avoid openly stating them. I will either keep quiet, or give a distorted explanation • I am pursuing my own goals, and I'll do everything I can to impose my position on the other party
Behaviour	• I want to even out differences in power by emphasising how important it is to me that we find a joint solution and confirming out commitment to working together • Right at the beginning, I will openly and truthfully state my feelings, interests, intentions and positions • When the other person is speaking, I will try to put myself in their position • I will not make promises as sweeteners, nor make threats to unsettle people • I will express any negative feelings in ways that do not cause hurt • I will remain calm when expressing any strong feelings (anger, impatience) (= hot conflict)	• I consciously emphasise power differences by emphasising right at the start that it doesn't matter if we reach a joint solution and by stating that I am not dependent on anyone else • At the beginning, I leave others unclear about my feelings, interests and intentions. I hold back, and let the other party come to me • I avoid putting myself in the other person's position. I think about where I can gain the upper hand • I start by making promises, but if that doesn't make the other party back down I switch to making open threats
	• I make it clear that my position is flexible • I adopt a cooperative approach, in order to create or to stabilise a cooperative relationship	• I hold back any negative feelings, but I reserve the right to use them later as targeted "payback" (cold conflict) • I make it absolutely clear that I cannot and will not move from my position • I appear cooperative in order to take advantage of the other party's cooperativeness in order to achieve my own goals

Fig. 22.12 Problem-solving and poker strategies (Berkel, 2002)

- The **poker strategy** is based on a conviction that every conflict must have a winner and a loser. It is a matter of imposing your will over the other party in order to come out as the winner.
- The **problem-solving strategy** assumes that every conflict represents a problem that can be solved, and that solving it together will bring benefits for both sides.

The Project Leader as a Mediator in Conflicts

Any manager who is asked to intervene in a conflict or to mediate must absolutely ensure that they only take on the role of a moderator, and that they do not themselves take on the conflict. The conflict parties are responsible for their own conflict.

The Adviser Role in Conflict Situations

When a team leader enjoys the confidence of their team members, they are often approached for help as a trusted person in conflict situations. The problem-solving strategy enables the team leader to offer advice in such a way that the employee who is caught in the conflict situation can systematically process their situation and is capable of taking action. In this adviser role, the team leader offers assistance to facilitate self-help.

Tips for Preventing the Most Common Conflict Situations in Projects

In project work, the most common sorts of conflict situations can be avoided or at least defused by carefully following the following suggestions. This applies in particular at the beginning of the project, but is also valid later on in the project.

- Make a conscious effort to perceive the various stakeholder groups and influences in the project on a differentiated basis.
- Clarify the different roles and responsibilities of the different parties and groups.
- Recognise and declare a person's different interests. These could be personal interests, or those of an interest group. These interests often arise from someone holding a particular post, or from a group's opinion. People in the project (especially project managers) often combine several different roles at the same time. So a role conflict can easily arise.
- Use preventive risk management so that the project is firmly embedded and well supported in the organisation.
- Clarify and clearly delimit areas of authority.
- Make sure that backs are covered, especially by management, which will provide the project managers with a certain amount of protection.
- As a project manager, constantly reflect with the team on the work processes, both on the hard skills issues and at relationship level. For example, undertake a group process analysis.

Helpful pointers for project managers in conflict situations

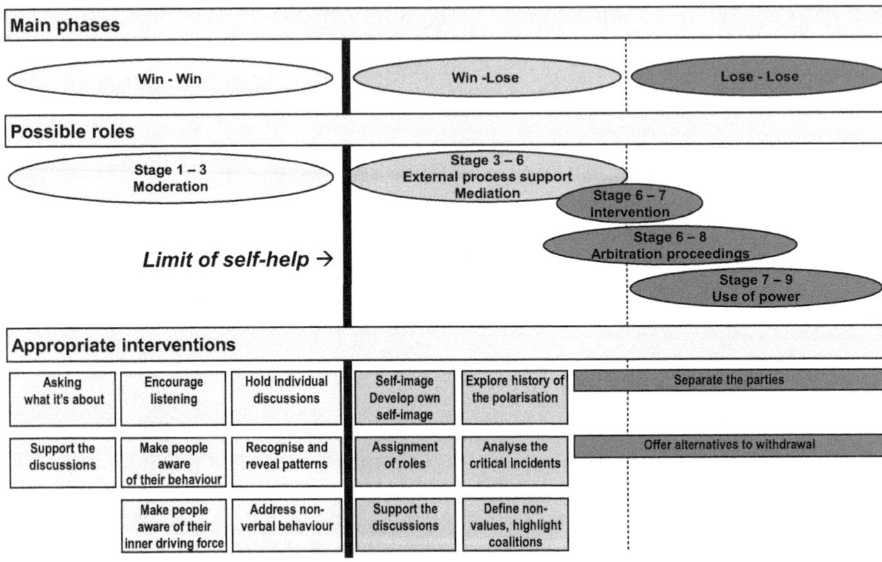

Fig. 22.13 Roles and appropriate interventions in conflict situations

For conflicts that are stages 1–3, project managers can ask the following questions:

- What is this conflict signalling to me?
- What does the conflict represent?
- What is the function of this conflict?
- What do I need to think about?
- What would happen if the situation were conflict-free?

Additionally, he might:

- Ask questions
- Actively listen
- Hold individual discussions
- Recognise and point out verbal/non-verbal behaviours

If the conflict escalates beyond stage 3, resolving it will need professional support for a trained specialist. The only effective intervention will be by a neutral person who is not directly involved in the conflict (Fig. 22.13).
At escalation stages 6–9, the only solution is usually decisions made by a higher authority. As the conflict progresses, the view of the parties in the conflict has become so limited and hardened that even if they were able to recognise their own destructive behaviour, the loss of face can no longer be corrected. At that point, they

work on a basis of "Better together into the abyss than to give any ground to the other party" So in this stage, it is simply a question of damage limitation. The power decision is normally pronounced by an institution, a court, a hierarchy or an even more powerful outside authority.

22.13 Requirements for Constructive Conflict Resolution

A project manager can only constructively try and resolve conflicts if they are in escalation stages 1–3 and the following requirements are met (Fig. 22.14):

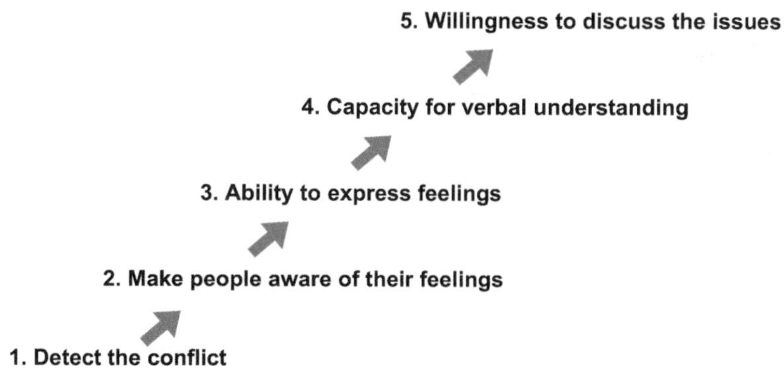

Fig. 22.14 Requirements for conflict resolution

- There must be enough time available
- The person who is decides to address the conflict should be in good physical and psychological shape
- There should be a social and economic basis present for everyone involved. If that is the case, the conflict can be constructively clarified and resolved, and it will be possible to move on to the next steps.

22.14 Guidelines for a Diagnostic Discussion

During a diagnostic discussion, the focus is on the following questions:

1. What issue lies at the heart of the dispute between the parties that are in conflict? What are the key issues?
2. How did the situation come about? What is currently happening?
3. Who is in dispute with whom?
4. How are the two parties connected? What is the state of their relationships with each other? What constraints is the organisation imposing? How are the parties informally managing their relationships with each other?

5. What do the parties that are in conflict want to achieve? Why are they in this conflict, and to what end? What do they want to gain from it? And what are they doing to achieve that? What is their fundamental perception of conflicts?

Identify How Willing the Parties Are to Change

If the conditions for resolving the conflict are in place, work on the initial situation can begin. When a change is necessary, this is often apparent from a vague dissatisfaction on the part of staff, customers or partners. Individual discussions and group discussions have proven to be a good way of finding out more about the viewpoints of those who are affected, and of understanding the reasons for the problem and people's willingness to change (Fig. 22.15).

What are the individual perspectives?	• How do the individuals see the situation? • What are people not seeing (blind spots)?
What are the relevant environments in the organisational unit?	• Who is linked to whom? • Who depends on whom?
What keeps the situation going?	• What patterns are there? • How can they be recognised? • What mechanisms impact on the system?
Where does the pressure for change come from?	• Where is the energy for change come from? • What are the positive perceptions of a "problem-free" future situation? • Which people and groups will be in favour of the changes, or opposed to them?
What will happen if things remain unchanged?	• Why is it good that there is a problem? • What will it achieve or prevent? • For who does the problem bring benefits / disadvantages?
How much time is available?	• with respect to the current market position • from the customer's perspective • from management's perspective

Fig. 22.15 Preparatory questions for conflict discussions

22.15 Procedural Checklist for Conflict Discussions

A discussion on a conflicting situation follows these steps (Fig. 22.16):

Step 1	Put the conflict on the table (confrontation). The seriousness of how much it disturbs you must be made clear to others. So don't beat around the bush, say it straight. "I" messages	"What bothers me..."
Step 2	Talking about your own goals clarifies your own needs	"I would like..."
Step 3	Identify other people's goals by asking direct questions. Control questions about the goal, active listening. Accept other people's goal as your own.	"What would you like?" "How do you see that?" "What do you think about that?"
Step 4	Seek things in common	"What do we both want?"
Step 5	Search for, accept, and evaluate ideas for solving the problem.	"What can we agree on?"
Step 6	Agreement: Concrete, detailed, clear deadline. After the agreed deadline, check whether the agreement can be implemented in practice and whether it works for us. If a solution is not possible: Postpone (agree a firm date) Identify the key question There are some problems that we cannot solve, but we can stop obsessing about them.	"What shall we agree?" "Do you think that would be useful?"

Fig. 22.16 Checklist for conflict discussions

23 Project, Program and Portfolio Management (PPP)

Companies whose success relies to a large extent on projects cannot generally avoid the need to introduce a system of project, program and portfolio management (PPP management). Thereafter, these must be constantly improved to suit the specific needs of the company's processes. This initiative is part of the corporate strategy. Project, program and portfolio management can be introduced relatively independently, and in practice it often takes place at different speeds. The PPP program manager is responsible for the PPP program life cycle.

Introducing PPP management in a company involves defining the best processes, methods, techniques and instruments and making organisational changes in the context of a continuous improvement process. This is highly likely to meet resistance. Standards and guidelines are useful tools when managing events that frequently occur in day-to-day project, program and portfolio work. Standardisation of terminology helps with shared understanding and forms a basis for contractual agreements between the parties involved in the project.

Close cooperation is essential in PPP management in order to ensure support from senior management for the projects, programs and portfolios. One of the biggest challenges in this process is connecting up the strategic and operational planning. Careful organisation of all roles in this extended context (PPP program manager, project portfolio committee, project officers, etc.) is key to the successful implementation.

23.1 Introducing Project Management in an Organisation

Results from various tests show that the success of a project depends not just on the professional work of everyone involved in the project, but increasingly also on the framework parameters and the extent to which the project is embedded in the corporate strategy. Whilst many relevant "surrounding systems" are known and are described as starting points in this book (such as portfolio management, project

J. Kuster et al., *Project Management Handbook*, Management for Professionals,
DOI 10.1007/978-3-662-45373-5_23

management standards), the integral design of all these components that is needed in order to bring about the required project culture is missing.
In the future, the focus of attention will move from individual projects to the contextual management of project work across the company. Some initial work on this includes various maturity models (e.g. Harold Kerzner) or the "Key 9 model" by Stephen Rietiker, which tries to integrate the regulatory, strategic and operational levels. At every level of the company, managers must recognise the support the right environment for the project work and the project manager, and provide the required support. However, these tasks cannot be defined in isolation from one another.

23.2 Implementing the Company Strategy

In most companies, the planning and implementation of strategies are not connected to each other. Formulating the strategy, an exercise normally undertaken by senior management, takes a lot less time than implementing the strategy, which is normally delegated to middle and junior management. To close this "strategy implementation gap", there is a need to carefully align three elements: "strategy", "structure" and "culture". The company's senior management is therefore responsible for providing the following:

Strategic Planning

Strategic planning is about identifying the available options for an organisation, and selecting the best one (doing the right thing, effectiveness). Even the best implementation cannot put right things that are missed during this strategic planning (Fig. 23.1).

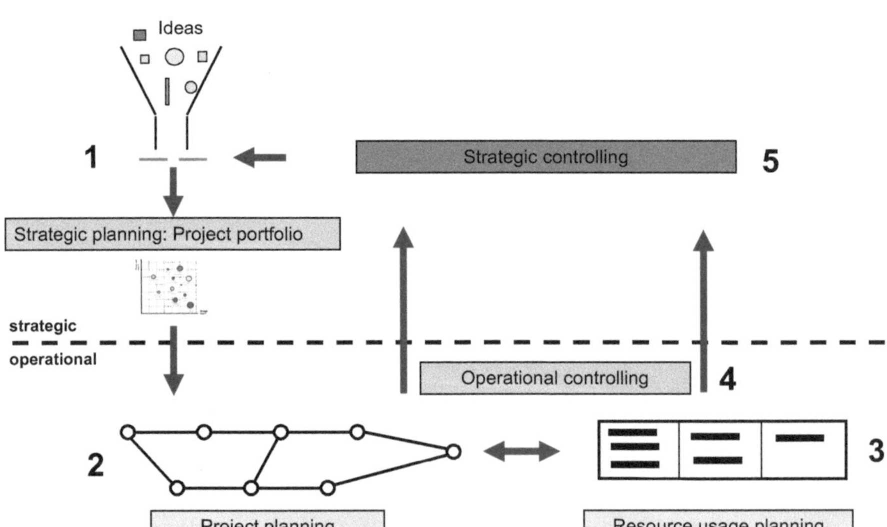

Fig. 23.1 Interaction between strategic and operational planning

Implementation of Strategy

Both before and during implementation of the selected strategy, care must be taken to ensure that the right methods and steps are defined (doing it right, efficiency). The strategic plan is put into reality by means of daily tasks.
This allows the project managers to design both the whole strategy cycle and the individual steps that it is made up from. In practice, the strategic planning process is often run as a project in its own right. In larger organisations, senior management appoint a strategic project controlling team or a project portfolio manager. This staff unit prepares the strategic project management and the reporting for the decision-makers.

Strategic Controlling

Strategic controlling (strategic project management, management of projects) ensures that the contribution made to the company strategy by the approved projects is maximised (Fig. 23.2).

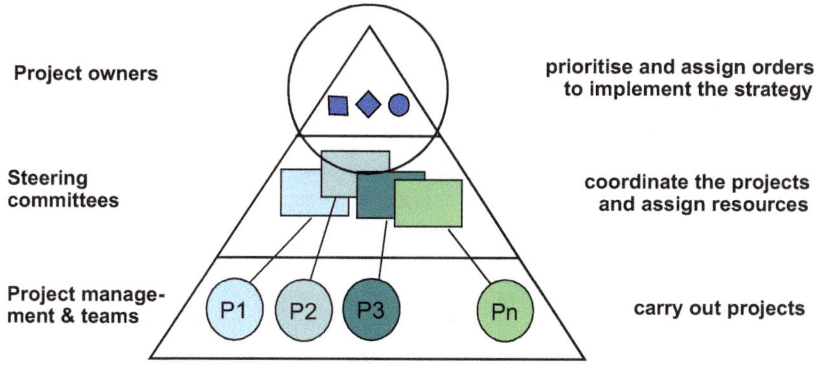

Fig. 23.2 Strategic controlling

Strategic Risk Management (SRM)

Identifies and monitors the organisation's risk profile. Big changes in this profile can lead to the strategic plan being revised or even completely changes, which has a direct impact on project management and on strategy implementation. It may even result in a new strategic version of the risk profile. Strategic risk management covers three areas:

- Strategic risk (the risk of not getting where you want to get to, or of clinging onto the old ways and going in a direction that is no longer desirable)
- Change risk comes from projects that are needed to take the organisation in a new strategic direction (project management is a tool for managing these risks)
- Operational risk, which covers the risks associated with the organisation's daily operations

23.3 The Project Portfolio

Organisations are always faced with more project ideas than they have resources available to implement. If a project fits with the formulated strategy and if it fulfils the framework conditions, the right place must be found for it in the strategic project portfolio. The portfolio is an effective means of answering questions about effectiveness: "Are we doing the right thing?" "Are we getting any closer to our vision?" The portfolio is intended to prevent too many projects being undertaken at once, which would mean that the projects would be competing for the same resources, which would hinder the projects.

Quantitative criteria compare costs, income, returns, whilst qualitative criteria examine the attractiveness or the potential of an idea. Organisations benefit from listing all the criteria that are important for them in evaluation tables and publishing these lists internally. For the quantitative criteria, an organisation should state which sums achieve which scores or weightings (e.g., a score of 0 for sums under 50,000 €).

A project portfolio is an overview of all current projects, presented either as a list or in graphic form, and which can be sorted according to various criteria. The company can choose how they distinguish between different project types, such as internal and external, short-term or long-term, highly complex and standardised, or customer projects and infrastructure projects. The important thing is that these different categories all call for different approaches and resources. That is the only way to establish priorities to make sure that resources are used in a way that fits in strategically.

The project portfolio enables periodic evaluation of planned initiatives using at least two predefined and adopted criteria. Depending on the nature of the initiative, it might be enough to simply compare using paired criteria, such as opportunity vs. risks, costs vs. benefits, etc. In many cases, several different views of this sort will be taken for a single project. But a project portfolio can also be multi-dimensional, using:

- Strategic criteria, e.g. a project score card
- Economic criteria (forecasts, market)
- Ecological criteria

- Ethical criteria
- Opportunity/risk criteria
- Mandatory criteria (new legislation, technological breakthroughs)

The normative, strategic management formulates and critically checks the company's corporate philosophy, vision, mission, models, goals, corporate strategy, business strategy, business models, corporate identity, culture, etc. They develop portfolio strategies, market strategies, product strategies and marketing strategies, diversification, cooperation. The project portfolio must be capable of classifying project ideas according to these criteria (Fig. 23.3).

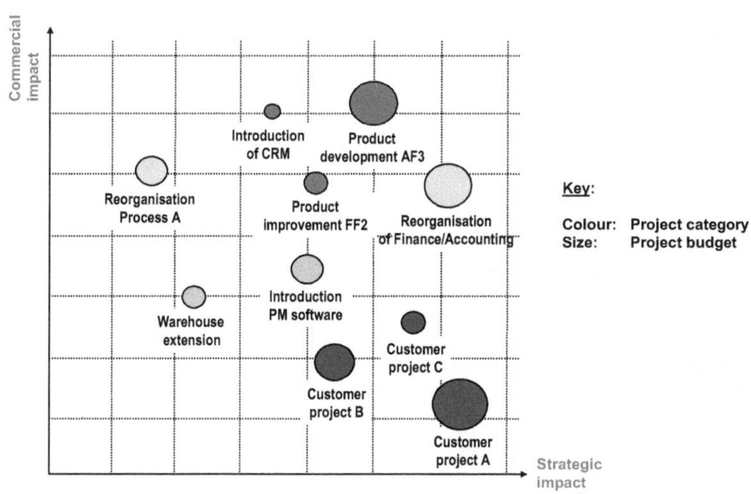

Fig. 23.3 Example of a project portfolio (strategy vs. commercial impact)

This project portfolio is managed either by the project office, or by a special Portfolio Management Committee (PMC), and includes a very wide range of skills:

- Strategic skills, provided by senior management
- Economic skills, provided by the strategic project controlling unit
- Methodology skills, provided by the portfolio manager
- Subject skills, provided by the "direct beneficiary" users

23.4 The Project Office

A lot of companies invest a great deal of time, effort and money in developing project management guidelines and instructions, in staff training or in project management software, only to find that the anticipated benefits often do not materialise.

The integration of a project office in the core organisation is one way of improving the company-wide support for running projects. A project office (also called a strategic project office (SPO), project management office (PMO) or project support office (PSO) is often the only visible or formal element for a project-oriented organisation, and it is typically responsible for the following tasks:

- Acting as a link between the strategic and operational levels (hence the term "strategic project office")
- Introducing project management processes, standards and methods
- Evaluation and rollout of software tools for project work
- Drawing up a project overview or project portfolio
- Responsibility for overall controlling
- Providing support, advice and coaching for project managers
- Providing a range of training and further training
- Development and qualification of project managers

Part IV

Tools and Instruments

This part describes a number of proven methods, tools and instruments that allow the project manager to select the perfect means for use in his situation. This extensive collection is supplemented with checklists, templates, forms and explanatory case studies. These make up the toolbox for operational leadership of a project.

Team Leadership Methods

24

24.1 Group Process Organisation

Definition

Group process organisation (GPO) provides a structured approach and rules for handling this topic in group work. It is drawn up jointly by all the participants before the group work starts specifically before they all become preoccupied with the group work. This makes the group work much more effective, more than compensating for the time required to complete this step. To function effectively, the group process organisation should be illustrated for all participants and should be visible for the entire duration of the group work (Fig. 24.1).

Contents

Who Performs Which Role?

Moderating

- Keeping the goal in mind and ensuring that the problem to be solved is understood by all group members and a common goal is made clear
- Structuring the work process with respect to procedures and timescale
- Leading without going off track
- Spurring all participants on to action and ensuring the same information level
- Listening to all parties, encouraging people to pursue their ideas without requiring immediate evaluation
- Helping to eliminate misunderstandings or rivalries between team members
- Not imposing own suggestions on the group
- Summarising and recording the current position

J. Kuster et al., *Project Management Handbook*, Management for Professionals,
DOI 10.1007/978-3-662-45373-5_24

Time Management

- Agreeing times for individual steps and the whole project
- Monitoring planned deadlines and intervening where necessary

Minute-Taking

- Writing down the results of the group work
- Making the work process visible to everyone (flipchart, slides, etc.)

What Needs to Be Achieved?

- Making the order or problem understandable to everyone
- Defining the goal or result

How Should It Be Achieved?

- Determining the necessary individual steps (including preparation of the presentation)
- Clarifying the use of layout and presentation techniques

When Will It Be Done?

- Estimating the time required for each individual step
- Setting all start and end dates
- Including breaks
- Scheduling in buffer time

Template

GPO Group process organisation

Who	Facilitation	
	Presentation	
	Minute-taking	
	Time management	
What	Task / problem	
	Objective / result	

How		**When**		
Individual steps	Technique(s)	Time required	from	to
Draw up GPO				
Buffer time				

Fig. 24.1 "Group process organisation" template

24.2 Kick-Off Event

The kick-off event is the official start of the project, but it does not only examine questions of content and organisation. The following checklist is intended to help plan all aspects of the kick-off.

Checklist

Relationship Level

This is probably the most important level in the start phase of a project. It is intended to define how people interact with one another:

- All the team members get to know one another
- The ground rules are defined and agreed by the team
- A good atmosphere is initiated
- Mutual trust is established
- Openness and critical review are allowed

- Mutual prejudices are eliminated
- Identification with the project and project team ("We" feeling) is established (by the meaning of the project, highlighting the special features of this team, etc.)
- The rules for communication have been addressed and allow mutual appreciation and active listening
- Informal platforms exist that allow an interpersonal exchange of views

Content Level

This level tackles the definition of problems and goals, and is used to sketch out possible approaches and methods for finding solutions:

- Participants are aware of the meaning of and background to the project
- They discuss the current progress of the project or order
- The intentions and goals are set out
- Team members have a common understanding of the task
- Perceptions of the goals are synchronised between team members
- The approaches to be taken are outlined
- The possible approaches are known and discussed
- The prior knowledge of individual members is disclosed
- Members state the times at which they are available and a binding agreement is made with their line managers

Organisation Level

This is the level at which working tools and work processes are agreed within the team:

- The existing project planning and controlling process are presented
- The links between individual assignments are explained
- Possible problem areas within the project are addressed and the team members with responsibility for resolving these problems are determined
- The project organisation is set out and the resources of individual members are known
- The communication and decision-making channels are defined
- The tasks, responsibilities and areas of authority are assigned to individual team member or groups
- The meeting intervals and minute-taking are defined
- The agenda is drawn up for the next meeting (using standard agenda items for every meeting, if necessary)
- The team members know what time is available to them, the project's priorities and their own role
- The available resources are known
- Every team member knows who to turn to in non-standard situations
- Perceived effort and deadlines have been discussed
- Expectations of the members have been set out
- Establish the content and format of reports and project documentation
- The ground rules have been agreed (team convention)
- Checks on the effort involved have been agreed (time, costs)

Timetable for a Kick-Off Event

A possible structure of a kick-off meeting can be (Fig. 24.2):

Time	Topic	Aim	Comments
08:00	Arrival, registration, coffee	Time for informal greetings and getting to know people	Coffee, fruit, etc.
08:30	Start, welcome from the project manager, introduce the kick-off process (with times), project manager introduces themself and explains why they are managing the project.	Start on time Orientation about the meeting Create an atmosphere of transparency and trust	Fllpchart
09:00	Introduce the project, give a quick history, explain the overall objectives	Make sure everyone has the information, introduce the project goal	Project owner
09:15	Let each project team member introduce themselves, address personal connections to the project: What do I personally want to get out of working on this project?	Getting to know each other, creating bonds to the project, discussing wishes, expectations, fears and relevant project experience. Start the networking process	Provide a framework for the introductions, so the process doesn't take too much time.
10:00	Break	Informal discussion	
10:15	Update people on the project status, or work together to establish it	Information	Reports, flipchart, pin-board, video, etc.
10:30	Draw up a project environment analysis, show the system boundaries	Acknowledge the environment and dependencies, get people to share their views on them	Pin-board, flipchart, mind map
10:45	Draw up a project force field	Identify which forces help the project, and which hinder it	
11:00	Identify any problem areas that are already known	Situation analysis, data for risk considerations	List any problems
12:00	Lunch		
13:00	Show the project's organisation in terms of working together and project processes (work packages)	Sketch out the project in organisational terms, acknowledge any bottlenecks, specify fixed deadlines, show where there is room to manoeuvre	Project plan, organisational chart
13:30	Outline details of working together Frequency of meetings Document management and minutes	Agree communications, ground rules	Flipchart
14:00	Discuss and agree the next steps	Create a clear work plan	
14:30	Each participant sums up their understanding of the next steps, of their role and function, and of the plans	Acknowledge engagement and resistance, develop commitment	
15:00	Closing words from the project manager or project owner	Show appreciation and reiterate the importance	

Fig. 24.2 A sample "kick-off event"

24.3 Moderating Work Processes

Moderation (lat. "moderari" means "to temper", "to steer", "to regulate") is a sophisticated art that demands a broad range of skills from the meeting chairman (who does not necessarily have to be the project manager):

- Neutrality
- Sensitivity to social issues
- Verbal skills
- Technical knowledge with respect to the topic and working methods
- Acceptance by the participants

Difference Between an Informal Meeting and a Moderated Meeting

Informal meeting	Moderated meeting
• Held regularly	• Generally for a specific situation or reason
• The participants know one another	• The purpose is to resolve problems or develop ideas
• There are agenda items	• Visualisation methods are used
• The participants sit at tables	• Facilitation methods are used
• The structures are set in stone	• The furniture can be moved

A Moderated Meeting...

- Utilises the benefits of moderation
- Uses moderation media such as flipcharts or pinboards
- Is chaired by a selected moderator
- Follows a structured timetable

The Roles of a Moderator

The moderator is the "primus inter pares" and assumes at least one of the following roles or functions:

- The gardener who creates the perfect conditions for growth
- The midwife who helps the group to resolve its problems
- The catalyst who speeds up the process without changing himself
- The transformer, transmitter and translator of information
- The director who coordinates the contributions to create a perfect whole

Moderation Steps

In terms of methods, moderation is made up of the three following phases:

Recording

- Bringing together all aspects of the topic
- Making them audible and visible
- Allowing and encouraging a broad approach to the subject matter

Condensing

- Developing the central topic with the group
- Channelling the subject matter

Firming Up

- Making sure that alternative solutions are considered
- Initiating specific steps for implementation

Rules of Conduct for the Moderated Meeting

To a large extent (but naturally not entirely), professional moderation can be described by the following rules of conduct:

- "Asking not stating"
- Moderator has a positive attitude towards people
- Working with the group, rather than fighting it
- "Social" disruptions have priority
- Distinguish between "perception", "suspicion" and "evaluation"
- "I" messages, rather than "one/people" expressions
- Look out for and investigate "non-verbal" signals
- Do not evaluate or judge
- Avoid being defensive or self-justificatory
- Do not discuss the methods
- Have two moderators if possible
- Analyse conversational practices
- Note body language
- Observe or even specify the seating arrangements

A Successful Moderated Meeting

A good moderator observes or plans the following points (Fig. 24.3).

- Outlines the meeting's purpose at the start
- Briefly introduced all participants
- Defines time frames
- Has time management in hand (possibly delegated)
- Defines ground rules and makes sure they are followed
- Explains his role, and the participants' roles
- Is neutral (with respect to the topic, and to participants)
- Visualises the goal setting process
- Can be both rational and intuitive
- Involves "quiet" participants
- Intervenes if people talk too much
- Pays attention to depth of detail
- Asks questions and follow-up questions
- Seeks to clarify statements
- Reduces the potential for conflict
- Makes sure specialist terms are explained
- Summarises things, and recaps the results
- Insists on concrete decisions
- Formulates what happens next
- Thanks everyone and wraps up
- After wrapping up, does not revisit things at content level

Fig. 24.3 Checklist "Successful moderation"

Sample Questions for the Different Work Phases

"He who questions, leads." This is particularly applicable to the moderator. He uses questions to steer the process through the various work phases.

Getting Started

- How well informed about the topic are you already?
- What is (particularly) important to you for today's meeting?

Gathering

- What information do you need about the topic?
- What needs to be discussed today?
- What still needs to be clarified today?

Selecting

- Which topics need to be given priority?
- Where should we start?
- What needs to be covered most urgently?

Elaborating

- What is (always) a problem?
- What can we do about it (first)?
- What do you suggest?

Planning

- What shall we now do specifically?
- Who is doing what, with what aim and by when?
- When shall we work on this again?

Closing

- How satisfied are you with this meeting?
- Have we achieved our goals?

Subject Moderation or Process Moderation

We distinguish between two types of moderation: subject moderation and process moderation. Subject moderation involves drawing up and recording all the aspects, opinions, ideas and objections relating to a particular subject. With process moderation, the emphasis is on the process. Different viewpoints are clarified and documented (Fig. 24.4).

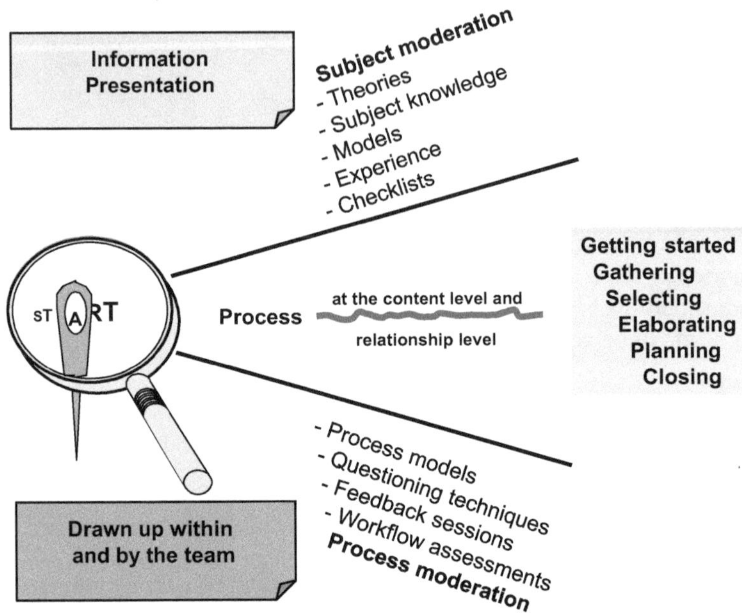

Fig. 24.4 Subject and process moderation

24.4 Running and Structuring Project Meetings

"So many meetings, so little achieved". A meeting takes up lots of resources, so it should first be clarified whether a meeting is the right tool for the job. To do this, we first need to know the goal of the meeting. It may be, for example, a:

- Coordination meeting (e.g. for resources)
- Information meeting
- Problem brainstorming and decision-making meeting
- Conflict resolution

Is a Meeting the Right Tool?

If the following questions can be answered with "yes", then a team meeting is probably the right format:

- Is the meeting really necessary? (Would it not be good enough to inform the people concerned by e-mail or letter?)
- Are the necessary specialists available (other deadlines, holidays, etc.)?
- Will the technical knowledge and personalities of all the participants allow them to contribute to the success of the meeting?
- Are there no more than ten people to be invited?
- Is it possible to put together a balanced group of participants made up of ideas people, organisers and doers?
- Can all participants be certain of being able to put forward their suggestions and ideas openly and without fear?
- Can the meeting be scheduled for an "off-peak" time?
- Will the meeting last no more than 2 h? (Two short meetings are better than one long).
- Is there a suitable room/infrastructure available?
- Does the invitation include the following information: location (with a map, if necessary), times (start and end), chairman and minute-taker?
- Are the topics and goals of the meeting clear from the invitation?
- Can the participants see from the invitation what is expected of them (receive information, come up with ideas, take decisions)?
- Will the participants receive their invitation with all the necessary documents at least 3 days before the meeting?
- Has the chairman prepared a script for the meeting (time for each topic or work step, how to handle it, materials required, etc.)?
- Have clear ground rules been defined to ensure that the meeting proceeds smoothly?
- Has time been scheduled in for feedback at the end of the meeting? (What went well? What needs to be improved in future?)

Preparing for Meetings

The effort involved in preparing for a meeting should not be underestimated. The corporate culture often has a massive influence on the way in which meetings are planned, run and followed up. This is often described as the "meeting culture". Serious planning for a meeting requires the chairman to carry out a degree of preparatory work in terms of both content and organisation and ultimately reflects the esteem he feels for the participants.

A checklist will be found very useful to avoid having to "reinvent" these recurring activities every time. After drawing up a personal checklist, it is a good idea to review, supplement and improve it on an ongoing basis (Fig. 24.5).

Topic of meeting	• Clearly described • Formulate in a value-neutral manner
Participants	• Team members • Subject experts • External people • Decision-makers
Documents for meeting	• Put together, put into order and select • Have sufficient copies available • Where appropriate, distribute in advance
Aim(s) of the meeting	• Define them clearly (time, contents, decisions) • At the start of the meeting, agree to have a concluding feedback round
Type of meeting	• Should suit the meeting's purpose
Agenda	• Agree with the responsible persons
Invites	• Topic / subject of meeting • Goal(s) • Date • Time (start / scheduled finish) • Location • Agenda • "Distribution list"

Fig. 24.5 Checklist "Meeting preparation"

Running Meetings

A checklist can also be drawn up for running the meeting. This acts as a meeting timetabling tool that takes the pressure off chairman and participants alike (Fig. 24.6).

Welcome	• "I" message
Introduction / topic	• Why are we here?
Introductions	• New people quickly introduce themselves
Absentees	• Who is not here? • Explain the impact on the meeting
Goals	• What do we want to achieve today?
Procedure	• Use the agenda to show the structure of the meeting
Leading discussions	• Who is managing the meeting, in terms of structure and content?
Ground rules	• How should we treat each other? • What is desirable? • Define the communication rules
Minutes	• Who will take the minutes during the meeting? • Who will write up the minutes, and by when?
Boundaries	• What is not up for discussion today? • What are the seams / interfaces to other projects or initiatives?
Structure	• What order will we discuss today's topic in?
Content	• Introduction to the topic • Make sure all aspects and viewpoints are covered • Visualise the knowledge that is acquired
Conclusion or decision	• What conclusions were reached? • What decisions were taken?
Summary	• What is the output from the meeting? • Ask for feedback from participants
Next meeting	• Agree the date of the next meeting
Thanks and close-off	• "I" message

Fig. 24.6 Checklist "Running the meeting"

Timetable for an Information Meeting

Preparing thoroughly for a meeting not only helps to make optimum use of the available time, but it also shows that all the participants who have made time for the meeting are appreciated.

Drawing up a script can offer additional certainty and guidance for an inexperienced project manager (Fig. 24.7).

Time	Contents, topic	Aim	Resources, comments
08:00	Arrival, registration, coffee	Time for informal greetings and getting to know people	Standing Coffee, fruit, etc.
08:15	Start, welcome Present the goals Framework for the meeting (with times)	Start on time Orientation about the meeting Aims of the meeting Create an atmosphere of transparency and trust	Flipchart, slides Handouts Work out a good seating plan
08:30	Quick questions to participants	To enable any specific information requests to be taken into account	
	Present the information in a structured manner	Quickly given an overview, and have time for main points	Flipchart Slides Handouts
	Conclude the information by looking to the future	Outline the prospects and provide information all round	
09:30	Round-table discussion	Open the floor for questions, comments	
09:50	Summarise Agree date and time of next meetings Thanks and close-off	Clarify any open issues, leave people ready for further activities	

Fig. 24.7 A sample "information meeting"

Evaluating Meetings

Thorough and self-critical evaluation after a meeting offers by far the greatest learning effect. The procedure for evaluating a meeting (ideally within the team) can be structured as follows (Fig. 24.8).

Overall	• Did the meeting work well? / Not well? • Why? • Improvements?
Aim of the meeting	• Achieved? / Not achieved? • Why? • Improvements?
Preparation for meeting	• Good? / Adequate / Not good enough? • Why? • Improvements?
Running the meeting	• Rules followed? / Not followed? • Why? • Improvements?
Chairing of meeting	• Good? / Adequate / Not good enough? • Why? • Improvements?
Participants	• Right participants? / Not right? • Why? • Improvements?
How did meeting go?	• From your own perspective? • From the participants' perspective? • Improvements?
Participants	• Engagement • Behaviour • Sensitivity
For next meeting	• Retain? • Improvements? • Avoid?

Fig. 24.8 Checklist "Meeting evaluation"

24.5 Visualisation and Presentation Techniques

People have five senses, which means that they perceive things via five channels. Nevertheless only one channel is used in most meetings: hearing. If we want to increase the retention and efficiency of a communication process, then we need to

address several senses at the same time. Things will never get moving if no pictures are created.
A visual appeal with appropriate images will reinforce the verbal messages (Fig. 24.9).

Fig. 24.9 Retention levels within the learning process

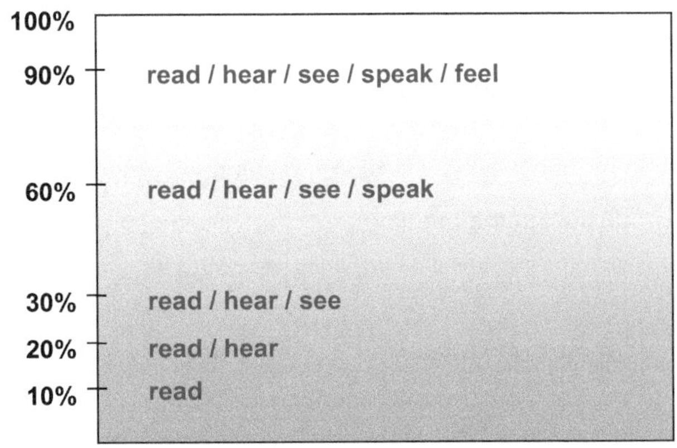

Visualised messages make it easier to interpret the entirety of a task or topic. Visualisation forces the team to specifically choose between important and unimportant information. It also avoids overloading the meeting participants' capacity to absorb information.
Visualisation allows messages, opinions, differences and results to be immediately represented in a visible form for everyone. The progress of the meeting can be seen at all times. It also becomes easier to document, interpret and summarise results. The participants identify with the result because they can make their own contributions and understand how the result came about. Visual aids should be kept available so that the group can start work immediately: questions, hypotheses, information, scenarios, etc.

Checklist "Presentation"

To prepare a successful presentation, follow these steps: (Fig. 24.10)

Preparation	• Define the target groups • Define the purpose of the presentation • Put together the presentation contents and slides • Establish roles for each member of the presentation team • Set the location and choose the room, then book it / reserve it • Work out a schedule (with breaks) • Draw up a catalog of possible arguments • Think about limitations / resistance points, and step through them with arguments • Do a test presentation (with timings monitored)
Scheduling	• Avoid clashes with other important events • Check and agree with other organisers at an early stage • Invitations with confirmation of date, remember that people are human!
Room	• Make sure the room is the right size • Arrange the tables so that they don't form barriers • Avoid disruptions (mobile telephones, laptops, etc.) • Air conditioning, lighting, technical facilities, temperature • Check the noise levels. Identify noise sources • Avoid interruptions (door sign, no telephones)
Media	• It is people that convince - not technical tools • Powerpoint is the curse of our information society, and can get in the way • Get equipment in advance, and check everything works • Flip charts, felt-tip pens, visual aids, handouts
Running the meeting	• Create a positive atmosphere • Keep watching the time • Monitor people's reactions • Address any questions, possibly hold over till later to keep things running • Develop the arguments raised by questions, ideally with examples from the participants • Make a note of any intermediate results
Conclusion	• Transition into discussion • Make sure the need for a decision is clear • Explain the next steps, draw up a list of activities • Produce minutes, and circulate them to all participants

Fig. 24.10 Checklist "Presentation"

A Successful Presentation

The following points must be observed for a successful presentation from Matthias Pöhm's book "Vergessen Sie alles über Rhetorik" (Forget everything you ever learned about rhetoric):

- Speaking is not about information. It is all about the feelings that are triggered by the information
- The effect of a presentation is determined 30 % by the text and 70 % by the packaging
- Stand in the centre of the room
- Deliberate body language that exudes self-assurance
- Look at the people, move your eyes around the room
- People first, then the matter at hand
- Rhetoric means creating images in other people's heads
- A speech will only be effective if the speaker is enthusiastic about the topic and is authentic
- Gestures start with the hands in the home position: hands touching lightly in front of the stomach, just beneath the navel
- Maintain the volume (made easier with gestures)
- Always give an example when you make a statement
- Show the visual aid for each topic
- Setting out the content at the start of a presentation is like revealing the murderer before the detective novel begins
- Speeches can be made fascinating by including elements that bring tension, curiosity, sympathy and fun
- A lecture should end if the listeners have heard everything they want to hear
- Short sentences, Verbs bring animation, Present, not past or future
- Any sentence that does not generate interest can be deleted without being replaced
- Words such as "dynamic", "flexible" and "innovative" belong on the rhetoric scrapheap
- Use a colon rather than a relative clause, two sentences rather than "and"
- Start by talking about the slide; short pause; display the image at the same time the text is presented
- It must be possible to take in the entire slide within 2 s
- Omit anything that takes time to read from the slide
- A statement is devalued if it is repeated on the slide

24.6 Analysis Within the Team

Process Analysis

Questionnaire

Answer the following questions from your own viewpoint in relation to your team. There are six possible responses to every question. If the statement on the left applies totally, then select the left-hand number (1); if the statement on the right applies, then select the right-hand number (6). If an expression falls somewhere in between, then select the most applicable number (2, 3, 4 or 5) (Fig. 24.11).

1. How did I feel in the team today?							
Uncomfortable and tense	1	2	3	4	5	6	Comfortable and relaxed

2. How were the objectives formulated today?							
Clear, understandable	1	2	3	4	5	6	Unclear, hard to understand

3. How did the team work today?							
Lazy, dissatisfied and superficial	1	2	3	4	5	6	Intensive, thorough, enthusiastic

4. How did the team discuss things today?							
Abstract, theoretical, unrealistic	1	2	3	4	5	6	Focused on the topic

5. What did the participants talk about?							
Focused on the content, addressed the issues	1	2	3	4	5	6	Development focus, personal feelings

6. What motivated the individual team members?							
Out to score points	1	2	3	4	5	6	Focuses purely on the matter in hand

7. Did people listen to differing views?							
No, they were not heard, or were not allowed	1	2	3	4	5	6	Yes, people discussed them and considered them

8. How did the team treat me today?							
I felt rejected, withdrawn, bored	1	2	3	4	5	6	I felt accepted, included in the group

9. How much help did the team give me today?							
My needs were not really heard	1	2	3	4	5	6	My needs were taken on board

Fig. 24.11 Questionnaire "Process analysis within the team"

Evaluation

Draw up a picture of the responses from all participants. Identify the questions or topics with the greatest spread of answers. Discuss these issues within the group.

The form shown above allows team members to anonymously reflect the process as they experienced it. If the summary of all the responses shows a scatter across at least three boxes, then it is a good idea to at least clarify or discuss the analysis. If the scatter ranges across four or even five boxes, then it is generally more than just a misunderstanding. Listening closely, asking questions and understanding what individuals want to express is then very important for project managers. In an open and trusting discussion, the individual's concerns or preoccupations are clearly expressed.

Team Portfolio

A recurring need for managers is to be able to assess success factors within their own team as rationally as possible. An end-to-end concept for measuring and improving the variables that influence a team's success was developed by Ruth Stock. These success factors can be assessed by team members using the following questionnaire (Fig. 24.12).

Team success		Not applicable					Applicable
		1	2	3	4	5	6
1	The contents of what the team has achieved are excellent						
2	I am convinced we can achieve more as a team than I can alone						
3	Every member of our team engages personally with everyone else						
4	I like all the team members						
5	The team leader takes the team members' concerns into account						
6	All the team members have good subject expertise						
7	All the team members systematically use problem-solving techniques						
8	There is total clarity about what is expected from team members						
9	The team leader encourages all team members to give their very best						
10	We often discuss personal issues						
11	We support each other if we have personal problems						
12	Personal disputes between us are very rare						
13	Information is circulated frequently and systematically						
14	We support each other with professional issues						
15	We have fair and open discussions about professional conflict						
16	Alternatives are explored and evaluated before decisions are taken						
17	The practicability of alternatives is assessed before decisions are taken						
18	Decisions are taken quickly						

Fig. 24.12 Questionnaire "Team success" (based on Stock, 2002)

The questionnaires are evaluated at the content level by the weighted mean of questions 6–9 and 12–18 and at the relationship level by the weighted mean of questions 2–5 and 10–12. These values show the assessed team's specific position in the following matrix for the relationship and content levels (→ team result). Depending on the position, the team leader can decide on suitable actions to develop the group into a real team (Fig. 24.13).

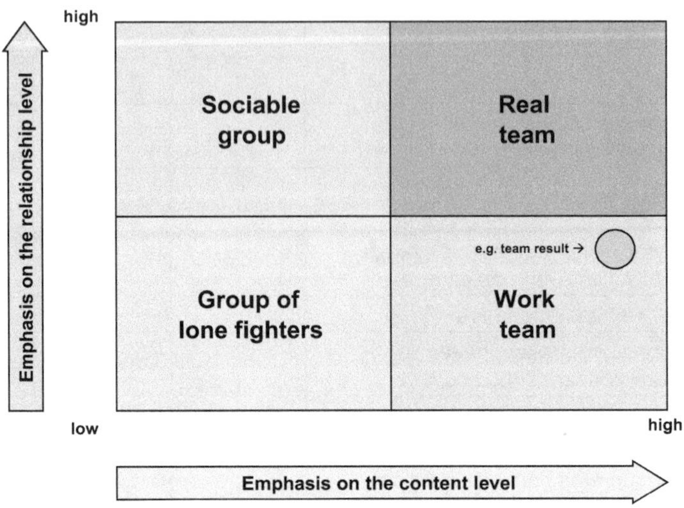

Fig. 24.13 "Team success" evaluation (based on Stock, 2002)

24.7 Intervision as a Problem-Solving Method

The psychoanalyst Michael Balint developed a method of discussing cases between analysts. The method is founded on the basic idea that, in problem situations, the input of other people is often readily recognised, while the analyst's own contributions to the problems are to a certain extent hidden (Johari window). Such contributions are quickly identified by neutral observers of a situation. They can offer valuable, critical feedback providing new ways of looking at the problem and reveal new approaches to finding solutions. The intervision method structures the way of handling a problem that a group member wishes to resolve with persons who are not present.

Michael Balint discovered that it is helpful to discuss the impressions of such outsiders without the presenting person immediately commenting on them. A Balint meeting thus has a crystal clear structure:

1. The Presenting Person Describes His Problem

The group listens without interrupting. The listeners make a note of what comes to mind during the description, what they feel and what they think of the presenting person (uncertain, nervous, arrogant, self-indulgent, etc.).

2. The Group Asks Comprehension Questions

The group does not ask any questions that contain assumptions, hypotheses, interpretations, etc. The moderator allows the question session to run until he feels that the group has sufficient information for the next phase. This sequence may be shorter or longer, depending on the problem.

3. The Listeners Describe Their Hypotheses

Thoughts, impressions, assumptions, causes, etc. are all permitted, but no solutions may be put forward yet. The participants may give free rein to their imaginations. Someone writes down the group's feedback in the form of keywords on a flipchart. The presenting person listens without making comment, even non-verbally. He must not join in, but may note down his thoughts, ideas, etc.

4. The Presenting Person Identifies the "Hits"

He goes through the list and says what is or is not applicable for him and why, or to what degree it is applicable. He crosses out the "false messages". His subjective view is the right one here! While the moderator strictly enforces the rules concerning speaking and intervention during the first three steps, he may allow more discussion in this phase.

5. The Group Draws Up Suggested Solutions

The presenting person or another participant notes down and displays the suggestions.

6. The Final Round

This gives the presenting person the opportunity to tell the group what he has taken from the meeting. The other group members consider what they have learned themselves. The more the presenting person has exposed himself personally, the more important it is for the other group members to now say something about themselves.

Timetable for an Intervision Meeting

A possible timetable for an intervision meeting can be (Fig. 24.14):

Phases	Individual steps	Questions
1. Show the starting point	The problem presenter outlines the situation	• What is it about? • What is the problem? • How did it come about? • Which people are involved? • How am I involved? • What have I already tried to solve the problem? • What are my thoughts and feelings now? • How do I see the goals?
2. Formulate questions to the group	The problem presenter asks the group questions, and notes the questions on a flipchart	• What issues would I like to clarify? • Where do I need your help? • Note question on flipchart
3. Clarification questions from the group	Members of the group then ask questions to clarify their understanding Then ask whether (and how) the question to the team might have changed	• What have I not understood? • How did that work? • When did that happen? • Where did that take place? • How many people did that affect? • Caution: No leading questions
4. Suggest speculative suggestions and thoughts (The problem presenter stands back from the group and allows himself to absorb the comments)	Members of the group express their assumptions, hypotheses and impressions The problem presenter notes these, without commenting	• What does that make me think, and feel? • What do I think could be possible causes of the problem? • How do I perceive you? • What are my wildest dreams?
5. Assessment by the problem presenter	The problem presenter comments on the accuracy of the comments	• Which hypotheses are correct, partially correct, or wrong? • What do these hypotheses make me start thinking about? • Has my perception of the problem changed? • If so, how? • Confirm / adapt the question on the flipchart
6. Develop ideas for solutions ((The problem presenter stands back from the group and makes a note of his own possible solution ideas)	a) Together b) Members of the group develop solutions and the problem presenter notes the ideas down	• What ideas or measures could the problem presenter implement that might solve the problem? • What does it bring to mind for me? • What solutions do we see in the light of the things we are hearing?
7. Select solutions	The problem presenter selects those solutions that could make sense, and comments on implementing them The problem presenter makes a note of the solutions on the flipchart	• What solutions shall I choose? • My next steps: Concrete measures? • What resistance should I expect? • How will I handle the resistance? • What might stop me implementing the solutions? • Where do I still need help? • Results on flipchart
8. Evaluation / reflection	Evaluate the results Reflect with all participants on methods used Thank the problem presenter	• How are you? • How did this experience feel for you? • What have I got out of this case? • What will I do with what I have learned?
	Final suggestions to the problem presenter	• One last thought for you to take away ...

Fig. 24.14 Timetable for an "intervision meeting"

24.8 Establishing Consensus

Differences of opinion often reinforce the battle lines. A dialectical approach is a useful way to establish a good consensus. The procedure described by Rupert Lay in "Dialektik für Manager" will help with argument and counterargument:

- "The alternative solution or a partial aspect of the solution then leads to the project goal if...". Use the flipchart to list as many conditions as possible that need to be fulfilled in order to affirm the project.
- Once the conditions have been described, the team defines all the terms that can be understood in different ways. The definitions belong to the list of conditions.
- All the conditions are then checked. Are they necessary, useful or sufficient? Alternatives will now be found for those conditions for which no consensus can be reached. One possible alternative is simply to delete the condition.
- The objective functions of the conditions are then also determined, if necessary: Does the condition help from the regulatory viewpoint in that it can be used to deflect serious damage to the common good? Does the condition help to bring about a better operating result? Or does the condition have an ethical effect in that it helps an employee to increase his own social and technical talents and those of his colleagues and employees?
- Finally it is checked whether the alternative fulfils all the conditions or allows them to be fulfilled with reasonable effort. It may not be possible to fulfil certain conditions until the project is implemented; these are intended to be used to assess and improve any pilot trials or to entirely reject the alternative.

Clearly then, with this method, the alternatives are given a "final polish" by virtue of the differences of opinion that are put forward.

24.9 Analysis of Conflict Styles

Just like a leadership style, a conflict style comes about because the parties to the conflict combine in their behaviour leanings towards their own concerns and to those of the opposing side in variable combinations. Karl Berkel has developed a questionnaire for this that may be helpful in practice.

Questionnaire

There are no "right" or "wrong" answers in the following diagnosis. Start by marking with a (I) the statement that applies most closely to your own behaviour. Then mark with a (II) another statement that may also apply to your behaviour. Make sure that you tick the variant(s) that you would actually do (... how you effectively respond), and not the solution that you think would be the best (... how you would like to be) (Fig. 24.15).

1. **Every conflict style is based on the feelings of the people involved. How would you describe your feelings in conflict situations?**
   a) I welcome an opportunity to get my feelings off my chest.
   b) I have concerns about conflict. I worry about what the other people are thinking and feeling.
   c) If get frustrated. Then I either become annoyed or resigned. So I don't really contribute to finding a real solution.
   d) I enjoy it, but I don't like the feelings to become too overwhelming.
   e) It often scares me. Open discussions always end up hurting other people.

2. **A friend has annoyed you. What do you do?**
   a) I tell him what I am annoyed about, and why. Then I ask him how he feels about it.
   b) What annoys me most is that I've allowed myself to get so annoyed by him. I keep out of his way until I have calmed down.
   c) When I'm angry, I explode, without asking many questions.
   d) I'm worried about getting angry. It could make me do something that I'd later regret. So I try to suppress my anger and do the exact opposite of what my anger is driving me to do.
   e) Getting really annoyed is good for us all occasionally, as long as nobody gets hurt.

3. **A meeting is dragging on and on because a colleague keeps reiterating his objections. What do you do?**
   a) I intervene so that he can set out his concerns. If he doesn't manage to convince people, he should accept the majority view.
   b) I try to find out why my colleague is seeing the problem in a different way to the rest of the group. Then we can all re-examine our own arguments from his perspective and better understand him.
   c) These differences of opinion weaken the group. I encourage everyone else to move on to the next item.
   d) The colleague is holding up our work. I would say so openly, and demand that if needs be, we carry on without him.
   e) I keep out of it when other people argue. Everyone needs to be able to explain their own opinion.

4. **Groups often need to discuss and coordinate their decisions with other groups. What are the points to consider when choosing a spokesperson for the group?**
   a) It should be the person who can best represent our opinions, but they should be flexible enough to revise our position in the light of arguments put forward by other groups so that we reach the best decision.
   b) They should skilfully represent our position, but should at all costs avoid us being trapped.
   c) They should be cooperative, friendly and reserved in order to avoid conflicts with other groups.
   d) They should be capable of touch negotiation, should make no concessions and should impose our positions as fully as possible.
   e) I would prefer someone who is willing to compromise from the start.

5. **Colleagues in your office are in open conflict. What do you do?**
   a) They should sort their differences out between themselves.
   b) I try and settle it by calming them down.
   c) I bring the parties together so that they can talk it through and resolve their problems.
   d) I suppress the conflict by directly managing them both. Conflicts should never be displayed so openly.
   e) I speak privately with each of them in order to gain an understanding of the problem, and to make clear to them that everyone stands to "lose" if there is conflict in the workplace.

6. **Your decision is challenged by colleagues. How do you react?**
   a) I repeat my decision in order to ensure that everyone clearly understands it. If they are still opposed to it, I explain that this is in everyone's best interests, even thought that might not be apparent to them.
   b) Once you start letting colleagues question you, it just encourages them to rebel and to refuse to obey. So my normal style is to say ".... that's my decision, and that's the end of the matter!"
   c) I explain things again, and if there are still questions then I step back from it in order to keep things peaceful, saying something like: "OK, do as you want!"
   d) I withdraw, or simply don't follow through on the decision.
   e) I explain how this decision was reached, and provide additional information that I hadn't shared before because I'd thought it unimportant, or information that was not previously available. Then I listen, so that I can hear any arguments that might change my decision.

Fig. 24.15 Questionnaire "My conflict styles" (based on Karl Berkel)

Evaluation

Transfer your responses to the sheet below and highlight the corresponding values (Fig. 24.16).

Question	Solution	Value (x/y)	Question	Solution	Value (x/y)
1	a	9/1	4	a	9/9
	b	9/9		b	1/1
	c	1/1		c	1/9
	d	5/5		d	9/1
	e	1/9		e	5/5
2	a	9/9	5	a	1/1
	b	1/1		b	1/9
	c	9/1		c	9/9
	d	1/9		d	9/1
	e	5/5		e	5/5
3	a	5/5	6	a	5/5
	b	9/9		b	9/1
	c	1/9		c	1/9
	d	9/1		d	1/1
	e	1/1		e	9/9

Fig. 24.16 "Conflict styles" lookup table

The numbers in the evaluation each represent a different conflict style, where:

9/9: Joint problem solving, creative collaboration, wanting to find the solution that is best for both sides, despite resistance and setbacks.
5/5: Compromise, everyone steps back from his maximum demands.
1/1: Escape, avoidance, disengagement, doing nothing, sweeping conflicts under the carpet.
1/9: Backing down, capitulating, relinquishing own goals, not exaggerating differences of opinion, smoothing over, creating harmony.
9/1: Having own way, compelling, You or Me, intimidating and using power, implementing the poker strategy.

Transfer your first (I) and second (II) choices to the following graph (the x values on the horizontal axis, the y values on the vertical) and identify the quadrant containing the most choices (Fig. 24.17).

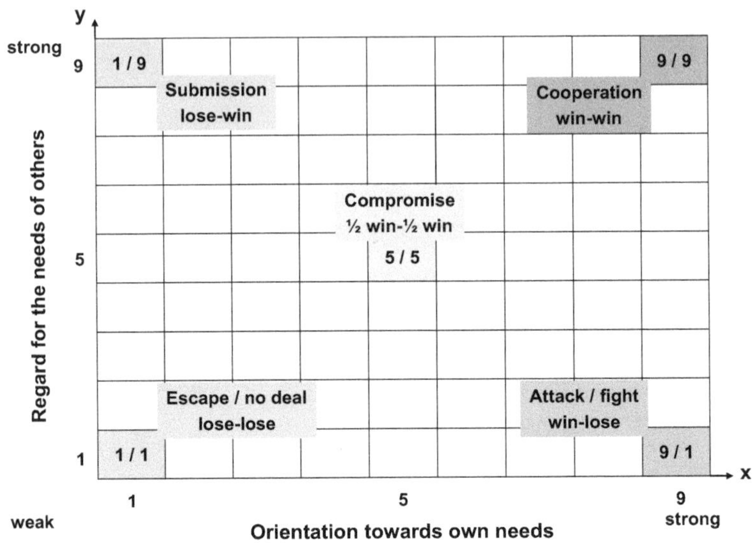

Fig. 24.17 Conflict styles

Notes

No style can ever be regarded as the only perfect style. Entirely different styles may be appropriate in different situations. A conflict can best be resolved if the participants are able to move flexibly between these styles.

Every person develops a sequence of conflict styles that are characteristic of his personality. He learns the best style with which to approach a conflict and which to switch to if the first is unsuccessful.

25 General Management Methods

25.1 Project Request, Project Order, Project Agreement

The project request is the basis on which decisions about a project agreement will be made. The term "project agreement" reflects an organisation's working behaviour. In the same way as a request for proposal is discussed and negotiated with an external customer, the internal responsible person agrees with the project manager what he can do to help achieve the company's goals. It is the incoming project manager's task to formulate the project request, which will be used as the basis for the project agreement. He does this through dialog with the project owner. As the person responsible for the project process, the project manager knows which points need to be discussed. He should write down the project owner's answers in order to ensure that both partners in the contract have the same understanding. In signing the document, the project owner confirms that the project manager has understood things correctly. He then gives a green light, normally for a project phase. The agreement is then in place.
Regardless of whether the project is internal or external, the following points should always be included in a project request or tender:

Starting Point

- The circumstances that led to the project
- The back history, factors that triggered the project
- Description of the current situation
- Problems and potential of the current situation

J. Kuster et al., *Project Management Handbook*, Management for Professionals,
DOI 10.1007/978-3-662-45373-5_25

Goals

Key points in formulating the goals:

- The goals should be defined in a solution-neutral manner, so as not to inhibit the problem-solving process too early. The goals can be defined either in terms of the required effects or in terms of avoiding unwanted effects.
- For many goals, creating a goal structure (or goal hierarchy) helps give an overview. Goals should be as verifiable as possible. For qualitative goals, however, it may not be possible to operationalise the goals.
- It should be possible to talk with the project owner about using existing knowledge or experience to modify or detail the goals.
- A distinction must be made between essential goals and desirable goals.

The following questions can help identify goals:

- What needs doing? What are we trying to achieve?
- What problems need to be eliminated?
- What situation needs to be improved?
- What needs to be avoided?
- What are the ideal type and scale of results?
- What potentials should be used?

Boundaries

This is where the limits are set for the project environment (project limits, project scope). These limits can be both positive and negative in nature: What does the project work cover, and not cover? The boundaries may relate to functions, data or organisational units, for example.

Dependencies and Influences

For example, there might be dependencies to:

- Other projects (content, time)
- External events (e.g. legislative changes)
- What are the key influences that the project is subject to?

Framework Conditions

Framework conditions are general requirements or restrictions that apply to the project (requirements to be taken into account, known restrictions).

Groundwork

What preparatory work or groundwork is the project based on? The groundwork might include:

- Studies, analyses
- Concepts, strategies
- Standards, norms
- Results from previous projects

Results

It is important that everyone involved in the project is clear about the end result and any important interim results.

- What should the end result look like? (presentation, document, system, etc.)
- What support can/will the project owner offer?

Project Costs, Benefits

- Have the relevant financial and human resources been budgeted for?
- What cost centre will the planned project costs be charged to?
- What quantifiable/non-quantifiable benefits or advantages will the implementation of the project bring?

Risks

- What are the existing, recognisable or likely risks, from today's perspective?
- What measures have been taken to reduce these risks?
- What are the consequences if the project is abandoned?

Approach, Schedule

- Project plan (=master schedule) with milestones
- When will the project end?
- Notes about any special approaches or procedures

Priority

- What priority does this project have?
- Is the project part of the project portfolio?

Project Organisation

- Project owner
- Contractor (project manager)
- Project team(s)
- Supervisory bodies (e.g. steering committee, policy board, etc.)

Information, Communication

This describes the information flow or communication with the project owner, supervisory bodies, users and other interested parties. This may be written communication, or verbal reporting. It should always be clear who produced the document and who the recipients are, and the frequency of the information:

- Progress report (status report)
- Project plan
- To-Do lists and decision lists
- Project newsflashes

Signatures

- Project owner
- Project manager
- Controlling (for large projects)
- Decision-making bodies (depending on financial literacy)

Appendix

- Project plan (master plan)
- Project organisation structure
- Detailed "cost effectiveness"
- Resource plan
- Planned labour costs

25.2 Benchmarking Shows How Others Are Doing It

Benchmarking is a planning and analysis tool that allows a company to compare what they are doing against "best in class" used by their competitors and against wider best practice used by companies in other industries. It is a process that compares the products, methods, processes and structures of one or more different companies' business functions in order to identify the potential for rationalisation or for improving quality or performance. It is an ongoing process, since practices in different sectors constantly change. One way of benchmarking is through sharing information with other organisations. Setting up a benchmarking group, and maintaining it, involves a major time commitment. It assumes that there is a readiness to learn from others. Professional benchmarking is structured in such a way that nobody has to disclose confidential information. Once a benchmarking group is up and running, the process continuously steps through the following loop:

- Set goals
- Analyse internally
- Compare with others
- Decide on actions
- Take actions, and check the effect that they have
- Set goals again

If the balanced scorecard concept is already being used in the company, it is worth building the benchmarking procedure around that. It allows up-to-date data to be quickly evaluated using a project scorecard, for example.
Good project management contributes in many ways to a company's market success. Project managers know where they are up to with their work, where the company's strengths lie and where there is potential for improvement. Benchmarking allows the user to compare their project against the industry average, for example. It can even be useful for comparing projects within a company, or within a department. Benchmarking enables project managers to see whether they are using the most appropriate modern project management tools to efficiently achieve their project's objectives.
Managers of multiple projects can evaluate individual projects in their portfolio, and compare them against each other or against industry leaders. Showing the analysis to project initiators, project owners and investors is a good way of helping them to form an impression of projects and service providers.
The European Foundation for Quality Management EFQM uses standardised questions to assess companies that seek to achieve the highest standards. The EFQM Excellence Award provides a way of comparing the assessments for the finalists and prize winners. Best practice can be assessed using the Business Excellence criteria, by industry and by country. The International Project Management Association IPMA presents its own IPMA Project Excellence Award, which is based on the EFQM model. This means that the criteria for excellence in project management are internationally comparable. To complete

the full circle and to bring things back from the organisational level to the personal level, IPMA also offers certification for project managers (see "IPMA Certification").

25.3 Change Request Management

Organising, managing and implementing changes in projects that are already running can be described like this (Fig. 25.1):

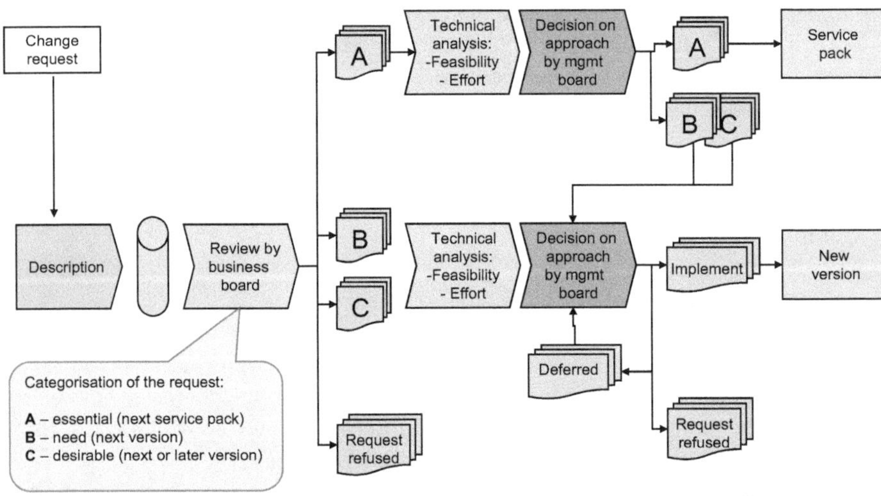

Fig. 25.1 Overview of a change request

It is essential that a systematic change management system must be set out in writing. This is the only way of ensuring that the additional demands can be checked at a later stage. The template outlined in Fig. 25.2 covers the key points.

Change request number:			
Project:	**Release**		**Priority**[1] 1 2 3 4
Sub-project: SET Classification	**Status**		
Work step:			
Relationship to other changes:			
Applicant:			

1. Request for change
Description of the problem
Result / scope, what needs changing
Reason for the change

2. Evaluation
Estimated effortSET RISK
What are the effects?
When benefits will it bring?

3. Design & report on implementing the change	
Analysis of effects of the change	
Proposal:	Result:
Discussed with:	Date:
Estimated effort	
"Lessons learned"	

4. Approval of the change

Item	QA measure	Approved by	Date

5. What has already been done (history)?

Version	Date	Person	Comments
1.0			
1.1			
1.2			
1.3			
1.4			
1.5			

Priority[1]

1 = Frequent function and business process, cannot be used	3 = Frequent function, but process can be performed manually
2 = Occasional function or business process is usable, but with difficulty	4 = Occasional function or process, can be performed manually

Fig. 25.2 A "Change request" template

25.4 Controlling

Benefits and Business Case

Organisations live from creating benefits. So the question of a project idea's value needs considering at a very early stage. In our economic system, organisations survive by having money in the bank. So each planned project must be checked to ensure that it increases the organisation's chances of survival.
Even before the project owner places an order for the project, the organisation must consider whether the project should be undertaken at all. This is not simply a question of quantifiable financial values. Strategic considerations, philosophies, resources and so on will also come into it.
Although the project owner will have to take the ultimate decision on whether or not to place the order for the project, all of these aspects must be transparent for the project management.
At the ideas management level, various questions about an idea's market appeal must be considered (some organisations call this the preliminary business case). The information below enables the organisation to assess the different project ideas:

- Monetary and non-monetary benefits for the organisation.
- Customer requirements, market segmentation (buyer groups) and market potential.
- Industry structure: Analysis of existing competitors, potential new competitors, threats from alternative products or services.
- Estimated investment in the project/operating costs for the solution.
- "Product" (customer needs): What will the idea look like?
- "Price" (cost-to-customer): What price can the idea be sold at?
- "Placement" (convenience): How will it improve service or customer support?
- "Promotion" (communication): What arguments work with customers?

Project Scorecard

The Balanced Scorecard concept is a measurement-based system for company management, but it can also be used for projects. The key benefits of a Project Scorecard are:

- Provides a direct link between company strategy and how the project is run
- Improved measurement-based monitoring of project progress
- Defined early warning signals help to reduce risks
- Project benchmarking, supported by a uniform assessment system
- Improves stakeholder engagement in the project
- A growing database for uniform project portfolio management

The extent to which a project scorecard will increase the administrative overhead for projects and thereby reduce the project's return on investment, or to which it can make a viable contribution to the success of a project, will of course depend on the type of project and on the company's project culture.

Review Techniques

A design review (the review) clarifies the question of whether the developed result meets the requirements, and shows up any defects at an early stage.

Principles

1. Four eyes see more than two
2. Overcoming blindness to deficiencies

The review technique was developed for complex innovation projects (e.g. research, space travel) in the 1960s, and quickly became established. It is now used in a wide range of industries for innovation projects. Examples of reviews include:

- Preliminary Design Review (PDR): After the definition phase, the draft is checked by internal and external experts who are not directly involved in the project, and by participants who represent different perspectives (users, interest groups, maintenance, general public, etc.).
- Critical Design Review (CDR): A decisive final check before a research project is started or before it is publicly announced.
- Peer review before publication of a scientific article.

Here's an overview of the "review" management method, using a design review as an example:

The Reviewer

- Must be competent in the subject area
- Must be independent (must not have been directly involved in the task)
- Must have a critical but constructive mindset
- Must study the documents (the project manager is responsible for making them available in good time)
- Must put thought into the best approach to take

Procedure

- The object of the review is selected, and boundaries set
- The reviewer asks open questions ("How have you ensured that. . .?")

- Using open questions often leads to the subject expert identifying the ideal solution because it helps key aspects to become clearer
- Checklists are useful tools during reviews. They must contain concise questions, suitable for the application and phase-specific
- Experienced staff should be involved in drawing up the checklists. They often know the stumbling blocks and identify with the result

The results of the review should be recorded in a report:

- The project continues as planned, but the To-Do list needs working through. Or the project is suspended, or maybe even broken off until certain conditions have been fulfilled
- The report shows who took part in the review
- The report must be signed by the responsible person
- The project manager is responsible for resolving the issues on the To-Do list

A review is a powerful instrument, with an excellent cost/benefits ratio. It should be used in a goal-oriented manner where the stakes are high and where the consequences of overlooking something would be significant (e.g. damage to the brand image, before launching a new service or product, before starting large-scale production). It should not be over-used, otherwise it becomes routine and gets dealt with in a superficial way, which is counter-productive. Companies that have defined their processes specify the milestones at which carrying out a review is a requirement. In the case of certified companies, the logs will be used during the audit process to check that the reviews have been carried out properly. A review that has been specified as part of the process at the start of a project can be taken out for commercial reasons, or an additional one can be introduced.

Milestone Trend Analysis

Milestone trend analysis (MTA) shows the project's development over time. In order to use MTA, the project being assessed must contain enough milestone dates. At regular intervals, the likelihood of achieving these dates is checked and entered into a system of coordinates (Fig. 25.3).

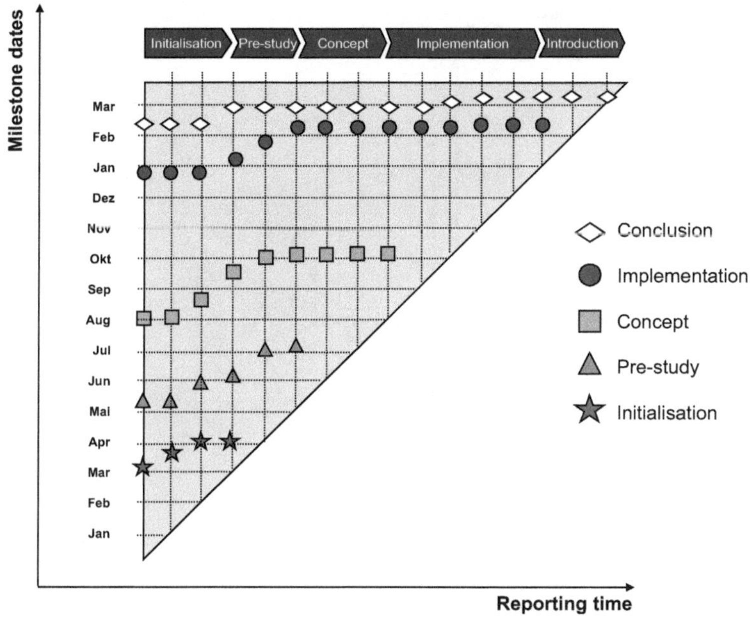

Fig. 25.3 Milestone trend analysis

The coordinates are then used to plot a trend line, which should ideally run horizontally (no deviation). By looking at the historical curve, an experienced project manager can forecast its future progress and anticipate the likely problems in good time. The particular value of a milestone trend analysis is that it is a combination of history of what has been achieved and a prognosis of what still needs to be done.

Cost Trend Analysis

In the same way as milestone trend analysis, cost trend analysis (CTA) uses the current cost situation to show how the project costs might develop in the future. Again, there must be enough meaningful figures available for the project in question (Fig. 25.4).

Fig. 25.4 Cost trend analysis

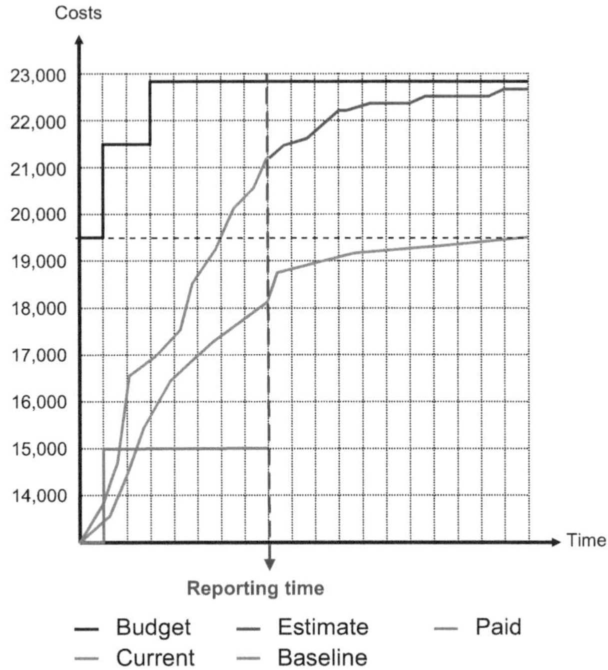

If it is regularly updated, cost trend analysis is an effective method of monitoring costs in a project. The planned cost profile (baseline) is compared against the actual costs (current). If there is a discrepancy between the two, that shows the cost divergence in the project. A forecast (estimate) can also be generated. This can provide a useful basis for an expenditure or cost estimate.

Earned Value Analysis

An earned value analysis involves identifying and calculating a project's key performance indicators. It tries to use time, costs and goals, the three elements of the "magic triangle", to assess the project. It does this by using the following variables:

- Planned value
- Earned value
- Budget burned (actual costs)

With the earned value method, the sum of all the planned costs for completed work is used as a measure of the value that has been created and of the status the project has reached. The planned cost of each work package is not counted until the relevant work has been completed (0/100 % rule). If the actual cost has been higher, that is not taken into account here. Ongoing work packages such as project management are counted pro rata, e.g. their planned cost per month. Half-finished work is not counted, as the value has not really been created at that stage (Fig. 25.5).

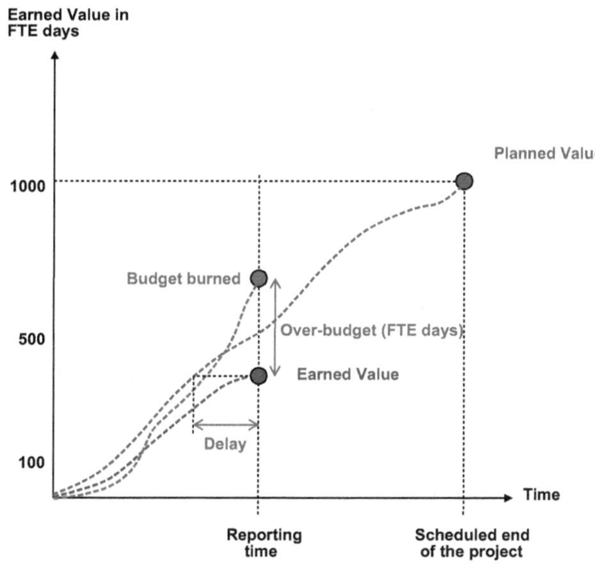

Fig. 25.5 Example of "Earned value analysis" (FTE days)

There are a few requirements before this method can be used: There must be functionally operational project management, project planning through to the work package level, a clear demarcation of work packages, good effort estimation, freezing of planned data as a baseline, regular status reports. The work packages should not be too large (2–20 full-time equivalent (FTE) days). The benefits of this method are that estimation errors during the schedule analysis are negligible. This method is conservative, and is fairly realistic in most cases. It can use either costs or FTE days.

Assessing the Economic Efficiency of Projects

Even before a project starts, the ideas management should include an initial assessment of an idea's economic efficiency. This assessment is always based on assumptions. It only becomes more reliable as the project gets underway. It should therefore be updated for each milestone decision. Something might also happen whilst working on a project phase that makes it worth carrying out an additional assessment of economic efficiency.

The assessment of economic efficiency estimates the future levels of cash reserves. With projects, money always needs to be spent (invested) at the start e.g. for resources such as personnel, equipment, raw materials, etc. The risks associated with this outflow of money are only taken on in anticipation of generating a larger inward flow of money at a later stage. In order to estimate the outward and inward flows of money, the effects on the system must be shown and then quantified in financial terms. The evaluation must only contain situations that can be changed in the future. For the assessment of economic efficiency, the key thing is not just the duration of the project (the project manager's horizon), but the lifetime including the utilisation phase and subsequent decommissioning (the project owner's horizon).

The assessment of economic efficiency answers the following questions:

- What benefits will this investment bring for the company?
- Does this investment ensure the company's ability to survive for longer?

An assessment of economic efficiency should be carried out before approval is given for each individual project phase, as a basis for the decision-making. Something might also happen whilst working on a project phase that makes it worth carrying out an additional assessment of economic efficiency.
One of the main purposes of the assessment of investment and of economic efficiency, and of the tools that they use, is to bring more transparency to the wide range of factual and personal problems through the use of a systematic approach. For real decisions, at least two versions must be available to choose from, or must be developed. The zero option (doing nothing) or the minimalist option (only doing the unavoidable) can count as versions for this purpose. The effects of the different versions must be made visible, so that they can be evaluated.

Assessing the Economic Efficiency of Projects

An economic efficiency calculation is a distillation of the financial figures, with the aim of being able to compare different variants. In most organisations, an updated economic efficiency calculation must be submitted for each phase decision (Fig. 25.6).

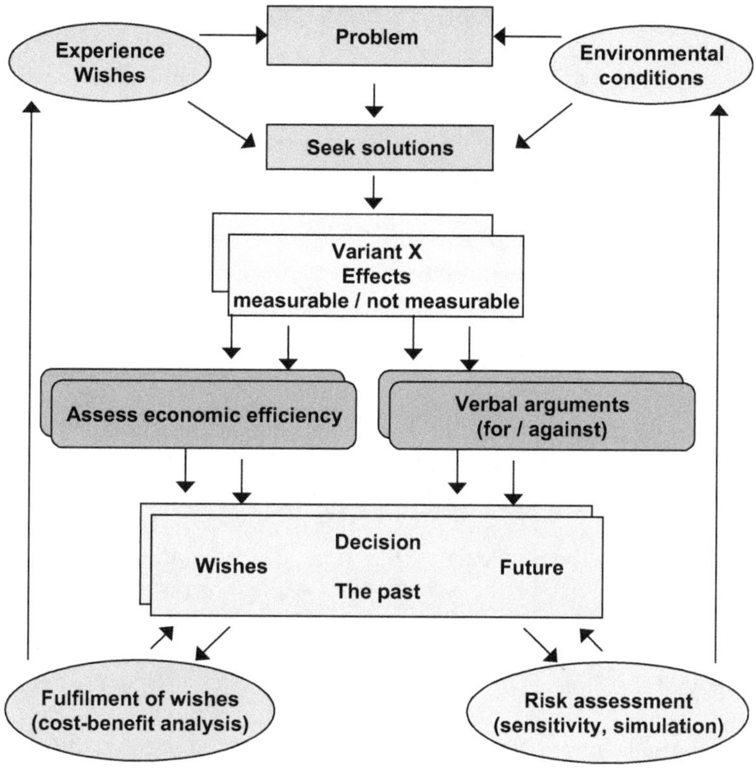

Fig. 25.6 Systematic approach to assessing economic efficiency

By compacting the information from the financial figures, some information is obviously lost. That represents a danger that the result could be influenced by the compacted elements.

Why is an economic efficiency calculation being carried out?

- To carry out a commercial assessment of the project.
- For budgeting and financial management of the project.
- For decisions in the preliminary study and concept phases.
- For setting priorities between projects.
- For evaluating the project on completion and to evaluate the system.

What does the economic efficiency calculation cover?

- Figures such as the total project cost, surpluses, losses, payback period, internal earnings and cost rates.
- Project costs by account class: Investments, personnel costs, external costs.
- Effects on the system: Income, savings, cost recovery, effect on the results.

Methods of Economic Efficiency Calculation

There are static and dynamic methods of calculating economic efficiency.

Static methods of investment calculation generally take just one period into account, and assume the same values for each period.

- A **cost comparison calculation** determines the costs of two or more investment projects, and compares them against each other.
- The **profit comparison method** looks at several investment opportunities and selects the one that offers the largest profit contribution. The main aspect that is compared is the break-even point (=fixed costs divided by the profit contribution per unit).
- The **return on investment method** calculates the rate of return (=100 × [profit or cost saving] per period divided by the average [additional] capital employed). The capital employed is equal to the sum invested plus the liquidation proceeds divided by two.
- The **payback method** calculates the time needed until the return on the investment has "repaid" the sum of the original investment. The payback period is calculated by dividing the amount to be invested by ([profit or cost saving] + depreciation). The shorter the payback period, the better the economic efficiency of the project.

Dynamic methods aim to determine the cash flows which occur at different times during the overall period of use.

- The **net present value** (NPV) method uses a discount table to calculate the present value from the difference between the incoming and outgoing cash flows. The higher the NPV, the better the economic efficiency of the project.
- The **internal rate of return** (IRR) method identifies which project ideas will generate the highest internal rate of return. The internal rate of return is the discount rate that gives a net present value of zero.
- The **equivalent annuity method** expresses the NPV as evenly divided annual inward cash flow (annuities). The aim is to achieve the largest possible annuity (>zero).
- The **dynamic payback method** (which looks at time values or returns) offers a helpful statement about preliminary investment analysis: "How long does it take until the total project costs have been recovered by way of discounted benefit?"

It is advisable even for non-profit organisations to undertake a careful preselection of ideas and possibilities through the use of a cost-benefit analysis. They also have customers, limited resources, and several alternatives with different risks and benefits. Instead of monetary values, they may look at benefits, effectiveness, ideals, material costs, and the effect on their image (Fig. 25.7).

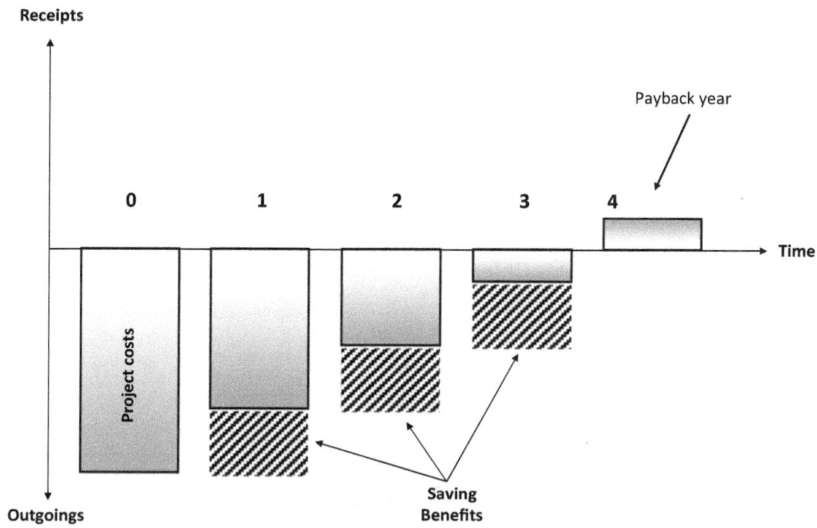

Fig. 25.7 Example of "Dynamic payback method"

25.5 Passing on the Results of the Project

User Training

The goal of the training is to create users who are able to use the new product or provide the new service. Further developing employees' abilities is one way to guarantee the company's survival.
Experience gained from the pilot phase or initial series can be leveraged to produce user-oriented documentation (e.g. operating instructions) and for training the users. The best training format will depend on a number of factors:

- Are there different target groups?
- Who is to receive the training?
- What information from the project has to be passed on?
- How extensive is the training?
- When is the training to be provided? And how?
- Do all users need to work with the new solution at the same time?
- Is the training short and intense, or would a slower, more gentle introduction be better?
- How will subsequent users be trained?

A new solution can be introduced in three ways:

- Abruptly: at a specified date

- In stages: one section after another
- In parallel: the old and new solutions are run simultaneously until the new version is up and working and can be used.

Whichever method is selected to pass on new knowledge or skills, it is important to consider the aim of the training. What do the participants need to know or be able to do after the course has ended?

The following options are possible, depending on the situation and requirements:

- E-learning
- Computer-based training (CBT)
- Instructions for use
- Direct online help
- Seminars and workshops
- Train-the-trainer, power user, snowball system

Pilot Trial, Initial Series

When new workflows or infrastructure are introduced into a company, it is a good idea to run a pilot trial or initial series within a limited organisational unit to ensure they will fit into normal practice. Only then should it be extended to the other organisational units, companies of the group or other countries.
Even though a prototype has been manufactured successfully, this does not yet mean that the manufacturing process is fully mastered. The production of an initial, limited series with the intended manufacturing process allows the processes to be optimised and any defects to be eliminated in good time.

Commissioning: From the Initial Series to Series Production

- Bringing the new system into service
- Preparing for and monitoring introduction
- Consolidation, elimination of defects
- Establishing and introducing organisational rules
- Handing over the new system to the users, starting series production

Concept for the Successor Organisation

- Maintenance and repair concept
- Organising maintenance, arranging warranty services
- Setting up a support organisation (with multiple levels, if appropriate)

25.6 Evaluating the Completed Project

Approach, Possible Solutions and Goals Achieved in Each Phase

A project evaluation can be done using the following template (Fig. 25.8):

Project phase:	
Question	Answers
How was the project (further) developed during this phase?	
What goals were achieved, what are the results?	
What successes should be noted? Why?	
What went wrong? How were those things dealt with?	
What direction did the (sub) project develop in?	
Are the specifications, original goals and procedures still appropriate?	
Process analysis within the team: Personal experience? Assessment of cooperation within the project team and with other relevant organisations? How sustainable is the workload for the project?	
Force field analysis: What forces enhance / inhibit?	
What risks and opportunities are there at the moment?	
Other negative points during this phase? How should they be handled?	
Other positive points during this phase? How should they be used?	
Consequences: A In general B For the current phase or next phase	
What should be kept the same?	
Who could do something more / differently / extra / or stop doing something?	
What measures should be introduced for similar situations?	

Fig. 25.8 "Project evaluation for each phase" template

Overall Assessment

To perform an overall assessment the following topics apply (Fig. 25.9):

Criteria	- -	-	+	++
Project initialisation Origination, intentions, how well it fits with company strategy				
Project agreement Goals and purpose of the project				
Project organisation / structure How does it fit into the company structure				
Project marketing Internal information, communication with external client				
Project phases Approach, milestones, steps, scheduling				
Planning / running the project Resource planning (people, money, time), planning of contents				
Project controlling Reporting, assessment of progress, milestone meetings				
Project support External people, consultants, cooperation partners				
Commitment of those involved Management, team members, stakeholders, departments				
Engagement Contribution by the relevant parties				
Professionalism Staff's subject expertise and project skills				
Social skills Cooperation and conflict management				

Fig. 25.9 "Overall assessment" template

25.7 Information and Communication

Successful Negotiation Using the Harvard Concept

The feeling and result with which I leave a negotiation greatly depend on the attitude and mindset with which I enter that negotiation. It thus may seem better in the short term to impose my will from my position and to emerge the winner using hard methods and attitude. The Harvard concept shows how to conduct a negotiation using the Win-Win principle.

However, negotiations generally take place between parties who will need to maintain a relationship in which they are able to resolve problems or achieve goals together or simply coexist. Thus the way in which, on the one hand, the negotiations are conducted and, on the other, the extent to which winners and losers are created will significantly affect future collaboration.

One commonly used negotiating concept is the Harvard approach (Fig. 25.10) which aims to achieve a reasonable result and is thus based on efficient and amicable practice. It enables both parties to the negotiation to win, hence it is also known as the "Win-Win method". It has four basic aspects (Fig. 25.11):

Features	Hard method	Soft method	Win-Win method (Harvard approach)
Mindset	The parties to the negotiation are opponents They mistrust one another	The parties to the negotiation are friends Trust / relationship is more important than the matter at hand	The parties to the negotiation are problem-solvers People and problems must be handled separately from one another
Goal	I want to win, to beat my opponent	I want agreement, I do not want to gamble with the relationship	I want a reasonable, efficient and amicably reached result
Strategy	Covert, tactical, pressure, insisting on own position	Disclosure of the negotiating strategy, one-sided concessions, eagerly-offered proposals, premature disengagement	Soft on people, hard with respect to the matter at hand, approach that has nothing to do with trust and mistrust, concentration on interests, only giving in to relevant arguments and not simply to pressure that is exerted
Problem solving	Looking for only one solution, uncompromising, not offering alternatives	Looking for the only answer that does not strain the relationship, avoiding appropriate and considered negotiation	Looking for possibilities that benefit both sides, considering as many options as possible, first looking together, then arranging, weighing up and finally deciding together
Decision-making criteria	Insisting on a solution that only I can accept	The solution that the other party, in particular, will accept	Insisting on objective, fair criteria

Fig. 25.10 The Harvard concept

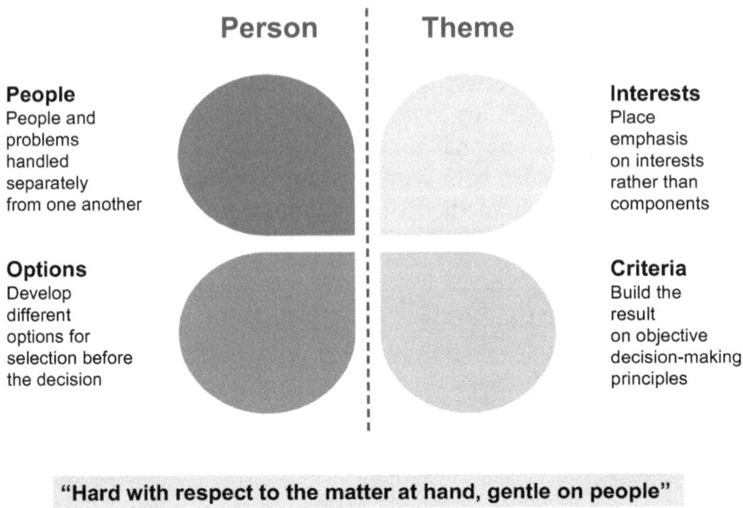

Fig. 25.11 An overview of negotiation methods

- People and problems are addressed separately
- The emphasis is placed on interests, rather than positions
- Various possibilities (alternatives, options) are developed before any decisions are taken
- The result is built upon objective criteria

People

Every negotiation has an object and a relationship level. It is thus important to be able to separate the basic interests in the matter at hand from the personal relationship, otherwise the two variables will be confused. The following approaches help:

- Putting oneself in the other person's shoes
- Making no assumptions about the other person's intentions. Asking instead
- Not attributing blame with respect to own problems
- The two sides put forward their presentations in turn
- Recognising, accepting and understanding emotions (including one's own)
- Observing the rules of communication (listen attentively, let the other person finish, argue intelligently, etc.)
- Tackling problems, not the people

Interests

Interests should take centre stage and not positions, although positions should first be specified as goals. Recognising interests has a motivating effect. Interests are the motivations behind the many different and confusing positions. Questions will generally help to move on from the positions to the interests. This often fails while the basic needs set out below are threatened or are not accepted to the same extent by the other party.

- Certainty
- Financial livelihood
- Sense of identification
- Feeling of recognition
- Self-determination

Alternatives

Different possibilities (alternatives, options) in negotiations allow scope with respect to the decision. There is often insufficient time or resources devoted to this aspect, however. The project manager wants to exit from the uncomfortable position of open negotiation as quickly as possible. A number of principles are helpful when developing different options for selection:

- Do not judge prematurely, encourage the process of identifying options and separate this from the decision
- Increase the number of options, rather than contenting yourself with the first available solution
- Be careful to consider the benefits of alternative solutions
- Develop suggestions that help to facilitate the decision

This does not mean haggling over ideas in order to find a compromise. Rather the aim is to develop creative ideas that will ultimately make it easier for both sides to reach a decision.

Objective Criteria

The solutions that are identified should, as far as possible, be measured against fair, objective and appropriate criteria. It is helpful for the assessment criteria to be determined at an early stage before any bias has developed for one of the alternative solutions. There is also considerable potential for conflict when defining the assessment criteria, so it is important to make sure at this stage that they concern the interests and that subjective positions are not permitted as measuring variables.

Management Summary

The statements in the management summary are given different scores or action levels. The "traffic light" approach is often used in practice (Fig. 25.12).

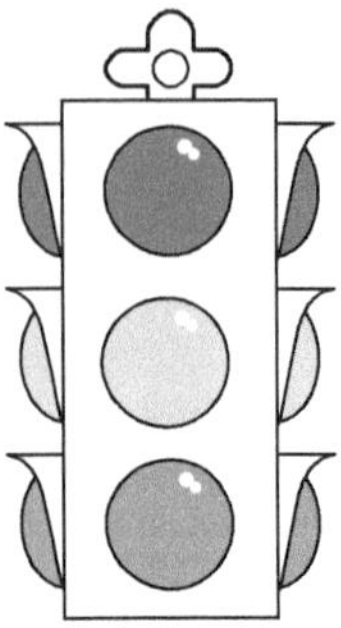

Management support required (alarm)
- ....
- ....

Pay attention (critical)
- ....
- ....

Going to plan (normal)
- ....
- ....

No.	Current situation	Person responsible Lead / participate
	General • Overall, the project is "in the green zone" • The problem list is being worked through, on the basis of prioritisation • FX / FY / AX need to urgently sort contracts, cost allocation between project / operations	
	FX • Plan contains training and rollout • Document layout to be finalised by end of month X	
	FY • FY migration completed • Work situation for help desk is stabilising • Some points on the problem list still need to be resolved • Define document layout by calendar week 40 • Prepare FY sub-project completion from Oct 20xx	
	FZ • First tests have been carried out, with good results • Second acceptance tests on dd.mm.yyyy • Third-party product B has been integrated • Document layout has been accepted by FZ • Training has been started in XX • Wyss subsidiary starting on dd.mm.yyyy • Rollout due to begin on dd.mm.yyyy	
	AX • Change management decision due by dd.mm.yyyy • Bring contract negotiations to a conclusion	
	NT • Performance has stabilised and is being further improved	

Fig. 25.12 A "management summary" example

Minutes

A sample of meeting minutes might look like this (Fig. 25.13):

Topic:	**Project committee project NT**
Date:	dd.mm.yyyy
Time:	hh:mm – hh:mm
Participants:	Y.N. (minutes) X.F. Z.S.
Document number:	P-6

12. **Information about the project's status**
13. Migration: ..... but overall, the project is progressing well.
14. The situation: .....

16. **Work lists**
17. Lists have been produced for planning.

19. **Critical points in the project**

22. **Prioritisation of activities**
23. The committee will check the prioritisation again if .....

26. **Actions / next meeting**
27. Monday, dd.mm.yyyy, 08:00 –09:00
28. Meeting room 22
29. By then, the following items will be completed:

Number	To Do	Who? (Lead / cooperation)	Deadline?
1.			
2.			
3.			
4.			
5.			

Fig. 25.13 Sample "minutes"

25.8 Network Planning Method

A network diagram highlights the dependencies between activities very effectively, although it does have the disadvantage of being large and rather cumbersome. Another illustration method is the bar chart (Gantt chart). The Gantt chart contains the same data as the network diagram. It is simply shown in a different format and has the advantages of being a compact representation, showing time relationships (duration, simultaneity, slack) and being easier to read.

Critical Path Method (CPM)

The scheduling for a project can be carried out using the network planning method. A network diagram is drawn up from the list of activities. It contains all the activities, their duration and the preliminary conditions that must be fulfilled in order to start the activity.
The critical path method (CPM) represents the activity by an arrow, and the event by a node. It is ideal for projects in which previous activities must be concluded before the next is started. It is simple to use, and the results are also easy for occasional users to interpret.
The activity and the duration are specified on the arrow. The earliest and the latest dates are specified in the node, for both the start node and the end node. If two activities are started at the same time, i.e. run in parallel, their durations may be different. It must therefore be possible to show the end date for every activity.
To make the results of both activities available for the subsequent activities, the two date nodes are joined to one another with a dummy activity. No work is carried out for this dummy activity, and its duration is zero time units (Fig. 25.14).

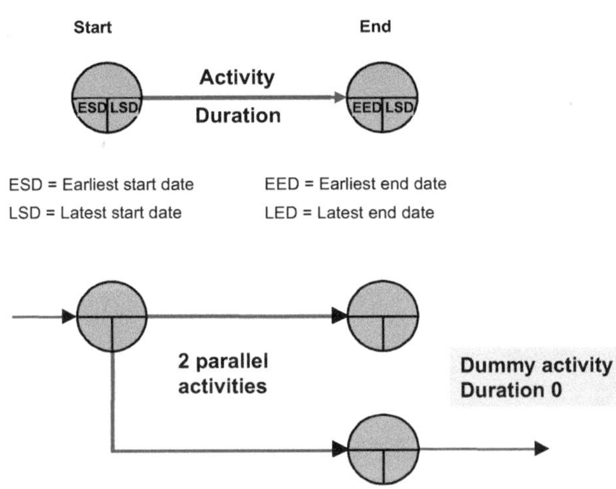

Fig. 25.14 Network diagram using the critical path method

The dates are determined in the network diagram using a simple arithmetical formula. The network diagram is first computed forward, giving the earliest dates. Where several arrows converge, the largest number is used. The next activity cannot start until the last of the predecessors has concluded. The network diagram is then computed backwards by making the earliest end date also the latest end date. Where several arrows diverge, the smallest number is used.
To check the result, the latest start date computed backwards must be equal to the earliest start date. The difference between the latest and the earliest date for an activity gives the slack. If the earliest and latest dates are the same, then there is no available slack. Connecting all activities without slack gives the critical path (Fig. 25.15).

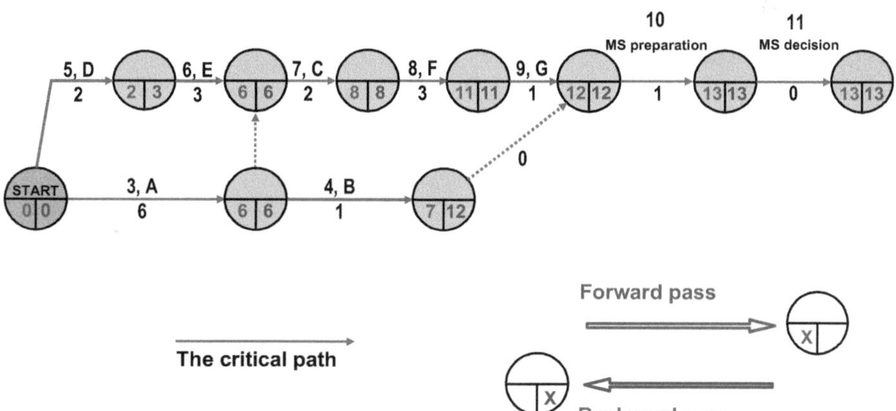

Fig. 25.15 An example of the critical path method

Metra Potential Method (MPM)

An alternative form of representation is the metra potential method (MPM). Here, the square node represents the activity with the same information: activity, duration, earliest and latest start and end dates. In this case, the arrows indicate the flow of data. Manual calculation quickly becomes time-consuming and error-prone when there are a large number of activities. With its clear rules, the network diagram method is perfectly suited for use with software (Fig. 25.16).

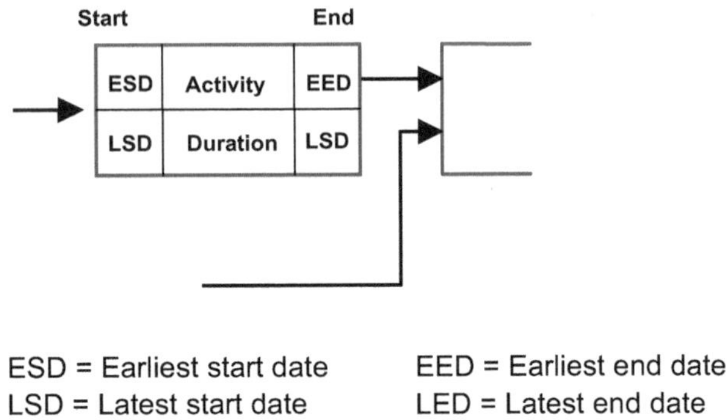

Fig. 25.16 Network diagram using the metra potential method

25.9 Setting Up a Project Portfolio

A degree of preparatory work is needed to set up a project portfolio. First all ongoing or approved projects actually need to be identified. This goal of itself can be quite a challenge in a large company. Ideas for projects can also be added to the project list, depending on the type of business (Fig. 25.17).

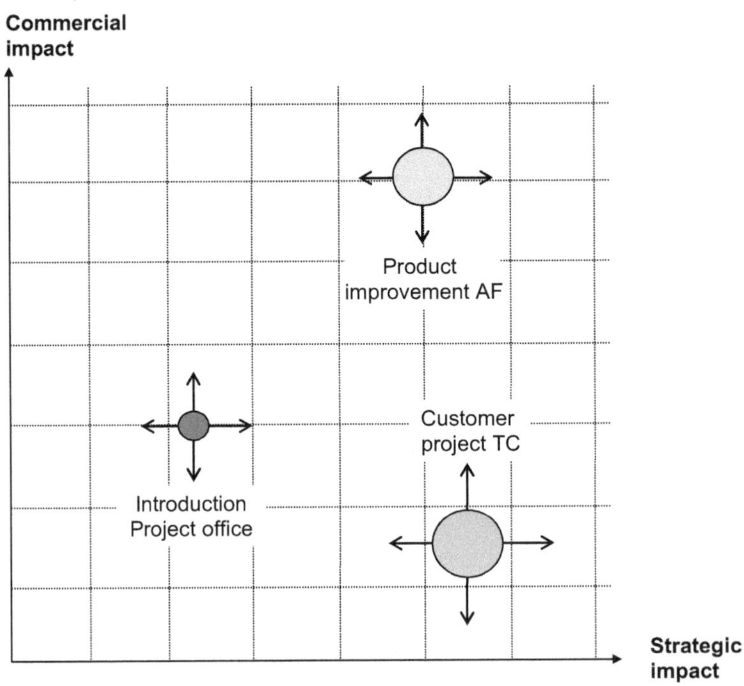

Fig. 25.17 A "project positioning and evaluation" example

Step 1: Identify the Criteria

Identify those criteria that make most sense for evaluating own projects. Depending on the nature of the initiative, it might be enough to simply compare using paired criteria, such as opportunity vs. risks, costs vs. benefits, etc. But a project portfolio can also be multi-dimensional. Possible criteria include the following:

- Alignment to strategy
- The appeal of the initiative: opportunities and risks (SWOT analysis)
- The project's potential: the organisation's ability and wish to implement the ideas
- Problem-solving potential: organisational improvements that are made possible by projects
- Economic indicators (ROI, NPV, IRR, profit contribution, etc.)
- Legal or statutory framework conditions (mandatory projects)
- Other principles (e.g. environmental, ethical, etc.)

Step 2: Evaluate

All known projects and project ideas (from the project list) are evaluated and positioned in the portfolio matrix with reference to these criteria. Here it is not the absolute position that is important, more the relative positions between the projects.
Established indicators may be used as the benchmark for assessing "economic importance", e.g. ROI, NPV, payback, etc. Quantitative methods are rather cumbersome and time-consuming for assessing the strategic importance. However it is sufficient to define the strategic importance using a purely qualitative approach.

Step 3: Prioritise

It is a very good idea to prioritise the projects, particularly with respect to the optimum deployment of resources. The labour available for the project work is first deployed for projects with both high economic importance and strategic importance (priority 1).
Allocation of priorities 2 and 3 then depends on whether the company assigns more importance to the economic projects than to projects that will improve the market position, for example. For priority 4 projects, it is questionable whether any resources should be made available to them at all. It is also sensible to define criteria that will cause a project to be abandoned. Companies often struggle with this step (Fig. 25.18).

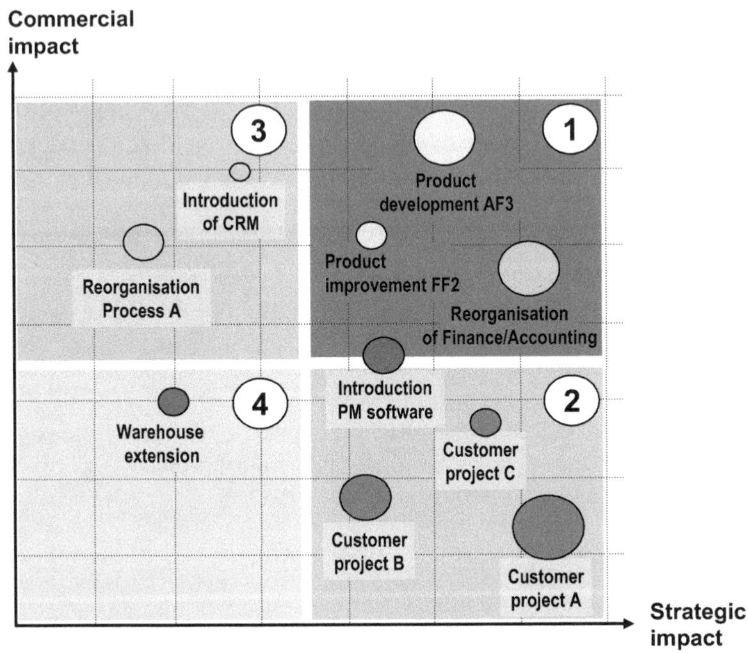

Fig. 25.18 A "project prioritisation" example (strategy versus cost effectiveness)

25.10 Self-management

Self-organisation

Location is increasingly no longer a restrictive factor in our lives. With all the communication options available to us today we can collaborate with other people at multiple locations, all at the same time. Management by objectives (MbO) is currently very popular, but one result of its use is that pressure from management to take on even more work has increased rapidly. In turn, we are all required to manage ourselves to a certain degree.

There is generally no management of restriction from an external source. In fact the normal situation is quite the opposite. We have to manage ourselves and decide where and how we wish to expend our energy. This means systematically aligning our professional, family, health and private spheres, bringing them into a favourable balance and reducing real or perceived sources of stress. Self-management thus demands self-awareness and self-esteem.

Lifeworlds

We often hear the term "work-life balance" used. Correctly speaking, it should rather be the life balance which is made up of both the "work life" and the "private life". Even more accurately, it relates to the balance between the three personal lifeworlds. This does not mean that all the worlds need to be the same size, but they must be present in a favourable ratio which will be different for every individual (Fig. 25.19).

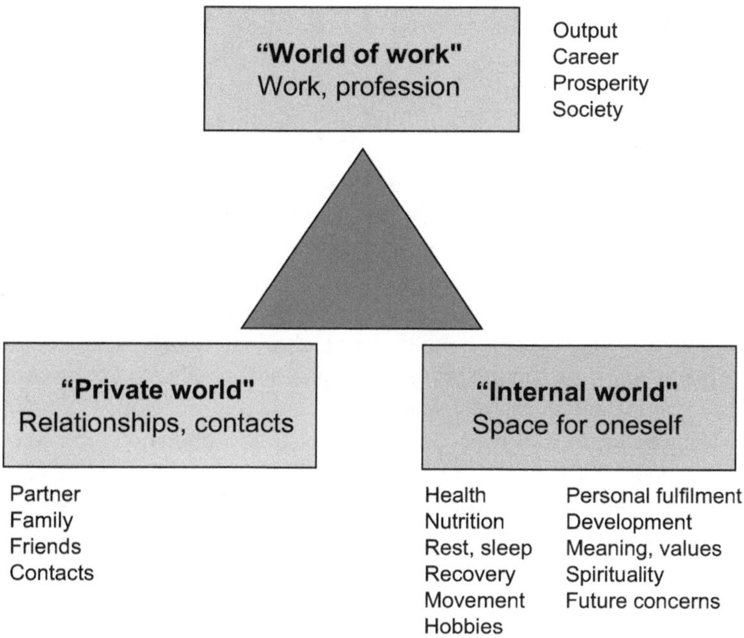

Fig. 25.19 My three lifeworlds (based on: "Men's work" department of the Evangelical Reformed Church of the Canton of Zürich)

Experience shows that, when we are overloaded, we tend to cut back on our own private world first. But this is the most important of the lifeworlds for personal regeneration, and its loss leads to the "burned out" feeling. If the pressure is maintained, the private world will start to suffer and will be replaced with a one-sided focus on the world of work. If balance cannot be restored in the long term, there is a risk of burnout.

Resilience

Rather than worrying how to prevent the risk factors and long-term stresses, it is also possible to focus on how this long-term condition came about or even to become yet more resilient on the basis of previous experiences. This will become

increasingly important as the pressure to perform rises (reorganisation, job losses, etc.). Resilience is the term used to describe the inner strength that allows us not only to overcome crises, heavy blows of fate, failures, difficulties and unusual trials, but to emerge from them strengthened and more mature and without lasting damage.
Resilience can be trained like a muscle. Resilience is also an ability we hold in reserve that prepares us for future difficulties. A resilient person is better able to cope with changes and upheavals. It also incorporates "elasticity", adaptability, ability to resist, "mental immune system". It allows people to bounce back to their normal state like an elastic band or stand up again like a "Weeble". In her book "Die Strategie der Stehaufmännchen" (The Weeble strategy). Monika Gruhl describes how resilience is essentially a question of fundamental attitude and the development of character. Her list of fundamental attitudes contains:

- **Optimism**. Seeing the good in the bad. Stop brooding. Positive self-image, self-confidence. Belief in own efficacy. Do not take failures personally. Realistic optimism (see the positive without ignoring the difficulties). "Difficulties hide opportunities". A belief that "This too shall pass".
- **Acceptance**. Accepting that misfortune, disappointments and adversity are part of life and that they cannot be avoided or eliminated without trace. Allowing yourself enough time, appreciating what has happened. Accepting ambiguity. Patience: giving the necessary space to become and develop. Confidence that things will change even if we do nothing. Trust in the bigger picture. Acceptance of what cannot be changed. Self-acceptance (strengths and limitations).
- **Focus on the solution**. "Anyone who wants something will find a way. Anyone who does not want something will look for reasons." Turn thoughts into solutions. Mobilise resources. Break free of "problem hypnosis". Develop options. Be open to new ideas and unusual perspectives. Flexibility: methods and behaviours can be changed when they stop working. Think creatively. Try new things. Create and organise structures within the chaos.

Resilient people take these basic attitudes and develop very specific strategies for the way they think, feel and act. The following characteristic abilities are important for this:

- **Regulate yourself**. Be aware of your feelings, but do not give in to them. Energise, reassure and cheer yourself on. Modify your moods, urges and responses (self-control). Build yourself up. Encourage yourself. Control your impulses. Express feelings and impulses appropriately. Apply effective stress management techniques. Intuition.
- **Assume responsibility**. Stop playing the victim. Identify what you can change yourself and act accordingly. Assume responsibility for your own feelings, thoughts and deeds. Let go of those things you cannot change. Stop playing

the blame game (with yourself and other people). Find answers to (take responsibility for) the current questions of life. Review and adapt/learn new attitudes and ways of thinking/behaving. Actively shape change.

- **Shape relationships.** Pleasant relationships and social networks create stability, affinity and support. Interest and empathy (for others and yourself). Social skills. Social commitment and public spirit.
- **Shape the future.** Have a meaningful picture of a future that is worth striving for (meaning, values, visions, dreams, inner picture, desire: these all give direction in periods of crisis). View the future as full of potential. Proactively take your life in hand: feel that you largely have responsibility for your own wellbeing. Know what you want to achieve in the long term. Recognise constricting presuppositions and dogmas. Work towards your lifelong dreams. Focus. Act.

Stress

Stress was described in the early 20th century by the biochemist Hans Selye, who defined it as an "acute state of tension in an organism that is forced to mobilise its defences in order to counter a threatening situation".

It is thus an adaptation response by the body and can be triggered by external factors such as cold, heat or danger, and also by inner threats such as fears, fantasies, subjective assessments, mental or physical underloading/overloading. The stress response is thus a physiological process that is highly complex and is broken down into three phases:

Phase 1 Alarm response	The hypothalamus triggers a cascade of chemical processes, including the release of adrenalin and noradrenalin. Energy is made available to the body for a possible conflict or to escape from the situation. The heart beats faster, blood floods into the muscles and brain
Phase 2 Adaptation or resistance phase	At almost exactly the same time, the body releases the hormone cortisol which has an anti-inflammatory effect and performs the function of effectively fighting possible injury. The threat is met and the "fight or flight" decision is taken and acted upon
Phase 3 All-clear	The threat is over, it was endured by fight or flight, the hormones and messenger substances are broken down once more and the body calms down. The body returns to its normal state

This physical process is an ancient one that is deeply rooted in us. It is a throw-back to the hunter-gatherer age, when man was exposed to many dangers and had to fight for day-to-day survival. The situation is often the same today. Only the type of threat has changed. Rather than facing down rival tribes, today we experience quite different moments of tension. For example there are critical and hectic road traffic situations or we need to get across the station quickly in the evening commuter rush while at the same time providing information about a complex interpersonal problem that has developed within the project. Or we may need to mask our fear

and hidden intentions in a tense negotiation meeting at the end of an 8 hour working day.
Stress is thus not fundamentally bad and something to be avoided. Quite the opposite. Stress gives us energy and gets us going. It causes us to act. We need it to adapt to the constantly changing challenges and environment. Stress researcher Hans Selye thus also stated that "The absence of stress is death".

Flow Theory

Another concept for describing healthy and unhealthy stress is the flow theory developed by psychology professor Mihaly Csikszentmihalyi (Fig. 25.20).

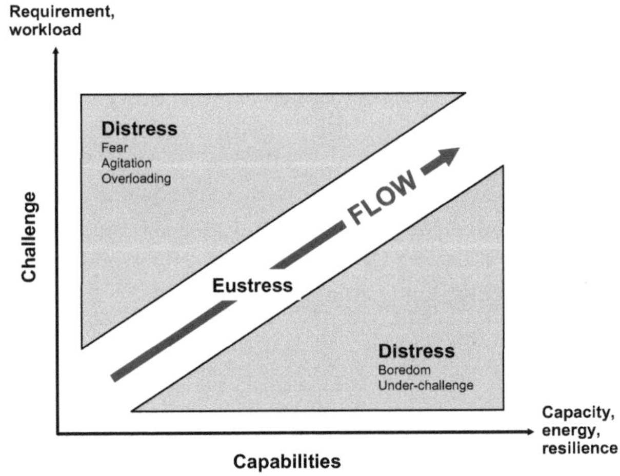

Fig. 25.20 Eustress and distress (based on Mihaly Csikszentmihalyi)

The boundaries between the three phases will differ according to the person and his abilities. The right level of stress creates a feeling of flow. Today this model is very popular for explaining motivation and performance, but medical research shows that it cannot be confirmed so easily. Indeed it is rather the duration, manner and intensity of the factors that trigger the stress that determine whether a form of stress is healthy or not.
In the world of work, heavy and long-lasting workloads have become a significant factor in health management. Stress is now the second most prevalent occupational health risk after back pain.
Chronic stress is the most dangerous. With chronic stress, the body is no longer able to move to the third "all-clear" phase (see Stress phases). The level of stress hormones in the blood remains high, either due to constant small micro-stressors such as criticism, denigration, contempt or because of constant overloading. The body is no longer able to settle down and the person is unable to relax. This total exhaustion is known as burnout syndrome.

Burnout

This form of burnout has been familiar for some time in the field of health care. For many years, this syndrome has been attributed, in particular, to the helping occupations with a high workload (both physical and mental), i.e. carers, teachers, social workers, child care workers. The burnout syndrome can now be found in occupations or activities that exert considerable pressure to perform (from the case worker through to the manager) or in people who make extreme demands of themselves (e.g. competitive sportsmen).
Even though stress is not bad per se, it is still a trigger for burnout. The burnout syndrome builds up slowly and insidiously as the stressor effect persists and more energy is expended. The burnt-out person attracts no attention and thus remains "undiscovered" for a long time. There are huge differences in how this process progresses in individuals, and the total exhaustion often trips in suddenly. Burnout is manifested in three specific and well-known forms in the advanced stage.

Physical exhaustion	Chronic fatigue, weakness, physical pain in the back or muscles, fluctuating weight, disturbed sleep, increased susceptibility to illness
Emotional exhaustion	Depression, helplessness, crying, uncontrolled emotional outbursts, irritability, emptiness, despondency, isolation
Mental exhaustion	Negative attitude towards self, work and life, increased feeling of inferiority, loss of self respect, increasing withdrawal from contact and refusal to communicate, cynicism and aggression

The affected persons are now unable to find their own way out of this state of burnout. As with other diseases, a long period of professional support is needed to return to a healthy normality. In addition to the actual recovery, a careful analysis of the various causes and conscious appraisal of the living situation (see Lifeworlds) for the affected person are needed so that they do not end up burned out again after a short time back at work.

25.11 Time Management

Time as an Economic Commodity

Time is capital, in short supply, cannot be purchased or duplicated. Time is life! Time management means a better overview, more creativity, less stress, more leisure time and helps us to achieve our goals. If it were not for one or two time thieves:

- External time thieves: telephone calls, visitors, meetings, deadline pressure, ongoing new orders with higher priority associated with a constant change of work topics

- Internal bottlenecks include the "tendency to put back", lack of priorities, excess paperwork, no delegation, inability to say no, unclear goals and lack of self-discipline

As in project management, clear goals have significant benefits for time management. They help to keep track of things, set priorities, use abilities, mobilise the subconscious and concentrate efforts. When identifying goals, personal goals, as well as professional goals, should be set out in specific terms, starting from a picture of the ideal life.
Professional goals can be broken down into long-term career goals and medium/short-term goals. Personal goals include health, partnership, family, friends, meaning of life and so on. It is helpful to define the greatest strengths and identify one's own bottlenecks in order to be able to set realistic goals.
The procedure planning method described by Descartes is known as the "salami tactic". It is notorious in politics, but less so in the organisation of work time: setting down goals in writing, breaking the whole down into parts, arranging into sub-tasks and completing according to priority. Last of all comes checking the result.
A written plan is the best aid for this. Planning saves time. Time planning helps to implement goals. The benefits of such planning are obvious to project managers as they make it easier to achieve goals, save time, give a clearer overview and reduce stress. As it is set out in writing, there is a clear overview. We do not have to rely on memory. It encourages participants to concentrate on the essentials and allows the results of each day to be checked, thus increasing success.

ABC Analysis

In the daily ebb and flow of events, it is an ongoing challenge to achieve one's own goals. How can we manage to do the right thing? With the ABC analysis, activities are listed over a specific period. Every activity is assigned to one of four quadrants. Important comes before urgent. Those activities that contribute to achieving the goals and demonstrate the greatest potential for success are important. This breakdown clearly shows which activities (Fig. 25.21):

- I need to do myself immediately: A tasks
- Could be delayed: B tasks
- Can be delegated immediately: C tasks (above a certain scope)

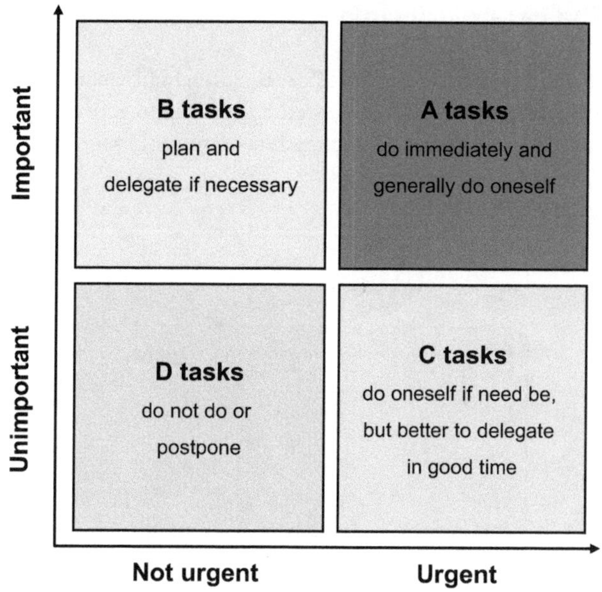

Fig. 25.21 ABC analysis

The following decision matrix is completed from left to right. It helps to do those activities that will achieve the goal (Fig. 25.22).

Responsibility	Importance	Time required	Urgency	Activity
Me	Important	Needs a lot of time	Can wait	Schedule immediately, or delegate
			Urgent	Involve somebody immediately
		Needs little time	⇨	Do it immediately
	Not important	Needs a lot of time	⇨	Straight to trash can
Someone else	Important	⇨	⇨	Pass it straight on

Fig. 25.22 Decision matrix

The matrix is designed to help you concentrate on A tasks, delegate B and get rid of C. In practice this means carrying out just 1 or 2 A tasks, plus 2 or 3 B tasks and then using any remaining time for C tasks.

The Pareto Principle

A small percentage of the effort is sufficient to achieve a large percentage of the result. This insight was first suggested by the Italian economist Vilfredo Pareto (1848–1923) at the University of Lausanne. It is commonly known as the "80/20 rule" (Fig. 25.23).

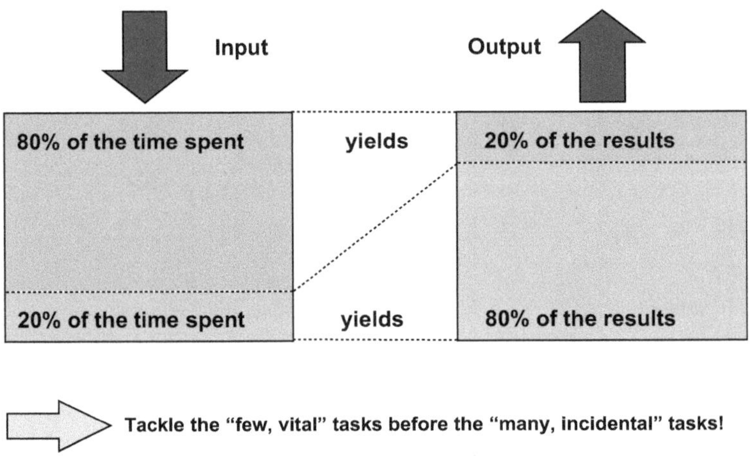

Fig. 25.23 Pareto principle

If the individual activities are classified by resources in and returns achieved, we can see that a small proportion of the activities contribute a large proportion of the result and, vice versa, that the large proportion of the activities return only a modest proportion of the overall result. This finding was published by Max O. Lorenz (1880–1962) (Fig. 25.24).

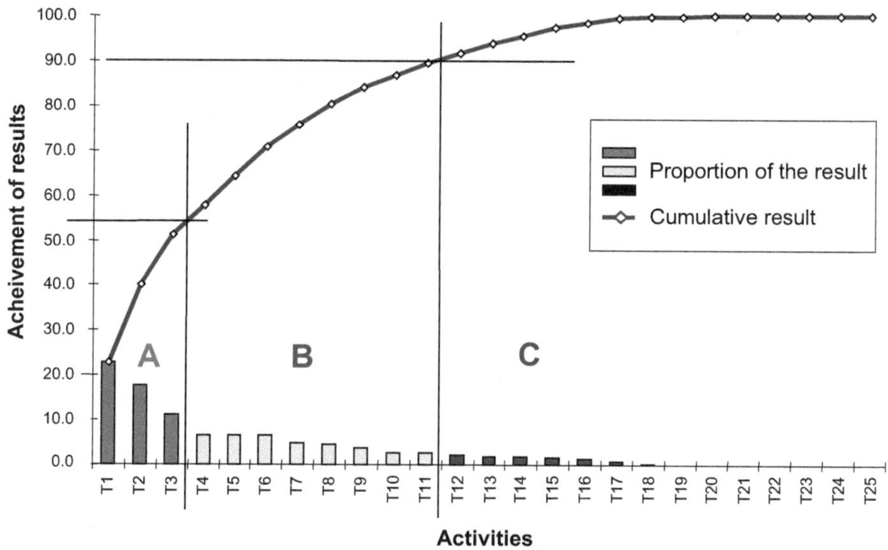

Fig. 25.24 The Lorenz curve

ALPEN Method

Time management specialist Josef W. Seifert developed another method for optimising normal work tasks. This is known as the ALPEN method:

- Assign tasks, activities and deadlines: personal To-Do list, tasks not completed the previous day, new daily activities, important deadlines, telephone calls and correspondence
- Length or duration of the activities: calculate time required, set time limit and eliminate disruptions
- Plan buffer time: roughly 60 % for scheduled activities, around 20 % for unexpected activities and around 20 % for spontaneous and social activities
- Establish decisions: priorities, cuts, options for delegation. Getting rid of excess tasks and giving one task clear priority will enable people to work more efficiently and effectively, making it easier to achieve the goal
- Review what was Not done, follow-up check, assign uncompleted tasks

It helps to follow the progress curve over the course of the day. For most people, peak performance occurs in the late morning, with an energy slump in the afternoon and an interim high in the early evening. The curve is very handy to use. Before starting work, check the day's plan from the previous evening, enter A tasks in the peak energy period (enter "do not disturb" time as an appointment with yourself in the day's plan, screen calls and interruptions and call people back later). Also schedule C tasks during the energy slump and at times that are prone to frequent interruptions, and B tasks during the interim high. Include breaks to restore strength and work at the following rate: 1 h of work, followed by a 10-min break. Keep drinking water and ensure you have plenty of fresh air. Check the day's plan before leaving for home, transfer any uncompleted tasks, draw up the next day's plan and finish the day calmly.

The Pebble Principle

Planning personal activities and assigning dates to the work calendar are very much like packing a car boot. It is important to pack the large items first. In terms of daily or weekly time planning, this means scheduling the large, whole-day or half-day events first and only then filling the marginal times with short appointments. This creates smooth transitions and less downtime and travelling (Fig. 25.25).

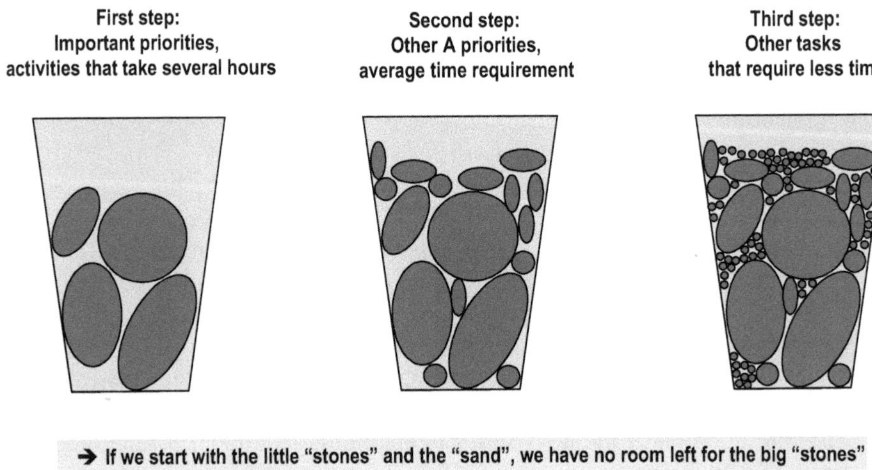

Fig. 25.25 The pebble principle

Delegation

The topic of delegation is also part of time management. Delegation saves time, utilises employees' capacity, helps people to progress and has a motivating effect. A person who delegates can give himself more time to work on A tasks and give employees opportunities to demonstrate their worth. Delegation should therefore be regarded as a management task.

Keeping on Track

These explanatory notes illustrate a number of aspects that will help to optimise personal time management. The best tool, however, is still a suitable time planning tool. This will provide an overview of all the tasks, and can be used to plan activities and dates and to check on events. It will include the calendar and system of reminders for all active tasks. A simple calendar may be used for this, or a time planner or PDA (Personal Digital Assistant) which can be synchronised with a computer. However, none of the available tools are a replacement for self-discipline. Consistent time management improves planning and organisation and makes it easier to achieve daily goals. Concentrating on what is important makes everything calmer and leads to positive successes.

26 Problem-Solving Methods

26.1 The Problem-Solving Process

The problem-solving process benefits from a systematic, structured approach, rather than knee-jerk reactions. It can sometimes be good to immediately embark on an "obvious" solution without actually fully exploring the current situation, or even without stopping to think about the goals and the possible range of solutions. This is a very common situation in practice, and is known as "jumping to a solution". But for projects that involve new territory, a different approach is more worthwhile (Fig. 26.1).

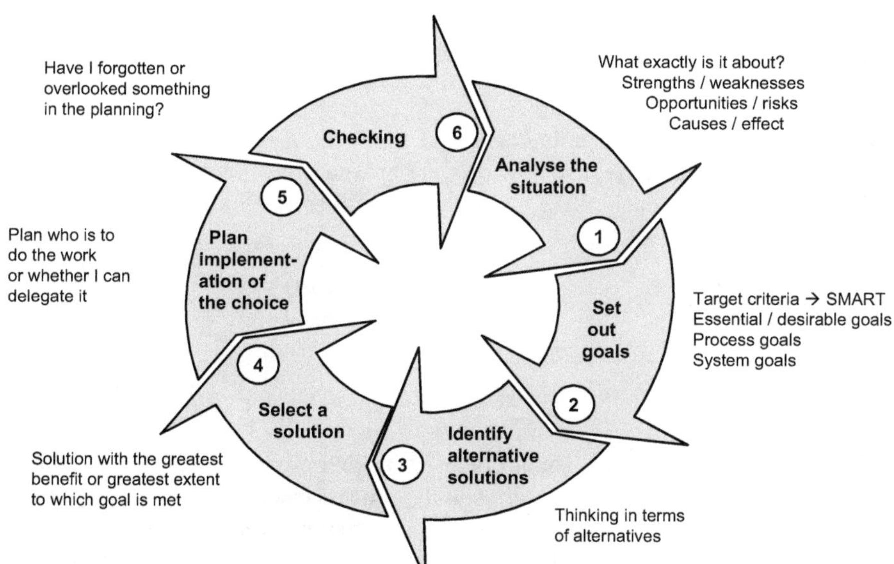

Fig. 26.1 Problem-solving technique

J. Kuster et al., *Project Management Handbook*, Management for Professionals,
DOI 10.1007/978-3-662-45373-5_26

The problem-solving process relates to the project's processes for designing products and services. It is a structured tool for solving task-related problems of any sort. The term "problem-solving cycle" is often used, in order to underline that in reality, finding a solution is not a linear process and that it may be necessary to go through the individual steps several times (iterative process) (Fig. 26.2).

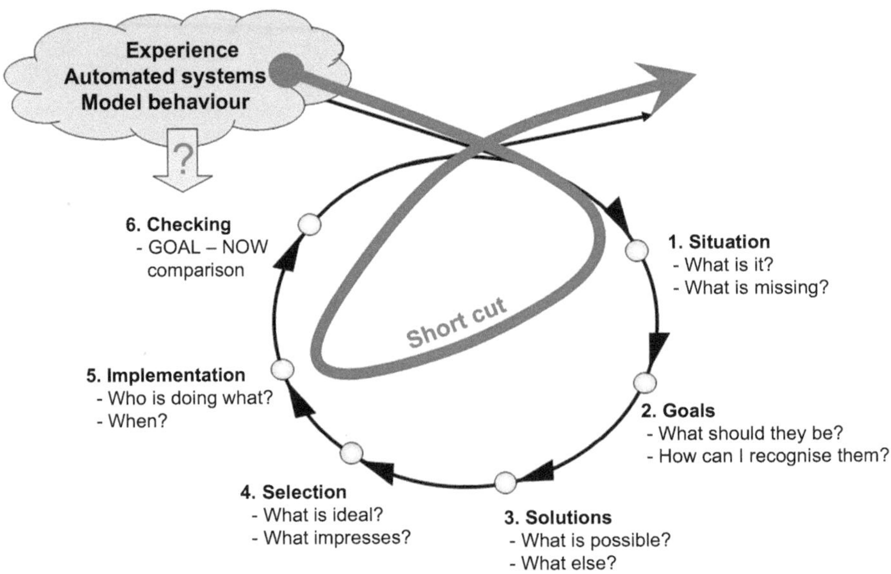

Fig. 26.2 Jumping to a solution

There are a wide range of problem-solving methods, and the main difference is usually the number of steps, and how they are broken down. The following method has three stages, with each of the main stages involved two individual steps:

Identifying goals	1. Situation analysis 2. Defining the goals
Identifying solutions	3. Solution synthesis 4. Solution analysis
Selection	5. Evaluation of the solutions 6. Decision

The problem-solving process should be used in each project phase. The emphasis on the individual steps in the problem-solving process changes during the course of the project. In the initial phase, the focus is on situational analysis and on formulating goals. But in later phases, the focus is more in analysis and synthesising solutions, and on evaluating possible solutions and making decisions about them. The amount of details also increases from one phase to the next.

The "situation analysis" and "defining the goals" steps, which occur at the start of the project, are of central importance for the way the rest of the project runs (Figs. 26.3 and 26.4).

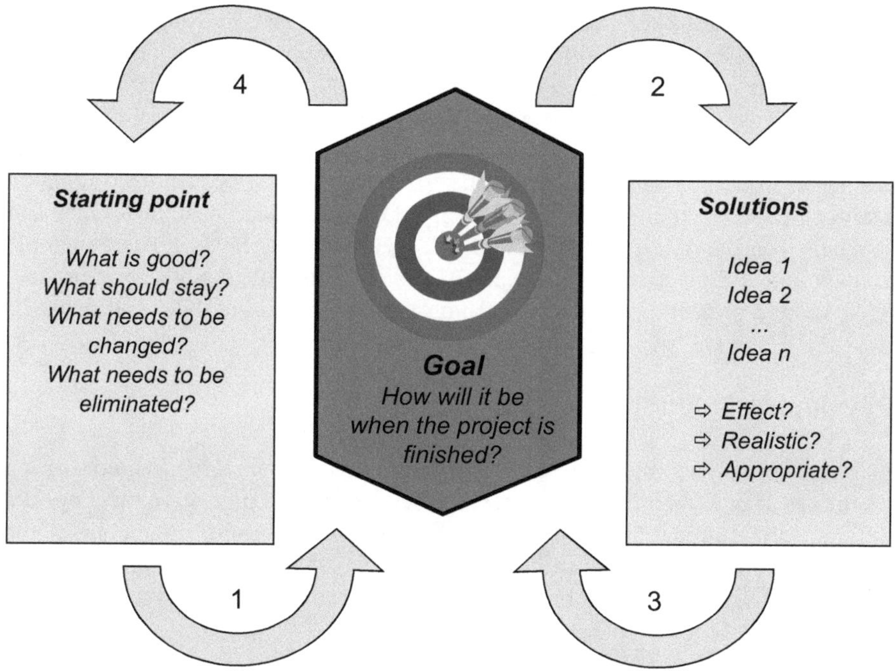

Fig. 26.3 The gradual approach to identifying goals and solutions

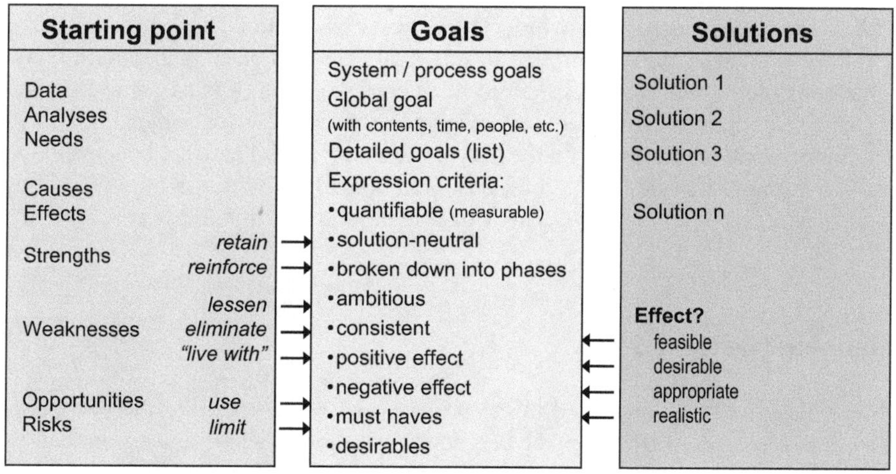

Fig. 26.4 Identifying goals from situations and possible solutions

Defining the Goals

The process of identifying goals generally follows on after the project agreement, which specifies the overall outcome and possibly some additional framework conditions. From experience, identifying goals is the hardest part of project management. It calls for good powers of imagination, creativity, and a sense of reality. Defining the goals forms the basis for developing solution principles and concepts, which follows on afterwards. It must therefore contain all the details that are needed when searching for a solution. In particular, this must include all planning data needed for the solution, minimal data structures and quantity estimates, which usually build on the structures developed in the situation analysis. Of course, this data is based on the understanding at that time, and depending on the nature of the project and on the starting level, it can be simple to draw up, or difficult. For example, it is easier to formulate goals for a product improvement project than for a research project or a project in the social field.

Developing Solution Concepts

In the concept phase, the focus of the solution cycle is on searching for and shaping a solution. It is a good idea to have the project team hold a creative meeting in a different and stimulating environment. An inspiring environment mobilises the team members' creativity. And having fun while they're about it also supports them in getting the best out of themselves intellectually and emotionally.

Making Decisions

At the end of the concept phase, the project owner must select one of the range of evaluated solutions presented to him. The project owner should be familiar with the solutions that have made it through to this final selection stage, and should know which one the project team consider to be the best solution. It is useful to evaluate the different options from the user's viewpoint, especially for projects that will require personal commitment on the part of the future user. There are benefits to the project manager carrying out a joint cost-benefit analysis with the future users, or with an accepted representative. This will help the solution to find better acceptance later on.

Planning the Next Steps

When the solution is presented to the project owner, the plans for the next steps should also be presented. When the best solution option is decided, decisions can be taken about the next steps and credit can be put in place for the next phase, which is generally the most expensive. The following points are generally included in the "next steps":

- Procedure plan for implementation
- Introduction concept and implementation concept
- Training approaches: learning alone or in teams? The project manager will build experience of adult education into the concept.

26.2 Alternatives to the Problem-Solving Cycle

The benefits of a problem-solving cycle are so substantial, and it is so widely used, that since the 1960s people have been refining and refreshing the approach under different names (Fig. 26.5).

Problem-solving cycle Goals, solutions, choices	**Kepner-Tregoe** Problem-solving and decision-making	**Deming circle** (PDCA cycle), Shewhart cycle, continuous improvement process	**Lean / Six Sigma (DMAIC, DMAEC)** Improving and optimising business processes
Situation analysis: What is it? What is missing? **Define the goals:** What should they be? How do we know that? **Alternative solutions:** What is possible? What else? **Selection and decision:** What is ideal? What impresses? What are the risks? Which alternative shall we choose? **Implementation and control:** Who is doing what? By when? Goal achieved?	**Situation appraisal:** Determine the overall situation **Problem analysis:** Define the problem and its boundaries **Decision analysis:** Identify goals, assess the different alternative solutions **Potential problem analysis:** Scrutinise against potential problems and take actions to minimise the risk	**Plan**: Analyse the current situation, and identify improvement potential **Do**: Try out, test, practical optimisation in small framework **Check**: Check and approve the results **Act:** Introduce as a standard on a broad front, regular checks (audits) for compliance	**Define**: What customer needs should the process meet? **Measure**: The process's key characteristics **Analyse**: Identify the causes of the deviation from the defined goals **Improve, Execute**: Find solutions for the identified problems, specify assessment criteria **Control**: Introduce and monitor improvements and new processes

Fig. 26.5 Alternatives to the problem-solving cycle

The project manager is confronted with a range of different tasks and problems. Depending on the situation, the approach and terminology of one model or another may be more suitable.

- Something new may be being created, something that has not existed before (new product or service). The emphasis may be on creativity and on new ground (e.g. pioneer projects).
- Something that fulfilled its purpose before may suddenly no longer do the job. That calls for an ability to spot what is not working, and to identify the causes.
- Existing processes might need improving or optimising.

Depending on the situation, one of the approaches might be easier to use and to implement than the others. We keep returning to tried and tested solutions, even with new names and slightly modified. And people tend to favour one approach over another because of their background, depending on whether they come from project management, operational management, or quality management.
In a different project environment, the procedures might be slightly different, or different terminology may be used. For example, research projects involve steps such as literature study, data collection, data evaluation, hypothesis, verification. Social projects have observation, hypothesis and intervention steps.

26.3 Identifying Goals: Obtaining and Analysing Information

The process of identifying the goals begins with a situation analysis. It helps clarify the problem situation, the relevant problem environment in the sense of an analysis of the situation and isolating the problem. Part of this clarification work has already been mentioned above: "First context analysis": Who is affected by the project idea?
In the preliminary study, a wide angle of view is needed when observing the starting point, and then it is narrowed down to the essential. Here, the analysis of the current situation is particularly important. The situation analysis is checked at each milestone, and can be repeated or expanded for certain aspects, if required. Depending on the nature of the problem and on the existing knowledge, a situation analysis may take just a few hours for an "ad hoc" analysis, or it could need detailed investigation by several people, possibly taking several months (typically for product development projects and investment projects), with the risk that the analysis may paralyse the project.

Compilation Techniques and Analysis

In order to understand the situation at the start of the project, information must be gathered (compilation) and then classified (analysis). When gathering information, ideas for future developments are naturally of interest. The compilation techniques shown above support this compilation work (Fig. 26.6).

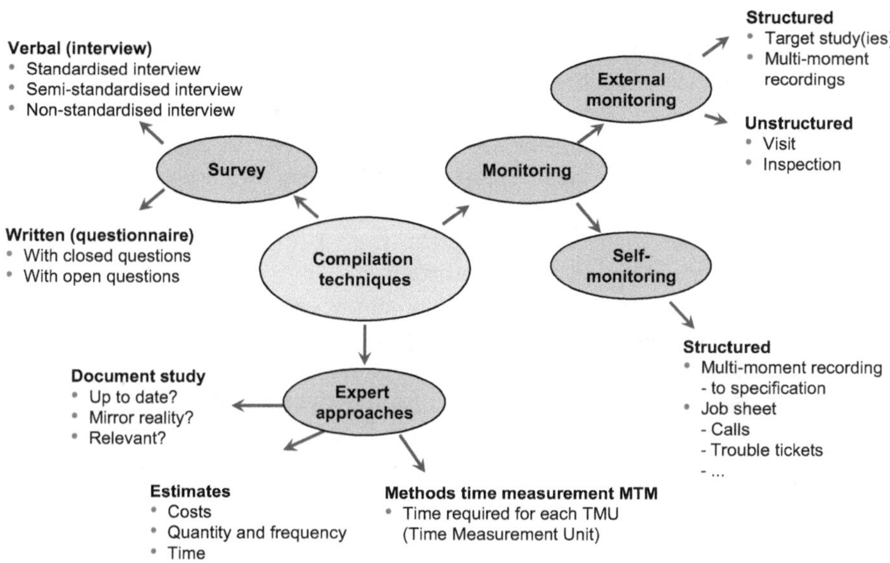

Fig. 26.6 Summary of compilation techniques

What Should the Analysis Show?

- There is often an external requirements document from a customer or an internal tender specification or specification sheet at the start of the project.
- Analysis is already an intervention in the organisation. It is important that the project instances, and in particular the project manager, know the processes that will need starting or initiating at the various different levels.
- Projects normally lead to resistance from the people affected by the project. In terms of economy of changes, the analysis should also show what must not be changed under any circumstances. Strengths should also be shown in the analysis.
- The analysis should not focus purely on internal issues within the company. The wider view must also show what is happening outside the company, what trends are important for the project.
- The information must be presented in a structured way. Possible classifications could include “strengths vs. weaknesses” or “opportunities vs. risks”.

Successful Analysis

The situation analysis can be broken down into three steps:

- Check the reasons that triggered the initiative, and add things to the project order if appropriate: If there are any doubts or question marks, the project agreement should be renegotiated with the project owner. Context analysis: Work out the boundaries and delimit the system and its environment, identify the various system aspects and relevant stakeholders. In other words, the system being observed and the associated project boundaries, for example by producing a bubble chart. The best approach here is "Set the boundaries inclusively and make sure they are wide enough": What does it include? What doesn't belong to it? What are the joins and interfaces with the surrounding environment?

- Analyse the current situation, ideally in the form of a SWOT analysis (strengths, weaknesses, opportunities and threats). Each of these four aspects is a starting point for later goals: What should remain as it is, what should be improved or changed? Processes and "bill of quantities" from an existing system can also be useful in understanding the current situation.
- Look to the future. The project may last months or even years, and the end result will be used for many years after the project has ended. From the very start, a period or "planning horizon" of at least 3 years will be needed in most cases: Think about implications right through to the expected end of the service or product's lifetime. When thinking about the future, it can be helpful to ask the following questions: "What influences between the system (project) and the changing environment around it need to be taken into account, and how will these influencing factors develop in the future?"
- A changed environment can have a significant impact on the project goals and on the subsequent benefits for the product. The more the influencing factors change, the greater their effect on the project. And the longer the time frames, the greater the effect of the change. The goal setting process must therefore be focused on the planning horizon (Fig. 26.7).

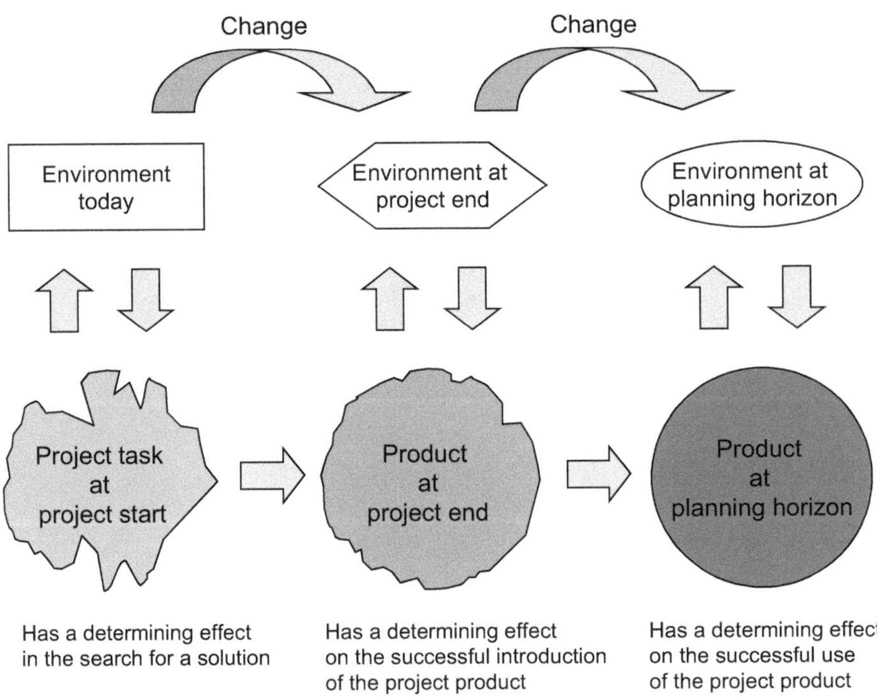

Fig. 26.7 Taking account of factors that will influence the future

Factors that might influence the project include:

- Natural influences (ground, climate, weather, etc.)
- Ecological factors (environment)
- Legal factors (laws, legislation, standards, etc.)
- Political considerations (protectionism, trade restrictions, etc.)
- Socio-political factors (trends, etc.)
- Economic factors (markets, the economy, etc.)
- Commercial factors (costs, profitability, etc.)
- Financial factors (capital, credit, interest, etc.)
- Staffing factors (headcount, personnel policies, etc.)
- Social factors (culture, society, etc.)
- Psychological factors (taboos, mentality, etc.)
- Technical factors (technological developments, etc.)

Group Simulation to Analyse the Initial Situation

For the project manager, it is well worth getting to know the make-up of the different stakeholder groups early on, during the initialisation phase. During the very early stages of the project, it is especially helpful if the project owner's vision of how the project will meet the various different demands can be made visible. The project manager can help guide the project owner towards giving a conscious explanation of his intuitive perceptions.

The project manager does this by using a "systemic structure" group simulation. It focuses on the solution and addresses several levels of symbolisation at the same time. The project owner decides which framework conditions and system interrelationships should be involved. People who have absolutely nothing to do with the project stand in for the stakeholder groups and consider how the individual system elements relate to the project. This sort of simulation can make clear what demands internal and external stakeholders will make of the project. It also helps give a feel for whether stakeholder groups are likely to passively oppose the project, actively support its objectives or whether they might actively seek to maintain the status quo. This orientation work helps show who the project needs to have on board, and how.

The systemic structure method can be used alongside classic methods. It offers a means of gaining an overview of complex structures within a short time, and reveals the background issues that normally remain hidden. This method has low overheads, which saves time and cost. This makes it suitable for use by non-experts as a means of diagnosing the starting situation.

After that, conventional methods can be used to run the project, taking into account the experience gained during the group simulation. The next section describes two methods of carrying out this constellation work – with representatives of a working group, or just with the project owner:

Variant with a Team

The person who had the idea (the idea generator), a moderator and around ten additional team members.

A solution-focused interview (the moderator asks the idea generator questions):

- "If this project was already successfully implemented, what would happen next?"
- "Who or which groups (teams, etc.) will play a role in this project, now and in the future?" Make a note of these people and groups on a flipchart, starting with "focus" (=idea generator).
- The idea generator picks people from the team to represent each person or group, adds their names to the flipchart, and perhaps gives each representative a name badge to help clarity.

Orientation work (the moderator guides the idea generator):

- "Go to the first representative, and hold them by the shoulders. Feel your breath, your contact to the ground and to the representative, and feel guided to where you want to position that person. Take the first step, and walk (with your eyes closed or open) to what you feel is the right place and the right position for them."
- Each representative will take on board what changes during this positioning exercise.
- Once all the representatives are standing, the moderator asks each of them in turn what they feel (feelings, physical perceptions) and what changed when new representatives were added to the mix.

Implementation work (led by the moderator):

- Capture this image in a bubble diagram on a flipchart so that the distances and angles between the different elements reflect the actual positions. Make a note of any special (+/−) connections if relevant.
- Show the project's boundaries by drawing a line around the core elements.
- Gather the feedback and tips that were given to the idea generator from the people playing the various roles. "In my view, we should pay special attention to . . ."

Variant with Just the Project Owner

The project owner has a moderator or coach available in the form of the future project manager. Instead of using representatives, they use moderation cards.

- The project owner asks the project manager to play one stakeholder role after another, and takes the project manager where his intuition guides him.
- The project manager places the card with the name of the represented system on the ground at his feet, with the edge of the card touching his toes. The card should be oriented so that it shows which way the element in question is facing.
- This process is repeated until all the cards have been laid out.
- Working with the project manager, the project owner checks the critical relationships (extreme size or small distances, direct confrontation or obvious aversion).
- Implementation work (bubble diagram): As in the "Team" variant.
- The project owner can then get a feeling for how the individual system parts stand in relation to the project, and he can give the project manager tips by standing in front of the various stakeholder group cards and exploring the constellation to the other elements. The project manager can change between the other positions and assume their standpoints.

26.4 Identifying Goals: Preparing Information

SWOT Analysis, SOFT Analysis

A SWOT analysis (or SOFT analysis) identifies Strengths, Weaknesses, Opportunities and Threats. It is a useful tool for exploring all the areas that are affected by a project. It analyses both the internal and the external environments. Strengths and weaknesses are controllable internal factors which can be altered if they are not acceptable in their current form. Threats and opportunities are generally uncontrollable external factors which cannot be influenced by the organisation. However, the organisation can take appropriate measures with respect to threats and opportunities.

A SWOT analysis can be used at the top level of an organisation to gain insight into overall performance. It can also be used at lower levels to run through the structure of individual projects and to concentrate on specific areas or departments. An organisation can also use a SWOT analysis for each individual project or individual service. To reduce the risks that are inherent in every project, an organisation should build on and develop its strengths, mitigate its weaknesses, take advantage of opportunities and prepare alternative options to project against threats and risks.

SWOT	Positive	Negative
Here and now (internal)	Strengths	Weaknesses
External and future (external)	Opportunities	Threats

SOFT	Positive	Negative
Here and now (internal)	Strengths	Faults
External and future (external)	Opportunities	Threats

Possible questions for members of an organisation:

Strengths

What's going well? What can we rely on? What are we satisfied with? Where do we get our energy? What are we proud of? What are our strengths?

Weaknesses

What is difficult? What disruptions are holding us up? What are we missing? What are we finding challenging? Where are our weaknesses?

Opportunities

What are we capable of? Where do possible future opportunities lie? What can we use in our environment? What is underexploited? What can be developed? What possibilities are open to us?

Threats

Where might threats lurk? What difficulties are we faced with? What do we need to consider? What are our fears?

Scenarios

Using scenarios as alternative views of the future helps to understand the effect of decisions in the future. The scenario method is particularly useful in identifying lasting solutions to problems. Scenarios are used in environmental research, in planning and in management.
Scenarios are used to develop and prepare a number of "If-Then" options. They can be used to examine the impact of individual measures on the organisation, or to explore the reliability of the forecast when predicting measures. The higher an anticipated action in terms of these two dimensions, the more attention should be paid to it, because reaction times and costs or investments play a very large role.

Cause and Effect Analysis (Fishbone Diagram)

This is a systematic method of representing complex causal relationships. It was developed in the 1950s by Kaoru Ishikawa in Japan. It is a simple technique for systematically determining the causes of problems. Today, the approach is also referred to as 6M, the fishbone method, or a Christmas tree diagram. The main causes are shown as arrows pointing to the main axis, and they can be applied to many situations: Man, method, machine, material, environment, metrics and management. Each of these main arrows also has smaller arrows pointing to it, which can show possible contributing causes.

The Ishikawa diagram is best drawn up in a moderated group. For this method to succeed, it is important that there are experts present to represent each affected area of the problem being analysed. This can sometimes mean involving external people (e.g. suppliers, customers).

Members of the team write the problem as clearly and comprehensibly as possible on a large sheet of paper (pinboard, flipchart) until everyone is happy with the formulation of the problem. The participants then write possible causes on cards during a brainstorming session. These cards are then sorted into an Ishikawa diagram using the standard blueprint. The use of horizontal and slanted arrows means that this method can drill down to many levels of possible causes (Fig. 26.8).

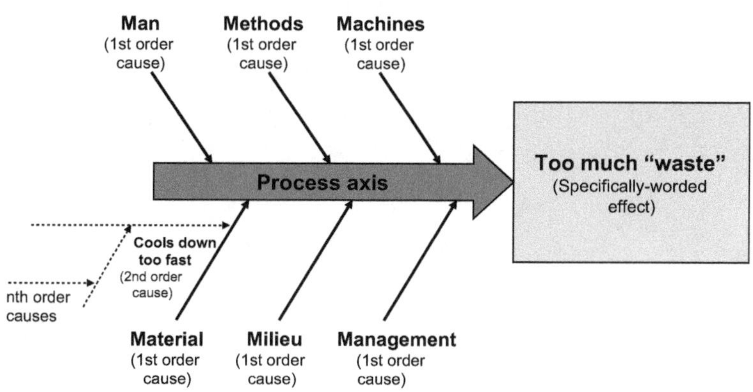

Fig. 26.8 Fishbone diagram

The dispersion method allows individual causes to be assigned to one of the main causes. Questions are asked about each individual cause: "Why does this cause (dispersion) occur?" The questions continue until the team cannot identify any more causes. A good approach here is the "five times why" method: Ask "why?" up to five times in order to identify the root of the problem.

The following questions might also help: "What causes this effect . . .? Why did that happen? What would happen if we changed this aspect? Who does this problem affect? How does that feel for you? Who suffers in this situation, this condition?"

The Ishikawa diagram can also be used to structure activities in processes, or to analyse processes. In that case, the tip of the main arrow points to the result of the process, and the individual "fishbones" contain the hierarchical activities.

26.5 Identifying Goals: Displaying Information

Context Analysis

It is good to strictly limit the number of surprises during initial consultations about a project. There are several different ways of establishing which parts of an organisation might be affected by a project idea. At the start of any project, there is normally only a very limited overview of the situation. The situational analysis is used to produce a relationship graph. This shows the elements that are affected by the project, and sets the boundaries for the project: e.g. the project team and the important elements around them that they need to communicate with (Fig. 26.9).

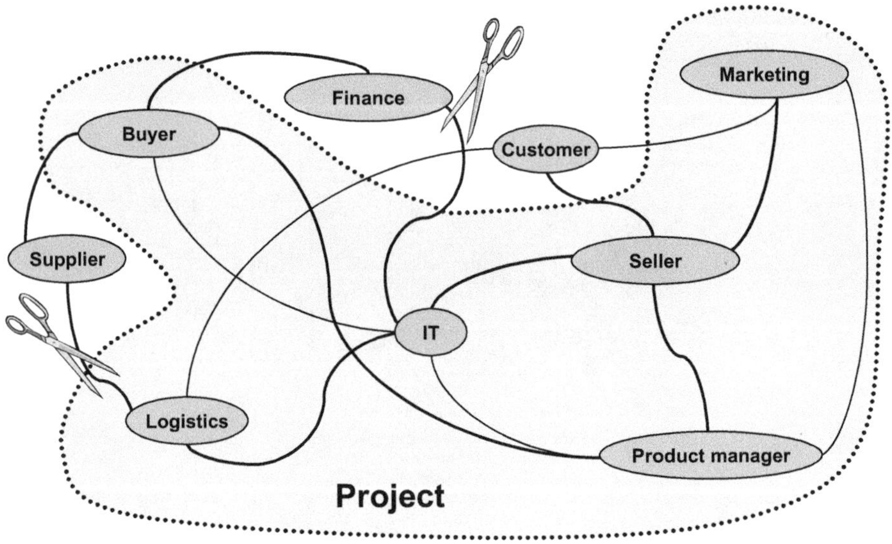

Fig. 26.9 Example of a relationship graph

Using Quality Tools to Support Systematic Analysis

As well as the fishbone method, there are also other methods available to support the continuous improvement process. They were developed in Japan in the 1960s, and show the results in graphical form. These methods are:

- **Tally sheets** or **assessment forms** are used to capture data on site. We can only assess something if it has been measured. Since the measured data is used as the basis for each improvement process, it is becoming increasingly common to measure the data electronically directly from the production process.
- **Histograms** and **bar charts** are used to show the frequency of events or faults by category (e.g. weekday, materials used, method, etc.). Histograms take their data straight from the tally sheets. Their effectiveness lies in their ability to group, so it is heavily influenced by the questions that have been asked.
- A **Pareto chart** is a histogram that has been sorted by frequency. It helps identify the main causes, so that successful results can be achieved as quickly as possible using the available capacity. As a troubleshooting process, it is particularly useful in guiding the effective use of capacity analysis and ensuring that energy is focused on dealing with "big fish".
- Graphical comparison, or **stratification**, is a primitive statistical tool for establishing the causes of a problem. The measured data is grouped according to suspected causes or by procedural perspective (e.g. day shift or night shift), and evaluated in the same way. The results diagrams (e.g. histogram of faults that have occurred) are compared against each other and checked for systematic differences.
- **Scatter plots** are used to check the correlation between measured data where a link is suspected. The strength of the link is determined by simple linear regression.
- **Control charts** are a complex instrument used for statistical process control (or SPC). Samples of a product or service are obtained according to specific rules, and the average value, standard deviation and extreme values are calculated. Continuous sampling and evaluation of the samples result in appropriate measures if an intervention limit is passed, which helps restore the appropriate quality standard.

As well as the classical quality tools, new methods are constantly being added to the "standard toolkit". These include, for example, process description flowcharts and regression analysis for correctly processing scatter plots. Besides the seven basic tools of quality control, which are used to monitor and improve ongoing processes, there are also three Quality Engineering methods that are used in more in-depth design and change processes:

QFD	Quality Function Deployment
FMEA	Failure Mode and Effect Analysis
DoE	Design of Experiments

26.6 Identifying Goals: Defining the Goals

Goals are statements about what the future solution should help achieve. So the question is:
What needs to be achieved, and not **how** should it be achieved?
In the business world, management by objectives (or MbO) has been used for many years now. By definition, projects are always goal-specific initiatives. Project goals show what is supposed to be achieved in the context of the initiative. They relate primarily to the functions that are being designed, and to their use (e.g. services provided, area of use, cost effectiveness, etc.).

Requirements for Goals

The global objective for the project order is differentiated during the preliminary study.

Project Goals Should Be

- Functions to aim for: What should be different when it is completed? How should it work?
- Wishes, hopes and emotions: What additional benefits should it bring?
- Effects to be achieved: What positive impacts will the project deliver?
- Criteria that the human senses can perceive, and which feed into future solutions: Quality requirements for the product, service and process.
- Always try to describe them positively: What should be, rather than what is not wanted.

Project Goals Should Not Be

- Measures or activities that lead to a solution.
- Suggested solutions.

Project goals are of central importance. They are the central guiding influence in the search for solutions. Every piece of work in the project relates to the goals, and they define what the project management team are working towards. That means the project goals must be formulated beforehand. Everybody involved in the problem-solving process must know and accept the project goals. The project's goals must always be formulated before the search for solutions starts, and never afterwards in order to justify a result. It is worth noting that in practice, people have different understandings of the term "goal", and use it differently.
On the one hand, when drawing up the project order, a goal is taken to mean just an outline goal or a global objective for the project. But on the other hand, the term can also be used at a more detailed level, to represent each individual detail target. And sometimes terms such as tender specification documents or specification sheet are

used instead of goals. These are simply detailed documents that are derived from the goals, and are normally used during the invitation to tender.

When formulating project goals, the following rules should be followed:

- Goals should be written as if they had already been achieved. This has a suggestive effect. For example: "Six months after the solution was introduced into service, the full project costs have been amortised."
- Goals should be formulated in a solution-neutral way: What can be perceived when the goal has been reached? If solutions are specified or described, there is a risk that other good solutions may be excluded too early in the process. Example: "When calling a customer, up to date information about the customer should be available." And not: "System XY provided up to date customer information on the screen."
- The goals should contain all the criteria that will later be used to evaluate or assess a solution or solution variant. Even from very early stages of the project, checklists ensure that nothing is missed. If no goal (criterion) is formulated for a particular aspect of a solution, that implies complete freedom in the choice of a solution (i.e. absolutely any solution can be chosen). That also offers insight into the following question: "How much detail should the goals go into?"
- As well as the goals, the framework conditions should also be made clear: What conditions need to be met? What must not be allowed to happen under any circumstances? For example, which safety aspects must be complied with?
- Goals should be formulated operationally: Simple, understandable, unambiguous, clearly measurable and formulated so that it is possible to assess the degree to which the goals have been achieved.
- Goals should be realistic, even if the solution is not immediately obvious at this stage. Realistic means that the people involved can have an active influence on achieving the goals. The goals may still be challenging, as this is a motivating factor, especially in highly innovative environments.
- Detailed operational goals should be formulated as soon as possible after the project order, and they should be formulated as precisely as possible. They may still be changed or added to later in the project.

At the start of the project, the goals should be set out in sufficient detail to ensure that anything where the specialists are not allowed complete freedom when searching for a solution is more or less specified in the goals.
The project owner sets guidelines but leaves scope for imagination, so that the team can find the best solutions. He should set hard limits, with no tolerance for deviation, where things are important to him or where something is absolutely required to comply with standards or legislation. He should go for open formulations with scope for innovation where the framework conditions have lower priority.

Goals Should Be

- Clear and precise
- Unambiguous
- Cross referenced to one another, where it is helpful
- Achievable with the available resources and framework conditions
- Support the company strategy
- Supported by the company's senior management
- Communicated to all interested stakeholder groups

Goals Should Be SMART

- **S**pecific
- **M**easurable
- **A**ttractive
- **R**ealistic
- **T**ime-bound

Identifying Goals Together

In theory, the global objective is specified by the project owner as part of the project order. In practice, the goals often have to be drawn out of the global objective and worked out in detail. Goals should be meaningful and ethical, and they should reflect the values of the parties who are closely involved with the project. For these reasons, the project manager never formulates the project goals alone. He should ensure that they are worked out in a group with the right expertise. This team should consist of people who can speak authoritatively about the results of the project initiative. They should have a direct interest in the project because they are either involved in it or affected by it, and because they will derive benefit from the project results. Goals should not just fulfil factual requirements. They should also be widely accepted. When formulating the goals, the people involved should:

- Be capable of thinking in terms of the big picture
- Be involved enough to contribute values
- Be able to express their thoughts in words
- Have above-average knowledge of the problem
- Be experienced in formulating goals

The group responsible for formulating the goals should be set up so as to ensure that all the relevant perspectives are heard. The following questions give some insight into which interests to take account of, and whether the project team needs extra people bringing in while formulating the goals:

- Which team members are future users?
- Which other important parts of the organisation will be affected by the project?
- Who else can provide information, and who else could either support or boycott the results?

Finally, when identifying the goals, the following points should be borne in mind:

- The detailed goals should be derived from the project order.
- There will be a wide range of goal categories, due to the different views of the involved and affected parties, or the company goals and company processes, or the project environment.
- For unclear goals, particular attention should be placed on ensuring they are measurable: How can we use simple tools to ensure that the goal has been reached?
- When talking of goals, we often use terms that can be understood in different ways. By defining the terms that crop up in the goals catalogue, we ensure that everyone has the same understanding.
- Goals are not limited just to the end of the project. Interim goals should be agreed as milestones. Which raises the question of how to recognise whether an interim goal has been achieved. If this question can be clearly answered, then that is ideal preparation for project monitoring.

Using a Goals Catalogue to Structure Goals

Formulating the goals for a project usually results in a long list of detailed goals. To help keep an overview, it is better to structure the list into subcategories. In the first instance, it is useful to divide them into system goals and process goals:
System goals are all the requirements and needs which will be met by the solution at the end of the project. They include the evaluation criteria for the project solution, performance goals and quality goals, deadlines, economic goals, etc. This includes understanding all of the project owner's perceptions of short-term and long-term impacts and benefits of the project result (Fig. 26.10).

System goals	Process goals
At least 5 bedrooms	Architect planning it around our lifestyle
Available budget: € 950,000	Roof finished before winter sets in
Detached house	Will do floors and plastering of walls ourselves
Full cellar	Moving date: dd.mm.yyyy

Fig. 26.10 An example of system goals and process goals

Process goals include all tasks or conditions that must be met during the project, but which are no longer relevant once the project ends. Process goals include fixed

milestones, the use of specific tools to carry something out, conditions for avoiding disruptions while the project is ongoing, etc.
The catalogue of detailed goals should contain mainly system goals. In terms of sheer numbers, there tend to be more system goals. To help keep an overview, a long list of individual goals should be structured. It should have some sort of hierarchy applied, either according to goal classes or goal groups and detailed operational goals (individual criteria). With this sort of structure, the "global goal" is at the top level, with "detailed goals" at the lowest level (Fig. 26.11).

Goals catalogue

Overall goal: In three months' time, our online services should be 40% faster than they are to day.
Maximum cost for the project: 800,000 €, Max. payback period 12 months

1 Custome rrelationship goals
1.1 Supplying customers is prioritised
1.2 There is direct feedback from / to external customers
1.3 Specific, individual services are agreed with customers
1.4 Relationships with internal customers are as good as with external customers (surveys)

2 Service development goals
2.1 All contributions to the overall service provision are complete
2.2 The service providers are kept informed about the status of the overall service provision
2.3 The process includes a completeness check
2.4 The process includes defect repair

3 Controlling goals
3.1 Each service has a current status showing how up to date it is
3.2 Deadlines for internal forwarding are being met
3.3 There is daily information about the output to services

4 Financial goals
4.1 The cost of providing the service will be 20% lower

5 Scheduling goals
5.1 Within three months of starting the project, all online services will be 40% faster

Fig. 26.11 An example of hierarchical structuring of goals

The global objective is the short, concise description of the project's purpose. It summarily characterises how things should be at the end of the project. It helps orientate project team members. The global objective should contain clear statements about:

What should be achieved (quality, functionality, scope)
Who should achieve it (people, groups)
When it should be achieved (deadline)
Resources to achieve it (costs)

The global objective is a vital part of the project order at the start of the project.

Goal Classes

These break the global objective down into individual "goal aspects", which in turn cover several related properties (detailed goals).

Detailed Goals

These are concrete (positive) requirements. They flesh out the goal class by explaining what is meant at an objective level. Where possible, detailed goals should always be quantified.

One useful form of hierarchical structuring is a goal schema with goal classes and detailed goals. The project manager draws up a suitable schema for each individual project. For some sorts of projects, standard checklists can also be drawn up. When formulating detailed goals in the goal schema, the principle of "Use the simplest and most understandable words to formulate and quantify" applies. The following questions should be clarified when the goals catalogue has been drawn up:

- Are the goals realistic?
- Can they be achieved with the available resources?
- What is needed to ensure the goals are accepted?
- What can be done to ensure the full commitment of everyone who is needed to achieve the goals?
- In case of open goals (e.g. research), are there at least target ranges such as "at least" or "no more than", or typical guide values?
- Is the goals catalogue ready to be weighted and ratified by the project owner?

Weighting Goals, and Making Them Binding

Even when detailed goals have been formulated so they are clearly described and quantifiable, the weighting given to the goals often causes a lot of uncertainty. The project owner must specify how important it is for each goal to be reached. In practice, the detailed goals are generally split into two categories:

- Essential criteria, or "must haves"
- Optimisation criteria, or "desirables"

Essential criteria, also known as "must haves", are conditions that absolutely must be met, even if it costs more or needs more time. That includes legal requirements, safety regulations and possibly standards. Goals that must be achieved in order for a solution to be useful or usable can also count as essential criteria. This precise definition of each essential criterion means that when the project is completed (if not before) it will be possible to clearly assess whether it has met the requirements. When comparing several different possible solutions, essential criteria are simply given a "yes" or "no" answer. If the criteria for a clear "yes" are not present, it may lead to difficult discussions about achieving the goal and about the worthiness of carrying out the project. If it cannot be clearly evaluated, then a goal cannot formally count as an essential criterion.

Another aspect that the project manager must check is the possible future consequences of an essential criterion. These consequences can sometimes be serious, or can lead to results that are technically incorrect. Where there is doubt, it should be made a heavily weighted optimisation criterion. Essential criteria should never be controversial, because that would bring about goal conflicts. Any essential criterion that pulls in the wrong direction should be deleted, or downgraded to an optimisation criterion.

Optimisation criteria are goals that are not essential, and they are often called "desirables". They may be mild wishes or strong wishes, and they can also be contradictory (for example, cost targets on the one hand and quality or performance objectives that cause costs on the other hand). So some sort of additional information is needed in order to assign weightings to the optimisation criteria. The weighting will be simplest to understand if it is expressed as a percentage. There are different ways of establishing a weighting:

Rankings compare and assign relative values to all weightings (comparing them against each other). Using pair comparison, all the optimisation criteria are looked at to determine which criterion is more important than the other.

Criteria	A	B	C	Position	Weighting (%)	Rounded (%)
A	(A)	A	A	3	50	50
B	–	(B)	B	2	33	30
C	–	–	(C)	1	17	20

A **preference matrix** determines which criterion in the table is the most important. The bracketed comparisons (A-A etc.) are ignored during this process. The points are then converted to a percentage.

Criteria	A	B	C	Position	Weighting (%)	Rounded (%)
A	–	A	A	2	66	65
B	–	–	B	1	33	30
C	–	–	–	0	1	5

Pair comparison can be time consuming, so an alternative is to assign points to the optimisation criteria: 4 = extremely important, 3 = very important, 2 = important, 1 = nice to have.

By dividing goals into "essential" and "desirable", the goals become clearer and the scope for people interpreting them differently is reduced. The direction of the search for a solution is specified. From the definition of the goals, it is also assumed that the goals catalogue will serve as a starting point for criteria used for the future evaluation of the solution variants. The optimisation criteria are clearer when comparing various different solution variants. They enable a graded evaluation of the individual variants, whilst the essential goals are either achieved or not achieved. Breaking the goals down into essential criteria and optimisation criteria with appropriate weightings calls for relevant shaping of opinion in the group that assigns the weightings, a group that is usually made up of the project owner and the steering committee.

Overview of Procedural Steps When Identifying Goals

- The project manager must understand the project order and must familiarise himself with the project owner's (spoken and unspoken) intentions.
- The project manager assembles a skilled project team.
- The project team analyses the current situation and the anticipated future development, and identifies factors that will influence the initiative.
- The project manager may choose to expand the team to form a temporary goal setting group. This helps integrate line managers, customers and affected parties into the team that sets the goals.
- The goal setting group or the project team critically review the project order and set clear boundaries for the system and the project.
- The project manager explains the methodology to be used for defining the goals, and suggests a structure for the detailed goals.
- The team formulates goals, ideally using moderation cards.
- The project manager structures the goal ideas.
- The team checks the goals for possible contradictions or conflicts.
- The project manager draws up the goals catalogue and passes it to the goal weighting group (project owner or steering committee).
- The goal weighting group assesses whether the catalogue is complete, and possibly adds or deletes goals. They then decide which criteria should be essential criteria and which should be optimisation criteria, and weight them. Any significant differences of opinion must be resolved.
- The project manager puts the results into a weighted goals catalogue. This is made available to all participants as a benchmark for use throughout the project.

Revising Goals During the Running of the Project

Projects are learning events, and they result in new knowledge. The problem-solving cycle is then repeated with the benefit of this new knowledge. Individual statements can then be corrected if necessary. The goals and framework conditions are checked against the selected direction of development, and can be adapted if necessary.
The customer also becomes more informed as the project progresses, and can make his wishes more precise. It is a good idea to agree right at the start of the project how changes will be handled. Until what point can clarifications be processed on a goodwill basis? When will other requirements with cost implications be possible? Who can accept change requests? How will they be evaluated and approved? Communications between the project manager and the project owner will be particularly intensive during the concept phase. At this stage, it is still possible to reshape things at a reasonable cost. All modifications should be listed in the tender

specification documents (claim management). This helps avoid conflicts, and enables the changes to be traced back.

26.7 Identifying Solutions: Creativity

The search to identify solutions follows on from defining the goals, and is the creative part of the problem-solving process. In this part of the solution-finding process, the focus is on finding a range of solution variants. The suitability of the possible solutions can be checked afterwards. The best alternatives are then systematically compared against each other and evaluated at a later stage. The solution-finding methods are mainly used in the concept phase, although they may also be used in the preliminary study (e.g. feasibility). The preliminary study should result in at least an initial outline solution variant, possibly with recommendations as to how to develop it.
During the creative solution-finding phase, the project team should consciously abstract themselves from the reality of the project environment and its framework parameters in order not to fall into the trap of choosing "safe and familiar" solutions.

Curiosity Is the Way to Creativity

Creativity refers to people's ability to have thoughts of any sort that are essentially new and which were previously unknown to the person who had the thought. In his book "Fühlen, Denken, Handeln" (Feeling, thinking, acting), the neurologist Gerhard Roth explains that creativity is more about intuition than intelligence (Fig. 26.12).

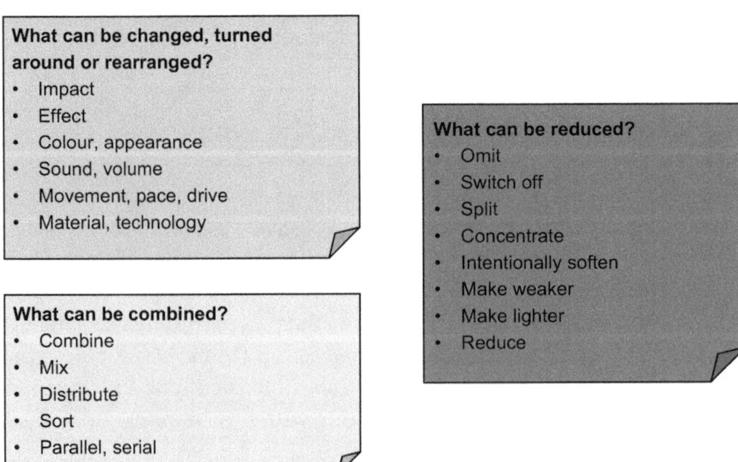

Fig. 26.12 Questions to stimulate the search for solutions

Many creative solutions are prepared unconsciously. During this phase, the moderator must create an environment that encourages the independence of the team members and supports their enjoyment of coming up with new ideas. The project goals must speak to the emotions, and thus appeal to the curiosity of the whole project team.

Ways of Being Creative

Creativity is often inhibited by blocks. These blocks can be caused or amplified by our own subconsciousness, or by external influences or unfavourable circumstances (Fig. 26.13).

Physical and enviromental blocks	Sociological blocks	Psychological blocks
Surroundings • Poor working environment • Poor or missing tools or resources • Noise, temperature • Calls, interruptions • Bad timing **Sensitivity** • Tiredness • Imperfect health • Burnout • Stress	**Cultural** • Lack of recognition • Taboos, morals, sacred cows • Cultural templates • Mentality **Group work** • Interaction • Conflict, tensions • Rivalry • High levels of competitiveness • Discouragement, criticism • Too many sub-groups **Management, politics** • Controls that are too rigid or too institutional • Regulation of communication • Bureaucracy • Failure to reward creative work • Unilateral activity • Too much routine work • Formality	**Closed thinking and intellectual rigidity** • Conformity • Functional fixation • Rejection of new ideas • Habits **Cognitive dissonance** • Dependence on authority • Fear of change • Risk aversion • Preconceived ideas • Perfectionism, looking for absolutes • Contradictory goals **Motivation** • Fear of total engagement • Disillusionment with work • Indecisiveness • Lack of self belief • Lack of curiosity • Lack of confidence • Psychological overload • Unchallenging goals

Fig. 26.13 Blocks to creativity

It is possible to overcome some blocks, and to have a positive influence on creativity:

- A capable team with minimum tensions
- A relaxed working atmosphere
- Clearly set tasks and goals
- The ability to postpone a verdict
- Different perspectives, views, standpoints
- Permission to stand things on their head
- Working with analogies to nature
- Allowing fantasy and images
- Visualisation of all contributions and ideas
- Make use of chance

Various creativity techniques have been developed in order to encourage creativity when developing ideas for solutions. They can be used when there are no routine solution methods. We make a distinction between intuitive creative methods on the one hand, and analytical systematic methods on the other. Depending on the level of detail, a problem-solving process generally involves using several techniques, one after another (Fig. 26.14).

Identifying goals				Identifying solutions			Selection
Information procurement	**Information compilation**	**Information representation**	**Wording of the goals**	**Creativity**	**Optimisation**	**Analysis of solutions**	**Evaluation/ Decision**
Information compilation techniques Workflow analysis Checklists Questionnaires Interview Information compilation plan Multi-moment recording Panel survey Questioning techniques Monitoring techniques Database systems Delphi methods Survey	Information preparation techniques ABC analysis Statistics Regression analysis Correlation analysis Black box method Input/output models Mathemath. statistics Random sample Forecasting techniques Projected forecast Exponential smoothing Saturated models Scenarios Trend extrapolation Cause matrix Probability calculation Networked thinking Influence matrix	(Information) representation techniques Workflow plan Block diagram Logic diagram Flow chart Assignment structures - Cause matrix - Cause-effect network - Influence matrix - Structure plan - Organisational chart Histogram Graph Graphics software Object breakdown structure	Operationalisation Goals catalogue Goals relational matrix Consideration of goals/resources Polarity profile (see also techniques under Evaluation/ decision)	Creativity techniques Analogy method Attribute listing Brainstorming Card deck Method 635 Morphology Scenario planning Synectics Problem-solving tree Cause-effect networks	Operations research Simplex method Linear optimisation Dynamic optimisation Sequential problems Simulations Monte Carlo method Allocation problems Competition problems Game theory Branch and bound Decision tree Decision theory Heuristic methods Queuing problems	Analysis techniques Disaster analysis Risk analysis Safety analysis Decision tables Fault tree Reliability analyses Value analysis	Weighting techniques Economic impact calculation Cost-benefit analysis Cost-effectiveness analysis Cost-utility analysis Weight measurement Criteria plan Score evaluation Sensitivity analysis Scaling matrix

Project management:	Network planning method Bar charts	Time/costs/progress chart Deadline trend analysis		PM system (computer-aided) Project structure plan
General:	Heuristics	Kepner Tregoe	PC software	Spreadsheet

Fig. 26.14 Overview of problem-solving methods (Daenzer, 2002)

A creative process has three phases. These can flow into one another. There will certainly need to be a warming up time, a logical phase, so that a team can get fully up to speed in the intuitive creative phase. After all, athletes always warm up first before a race (Fig. 26.15).

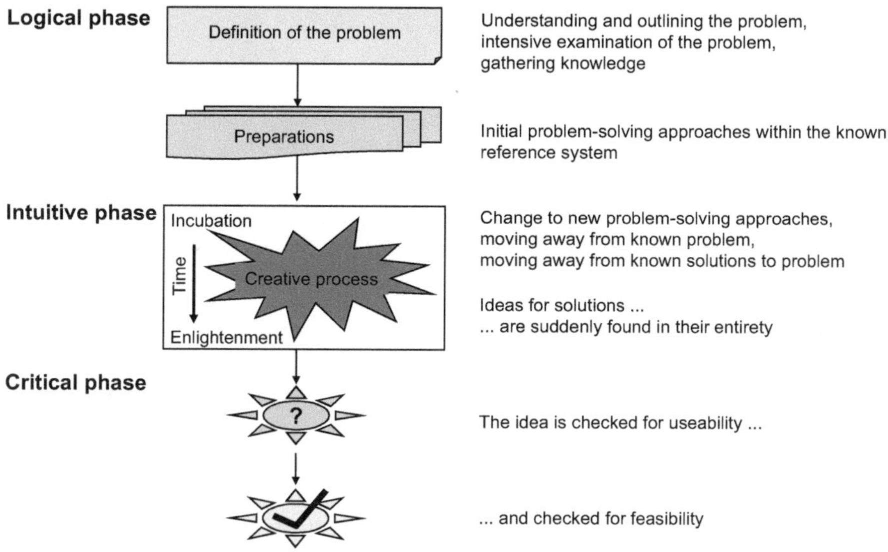

Fig. 26.15 A creative process

Brainstorming

Brainstorming was developed in the 1940s by the American A. F. Osborn. Its aim is to productively and effectively capture a flow of ideas during a problem-solving session without restricting creativity. It separates the creative process from discussion of the suitability of each idea.

Preparation

- Assemble a heterogeneous group that represents the system (5–12 people)
- Appoint a moderator and a "secretary" (to write the ideas down)
- Decide on the topic
- Create a productive working atmosphere (environment)
- Agree a time frame (20–30 min)
- Explain the rules: No criticism (even non-verbal), evaluation will come later, quantity is more important than quality, let your imagination run free

Running the Session

- Clearly formulate the problem
- Write down each idea so that everyone can clearly see them
- Let people spontaneously come up with ideas. Don't discuss them, don't criticise them. Intervene if anyone criticises
- Welcome any unusual comments, and anyone building on other people's ideas
- Give the group new impetus when they start to run out of ideas

Evaluation

- Group the ideas
- Evaluate the ideas, and eliminate unusable ones
- Firm up usable ideas (morphological box, for example)
- For challenging topics, share the ideas with the subject experts for them to develop
- Give the group feedback, tell them what happened to the ideas

Brainwriting, Method 635

Brainwriting, which is also the 6-3-5 method, is slightly different to brainstorming in that it is a written method. Six participants each think up three solution ideas for a problem, or for up to six sub-problems (functions). They write the ideas down on a worksheet, and pass them to the next participant, who reads the ideas and uses them for inspiration to write down three more ideas. Each worksheet is passed on five times to stimulate three new ideas each time. This results in 108 ideas in a short space of time. The moderator draws up a worksheet using the following template (Fig. 26.16).

Issue / problem	

	1st solution idea	2nd solution idea	3rd solution idea
Person 1			
Person 2			
Person 3			
Person 4			
Person 5			
Person 6			

Fig. 26.16 Brainwriting template

The moderator writes the question on the sheet, and makes copies of it. If there are sub-questions, these should be entered after the master template has been copied. The blank worksheets are then handed out, and the participants all fill in the first line of their worksheet. The worksheets are then circulated until they are completely filled in. The worksheets are then posted up clearly visible. Any repeated ideas should be crossed out, to make the weighting process easier. The group then gives points to each idea, to weight them. The ideas with the most votes are then followed up.

Analogy: Bionics and Synectics

Analogy methods use recognisable similarities in form, characteristics or function between two phenomena. Analogy is somewhere between identity (completely the same) and diversity (completely different). Bionics looks for solutions by identifying models in nature and imitating them. Synectics attempts to increase the intensity of the search for a solution by establishing imaginative analogies (Fig. 26.17).

Step 1	Specify the required property or function
Step 2	Search for existing models that have similar properties or functions
Step 3	Examine the system that features these properties or functions
Step 4	Explore whether and how its way of working could be copied

Fig. 26.17 Procedural steps for the analogy method

Producing the Solution

In the information technology sector, producing the solution can mean coding the program. For a construction project it would mean the diggers getting started, and in the consumer goods sector it can be acquiring the means of production. These three examples show that human work is involved in making the project progress in this end phase.

It is often useful to implement the solution in the form of a prototype in the first instance. In that case, the functional sample is turned into a prototype. Then, in a further enhancement of the prototype, checks can be made to ensure that the solution works perfectly in technical terms, and people can familiarise themselves with using it. Production of the prototype or initial series shows whether the production process is working properly. Where projects require users to change their behaviour, this sort of trial can reduce resistance. People can earn how to use the new solution by practising with it, and they can experience the benefits.

Testing the Solution

Before a solution moves to series production it must be thoroughly tested, for example by means of a prototype or initial series. The solution is measured against the goals catalogue:

- Does the solution fulfil the specified goals?
- Does it also meet the specified requirements?
- Some requirements can only be checked once the solution is in operation. Have those requirements been met?
- Is the company's production process working properly?

An isolated test using individual parts of the solution does not guarantee that the solution will actually work in practice. In order to test complex systems, they must be put into a meaningful sequence during the integration phase, and then put into operation. The subsequent troubleshooting or debugging often takes longer than originally anticipated. The way the individual functions work together must be checked, as must the way the users interact with the new system. Feedback from critical users and from users "with two left feet" is particularly helpful: The results from the pilot plants and the initial series are then carefully evaluated, in order to ensure high-quality deliverables.

26.8 Identifying Solutions: Optimisation

Idea Management

The unfortunate thing about ideas is that they are easily forgotten. For all organisations, it is good to use active idea management to channel the good ideas. Exactly what sort of idea management is required will vary significantly from one industry to another. In the chemicals sector it involves a good understanding of the product components and of the application's requirements. On the other hand, service sector companies need to recognise how they can deliver greater perceived benefits to their customers. In each industry, the company strategy will determine the business sectors. A business sector describes the different combinations of products and services with the market. Search fields are used either within a business sector or across business sectors to describe the customer groups' concrete need for action.
The questions here are: What are the unique aspects of this target group? What can we improve for them? Search fields make it easier to work on a focused basis when gathering, preparing and evaluating ideas. Ideas which receive a positive evaluation can then be pursued further.

Investigating Alternatives

A key characteristic of projects is that they are not simply a linear progression towards a single solution. They always involve choosing between several different alternatives. The various different ideas should be condensed into different variants. Suitable solution variants can be designed by combining known partial solutions. The project team should use their expertise to limit the possible number of variants. The aim of having different variants or alternatives is to extend the scope for solutions. This involves improving unsatisfactory partial solutions, optimising solutions according to specific criteria (e.g. cost, weight), ensuring that "all" conceivable solutions to a problem have been identified and that decisions have only been taken once all solutions have been explored.

During the concept phase, the higher-level solution concepts from the preliminary study are developed into several solution variants. For large projects, it is a good idea to get the project owner to confirm rough solutions before moving on to prepare the necessary detailed solutions for the implementation phase. As markets become more global and more transparent, the design of products and processes is becoming increasingly customer-specific. In many cases, different variants of products are offered. Improvement management involves taking an additional look at the product life cycle, at the way the product is used (e.g. knowledge gained from handling complaints) and at the production contexts (products, processes and equipment), in order to incorporate it into the variants.

26.9 Identifying Solutions: Analysing Solutions

The ideas for solutions to the individual functions are only partial solutions, because the individual functions are only partial problems. New overall solutions cannot be identified until the solution ideas for the individual or partial problems have been combined.

Using creativity methods leads to a large number of possible solution variants, although many of them will of course be unsuitable or not possible. It won't be feasible to take all the idea combinations forward for development as solution variants. Some form of initial filtering is needed, a first selection of a limited number of good, usable variants. Possible barriers for individual variants might include:

- A variant clearly fails to meet an essential criterion
- A variant is clearly too expensive when compared to the target cost
- Implementing a variant will take too long for the required deadline
- A variant is psychologically or politically inappropriate or unusable

For a meaningful selection of solutions there must be at least two variants, although one of these could be the existing solution.

Failure Mode and Effect Analysis (FMEA)

An FMEA (Failure Mode and Effect Analysis) is carried out at the end of the development phase, once the result of the innovation project is available on paper. All potential failures that might occur in a wide range of situations are listed. Then preventive measures are taken, focusing on the most important, and the effectiveness of the measures is checked (Fig. 26.18).

Failure Modes and Effects Analysis (FMEA) **Threshold for risk priority number: 250**

Subject: New bone implant AD45a Created by: WE Initials: WE, dd.mm.yy Approved by: TR, dd.mm.yy

				Current status							Improved status			
System, process	Potential fault	Potential conseq. of the fault	Potential causes of the fault	W	T	E	RPZ	Recomm-ended action	Respon-sible, date	Action taken	P	T	E	RPZ
Bone implant	Surface not adhering	Medication >1 year	Careless surface treatment	8	2	8	128							
	Shaft breaks under load	New OPS required	Cavities in the shaft	8	9	6	**432**	100% X-ray US	WE mm.yy	Equipment in prod. process	8	9	1	**72**
	Cartilage attaches	Endoscopy every 3 yrs	Surface roughness in places	4	6	2	48							

Key: W = Probability of occurrence
T = Scope of consequences
E = Likelihood that the failure will be detected by the company

Fig. 26.18 Example of an FMEA analysis

FMEA Procedural Steps

List all conceivable failures or problems

1. Describe the possible consequences of the failure or problem
2. Formulate possible causes and corrective measures
3. P = Probability. The likelihood of failure occurring [1–10]
4. S = Severity. The amount of harm that the failure will cause [1–10]
5. D = Detection. The likelihood that the failure will be detected by the company [10–1]

The values for P, S and D should be printed out using a clearly defined scale. The product of these three variables is the risk priority number.

$$P \times S \times D = \text{Risk priority number (RPN)}$$

The higher the risk priority number, the more important it is to take preventive measures, especially if the cost-benefit ratio is still positive.

The company defines an upper limit for the risk priority number, to match their quality expectations. If this limit is exceeded, preventive measures must be initiated and their effectiveness must be checked. A new risk priority number is then calculated after the measures have been taken, and this new number must be below the threshold value.

Morphological Box

After the initial selection, we are left with usable parts for a new solution. One effective method of coming up with surprising combinations was developed by the Swiss astrophysicist Fritz Zwicky: A morphological box offers a means for numerous ways of combining the partial solutions into possible overall solutions. The diagram below shows an example to give an idea of this method (Fig. 26.19).

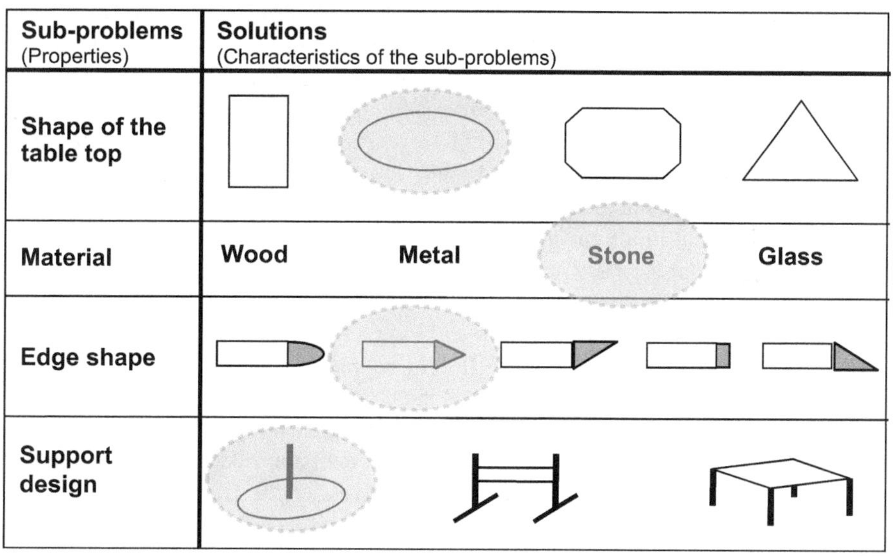

Fig. 26.19 Example of a morphological box

The partial problems, characteristics, functions or partial objects (called parameters) are noted in the left column. For each partial problem or partial object, solution ideas (partial solutions, or solutions to part of the problem) are listed or represented by a sketch along the horizontal axis. Setting up a morphological box is time-consuming, but it is one of the best ways of developing new solutions. This systematic written matrix also allows the creative process to be traced back at a later stage, which helps optimise the individual steps into the best overall solution.

26.10 Selection: Evaluating Solutions

Value-Benefit Analysis and Cost-Effectiveness Analysis

For most people, a situation where they have to make a decision is often a barrier. To decide means giving up various alternatives in favour of one selected option, means separating oneself from the other options. And even after a careful situation analysis, other viewpoints in the company, viewpoints that may be outside the project management team's awareness can exercise considerable influence on the decision-making.

All assessments and decisions are subjective. Decisions can be made consciously, or unconsciously. In his book "Fühlen, Denken, Handeln" (Feeling, thinking, acting), Gerhard Roth states: "All our conscious decisions, even those that we take after long reflection and after weighing them up carefully, are prepared and taken by unconscious processes. When it comes to decision-making processes, emotions have the first and the last word. There is no such thing as a purely rational decision. When it comes to decisions, understanding and reason are simply advisers. Our mind always decides simply on the basis of anticipated rewards. Even the avoidance or reduction of unpleasant circumstances is a form of reward. Expectations of reward are transmitted by feelings. The emotional experience system is responsible for generating hopes, intentions and plans. And it decides whether what we have planned should look like this, rather than something different. That guarantees that in all our negotiations, we are guided by our past experience. However, it does not stop us from making mistakes. Understanding and reason are necessary for evaluating complex, detailed situations, for accessing and applying expert knowledge, for gauging medium and long term consequences (especially in a social context) and for weighing up different courses of action. But understanding and reason do not make decisions. That is the emotional system's job."

For project management, this means that decisions that call for a different approach must be emotionally prepared for, so that the decision-makers and the people affected by the change feel that they will derive benefit from the innovations. Decisions must be understandable in emotional terms.

Decisions can be made understandable using systematic methods. By far the most common method for decision-making processes in practice is the value-benefit analysis. This method looks objective, but also takes account of subjective assessments (Fig. 26.20).

Criteria / detailed goals		Variant 1		Variant 2		Variant 3	
Essential criteria			Fulfilled		Fulfilled		Fulfilled
Optimisation criteria	Weighting W	Score S	W x S	Score S	W x S	Score S	W x S
Overall utility value							

Fig. 26.20 Value-benefit analysis template

With a value-benefit analysis, a points value is assigned to each possible solution. This points value is an indicator about achieving the project's goals. There are five steps to a value-benefit analysis:

1. Establish the goals and weight them (before identifying solutions)
2. Assign values (points) to each variant
3. Multiply the values by the weighting
4. Note the overall weighted values
5. Analyse the sensitivity of the results

How Should the Solutions Be Evaluated?

Selecting the definitive solution involves comparing and evaluating the different variants. The evaluation criteria should be the detailed goals that were defined when drawing up the goals. Supporting criteria derived from these detailed goals may also be used (e.g. cost components for a cost goal; however, note that financial aspects really belong in the cost-effectiveness analysis, which builds on the value-benefit analysis). Assessing solutions means evaluating the degree to which they fulfil the detailed goals. The first check is to ensure that all of the solution variants meet all the essential criteria. If a variant does not meet an essential criterion, it is unusable. Solution variants must be developed in sufficient detail (usually during the concept phase) to make it possible to check that they definitely meet the essential criteria.

The next step is to assess the degree to which the usable variants meet the optimisation criteria. Each variant is given a points value for the degree to which it meets each optimisation criterion. For this process, a clear, easily recognisable measuring system should be used (e.g. 1–10, A–F). It might be important to check whether a neutral value can be obtained (1-2-**3**-4-5). This may influence whether to choose an odd or even scale. The weighting multiplied by the value gives the net value (or utility value) of the solution variant for that specific assessment criterion (or in the case of a sub-goal, the partial utility value). In reality, it is often hard to assign any (numerical) value at all to a criterion. Example: "How user-friendly is an IT solution?" In this case, the group has to draw up a framework or table to facilitate objectively assigning scores to non-measurable criteria.

If the value-benefit analysis is carried out by a team, the weightings should ideally be agreed by consensus.

If the evaluation of the variants results in values of benefit that are very close to one another, there must be questions about the randomness of the results, and thus of their ranking. A sensitivity analysis can be helpful in such instances, i.e. the weights or values are slightly altered, and the impact on the ranking or order is observed. If this reverses the rankings, the variants in question are probably within the weighting or evaluation's scatter band, and should therefore be viewed as being the same. This sensitivity analysis determines how sensitively or insensitively a valuation result reacts to changed assumptions. Where there are any doubts about the results or the order, the completeness of the optimisation criteria and the weightings should always be checked immediately. In practice, entire goal classes such as psychological goals or political goals are often completely overlooked as evaluation criteria, which can lead to serious defects in the evaluation (Fig. 26.21).

Cost-utility analysis for a house purchase										
Goals		Alternative A			Alternative B			Alternative C		
Essentials		Info	Yes / No		Info	Yes / No		Info	Yes / No	
Cost/month <2500 €		€ 1900	yes		€ 2400	yes		€ 2800	No	
At least 3 bedrooms		3	yes		4	yes				
Desirables	W	Info	S	WxS	Info	S	WxS	Info	S	WxS
Central position	25	10' by bus	4	100	5' on foot	9	225			
Close to schools	20	20' by bus	4	80	15'onfoot	8	160			
Comfort	45	4½ rooms	2	90	6½ rooms	9	405			
Style	10	Pre-war	3	30	Concrete	6	60			
Total utility = Sum of all WxS				300			850			
Extent to which goal is met				30%			85%			

Fig. 26.21 Example of a value-benefit analysis

Financial goals should only be used as essential criteria in a value-benefit analysis. The overall benefit of a variant must be determined independently of the financial goals (using points, or in %). A cost-effectiveness analysis is then used to calculate how much a point costs, or how many points there are for each monetary unit for each variant (Fig. 26.22).

Example: Cost-effectiveness analysis of buying a house (direct continuation)			
Cost / points in €	6.33	2.82	
Points / monetary unit	0.16	0.35	

Key: G = Weighting / W = Assessment / Value (maximum value = 10)

Fig. 26.22 Example of a cost-effectiveness analysis

As well as the goals, additional criteria such as the results of a risk analysis may also be taken into consideration. If the risks vary significantly from one variant to another, these factors should also be incorporated into the evaluation.

Who Should Evaluate the Solutions?

The evaluation method described here is the one that is generally used for evaluations. It is also clear that any such grading system is to a greater or lesser extent subjective, and should therefore not be carried out by just one person (unless that person is the project owner, who in any case has the ultimate say about the definitive solution). In projects, evaluating solutions is simply another form of checking that the goals will be achieved. It is therefore a key task for the project manager, because they are responsible for achieving the goals.

So it is down to the project manager to assemble a suitable evaluation team (sometimes also called a "review team" or "jury") and to work with them in evaluating the different solution variants. The evaluation team should consist mainly of representatives of the future users and key stakeholders. So the project manager should be thinking primarily about the goal evaluation group (possibly with some changes or additions). This group is particularly suitable as its members are familiar with the problem, and have already been involved in deciding on the goals and the weightings. They will also not usually have been involved in the search for solutions, so they should be fairly neutral. Where there is already a review team or steering committee in place, it is useful to involve that body for the evaluation.

To enable this team to give a quick, informed evaluation, the benefits and drawbacks of each solution variant must be explained to them in relation to each of the criteria. Additional specialists who know the solution variants may be brought in to do this, in order to explain the variants and to answer the team's questions. Alternatively, the team may simply set out each variant's characteristics, benefits and drawbacks in a written report.

Overview of Procedural Steps When Evaluating the Solutions

- The goals must be defined, with weightings assigned.
- In order to carry out a meaningful evaluation, there must be at least two solution variants that meet all the essential criteria.
- Organise an evaluation team (review team, committee, jury) and, if necessary, subject experts to explain the variants (see goal weighting team).
- If possible, draw up an evaluation schema with the evaluation team and agree a scale or grading system.
- The team must check that each variant meets all the essential criteria.
- Individually evaluate each variant against the specified optimisation criteria.
- Calculate the benefit and the cost effectiveness of each variant, and rank them.
- If there is any doubt, check the completeness of the criteria (detailed goals) and the weightings, and possibly carry out a sensitivity analysis or risk analysis.
- The proposed best solution variant must then be confirmed by the project owner.

Even if he is a generalist and was not therefore directly involved in drawing up the solution options, it is essential that the project manager is involved in this evaluation process, and that he ensures it runs correctly. This also puts him (or her) in the best position to identify with the proposed solution. This transparent evaluation procedure is the best option in cases where the solution variants are relatively similar and where sorting them into a ranked order is not immediately obvious. Experience also shows that it leads to better acceptance of the solution, both by the project owner and by those who may be negatively affected by the solution.

Certification in Project Management

1. Certification Systems

Project managers have two means of personal development open to them. They have a corporate career path, and they have individual certification. "Project manager" is not a protected professional title, which is one reason why individual professional certification has become very important. These qualifications are a reflection of the project manager's knowledge, ability and experience, and they increase the market value of the individual in question. At the same time, some form of common language and common approach is needed, particularly where an organisation is working towards the same certification standard.

There are several certification routes available today. The leading organisations are the IPMA (International Project Management Association), which is recognised internationally and especially throughout Europe and Asia, and PMI (Project Management Institute) which was originally an American organisation and which has international members but a core base in the USA. The IPMA standard is based more on a generic model, with an emphasis on soft skills. The PMI standard features clearly defined terminology, with more emphasis on methods and tools.

The two systems offer the following certification levels (Fig. 1):

J. Kuster et al., *Project Management Handbook*, Management for Professionals,
DOI 10.1007/978-3-662-45373-5

PMI		IPMA	
PgMP®	Program Management Professional	IPMA Level A®	Certified Projects Director
PMP®	Project Management Professional	IPMA Level B®	Certified Senior Project Manager
		IPMA Level C®	Certified Project Manager
CAPM®	Certified Associate in Project Management	IPMA Level D®	Certified Project Management Associate
			Basic Certificate in Project Management (GPM – Deutsche Gesellschaft für Projektmanagement)

Fig. 1 Comparison of IPMA and PMI certification levels

As well as these two main systems there are also a wide range of other standards. These tend to expand on the main systems, or to be of a complementary nature. For example, it can be beneficial to add one of the following additional qualifications to a PMI or IPMA certification:

HERMES: This is a procedural model that was developed in Swiss government circles, and is used mainly in ICT projects. There are two different certificates:

- HSPTP HERMES: Swiss Project Team Professional
- HSPM HERMES: Swiss Project Manager

PRINCE 2: A process-oriented approach, developed in Great Britain as a government standard for ICT projects and used mainly in the UK.

SCRUM: Two role-specific certificates for this agile approach in the software development field.

As well as certification schemes for individuals, there are also certification schemes for organisations. These include CMMI, a maturity and improvement process for business project management.

2. IPMA Certification

Title	Capabilities		Certification Process: Stage 1	Stage 2	Stage 3		Validity
Certified Projects Director (IPMA Level A)®	Competence: • applied knowledge • relevant experience	A	• Application • Curriculum vitae • Self assessment • References • Project list	Program-/ Portfolio-description	Projects Director report	Interview	5 years
Certified Senior Project Manager (IPMA Level B)®		B		Project description	Project report		
Certified Project Manager (IPMA Level C)®		C		Written exam	Short project report		
Certified Project Management Associate (IPMA Level D)®	Knowledge	D	• Application • Curriculum vitae • References	Written exam			

Fig. 2 The IPMA certification process

There are two leading international certification organisations in project management, the IPMA (International Project Management Association) and the PMI (Project Management Institute).

Because the IPMA is more widely used and has greater acceptance in Europe, we decided to produce a reference list of all of the topics that are required and tested for IPMA certification, with links to the relevant pages in this book. IPMA's four certification levels are described in the image above (Fig. 2).

3. The IPMA Eye of Competence

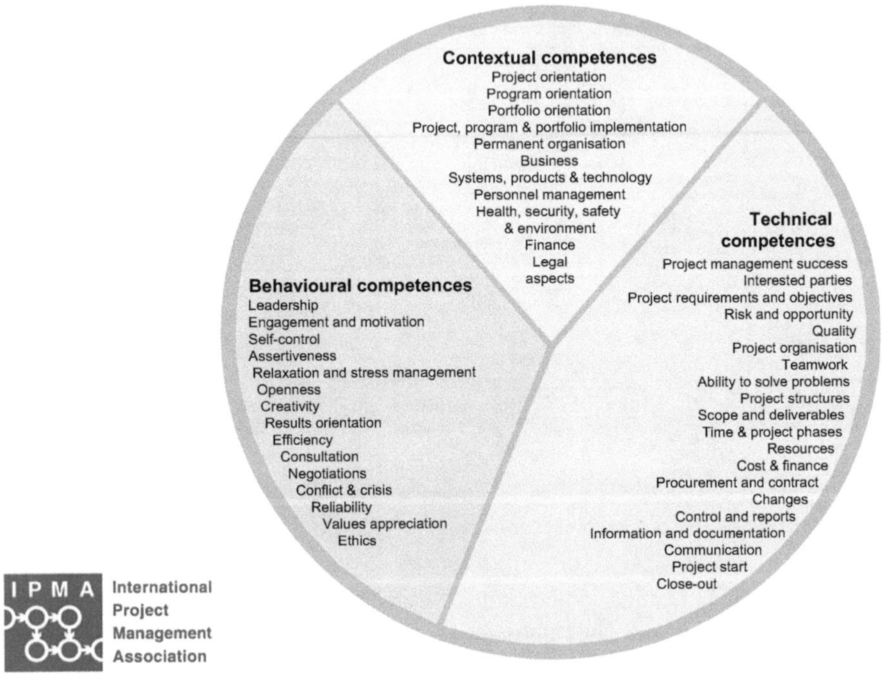

Fig. 3 The IPMA eye of competence (based on the Swiss National Competence Baseline, Version 4)

4. Reference List to IPMA Competence Baseline

The first step towards certification in project management involves a self-evaluation. This helps assess your own knowledge and experience of the topics discussed above in the project management field. The following reference list is based on the IPMA Competence Baseline (ICB Version 4), and provides links between the IPMA competence elements and the relevant pages in the book.

4.1 Technical Competence

No.	Subject	Reference (Chapter)
1.01	Project management success	15, 17
1.02	Interested parties	11, 14, 18
1.03	Project requirements and objectives	9, 13, 26
1.04	Risk and opportunity	17
1.05	Quality	16, 17
1.06	Project organisation	3, 10, 15
1.07	Teamwork	19
1.08	Problem resolution	26
1.09	Project structures	10, 11
1.10	Scope and deliverables	11, 15, 16, 25
1.11	Time and project phases	PM compass, Part II, 16
1.12	Resources	Part I, 16
1.13	Cost and Finance	16, 17
1.14	Procurement and contract	17
1.15	Changes	17, 25
1.16	Control and reports	17, 18
1.17	Information and documentation	18, 26
1.18	Communication	18, 25
1.19	Start-up	19
1.20	Close-out	13, 18

4.2 Behavioural Competence

No.	Subject	Reference (Chapter)
2.01	Leadership	19, 20
2.02	Engagement and motivation	19, 20
2.03	Self-control	20, 25
2.04	Assertiveness	19
2.05	Relaxation	20, 25
2.06	Openness	15, 19, 24
2.07	Creativity	19, 26
2.08	Results orientation	10, 11, 12, 20
2.09	Efficiency	20, 25
2.10	Consultation	15, 19, 21, 24
2.11	Negotiation	14, 20, 25
2.12	Conflict and crisis	17, 22
2.13	Reliability	19
2.14	Values appreciation	19, 20
2.15	Ethics	21

4.3 Contextual Competence

No.	Subject	Reference (Chapter)
3.01	Project orientation	3
3.02	Program orientation	7
3.03	Portfolio orientation	7, 15, 25
3.04	Project, program and portfolio (PPP) implementation	15, 25
3.05	Permanent organisation	15, 18, 19
3.06	Business	21, 25
3.07	Systems, products and technology	14, 15, Part IV
3.08	Personnel management	8, 19
3.09	Health, security, safety and environment	20, 21, 25
3.10	Finance	16, 17
3.11	Legal	14

Bibliography

1. Basics

Daenzer, W. F. (Hrsg.). (2002). *Systems engineering*. Zürich: Verlag Industrielle Organisation. ISBN 3-85743-998-X.

Gandolfi, A. (2001). *Von Menschen und Ameisen: Denken in komplexen Zusammenhängen*. Zürich: Orell Füssli. ISBN 3-280-02669-5.

Ninck, A. (2001). *Systemik, Integrales Denken, Konzipieren und Realisieren*. Zürich: Orell Füssli. ISBN 3-85743-994-7.

Schein, E. (2010). *Organisationskultur*. Humanistische Psychologie. ISBN 978-3-89797-014-4.

Staehle, W. H. (1999). *Management, eine verhaltensorientierte Perspektive*. München: Vahlen. ISBN 3-8006-2344-7.

Steiger, T., & Lippmann, E. (2008). *Handbuch angewandte Psychologie für Führungskräfte*. Berlin: Springer. ISBN 978-3-540-76339-0.

Thommen, J.-P. (2004). *Managementorientierte Betriebswirtschaftslehre*. Zürich: Versus. ISBN 3-03909-000-3.

Vester, F. (2002). *Unsere Welt – ein vernetztes System*. München: Deutscher Taschenbuch. ISBN 3-423-33046-5.

2. Project Management in General

Becker, M. (2004). *EDV-Wissen für Anwender*. Zürich: Verlag Industrielle Organisation. ISBN 3-857-43700-6.

Burghardt, M. (2002a). *Projektmanagement*. Erlangen: Publicis MCD. ISBN 3-89578-199-1.

Burghardt, M. (2002b). *Einführung in Projektmanagement*. Erlangen: Publicis MCD. ISBN 3-89578-198-3.

Buttrick, R. (1997). *The project workout*. London: Pitman Publishing. ISBN 0-273-62680-9.

DeMarco, T. (2005). *Der Termin, Ein Roman über Projektmanagement*. Wien: Hanser. ISBN 3-446-40165-2.

Dobiey, D. (2004). *Programm-Management: Projekte übergreifend koordinieren und in die Unternehmensstrategie einbinden*. Weinheim: Wiley. ISBN 978-3-527-50121-2.

Doppler, K., & Lauterburg, C. (2002). *Change management*. Frankfurt: Campus. ISBN 3-593-36819-6.

Gareis, R. (2006). *Happy projects*. Wien: Manz'sche Verlags-Universitätsbuchhandlung. ISBN 978-3-214-08438-7.

J. Kuster et al., *Project Management Handbook*, Management for Professionals,
DOI 10.1007/978-3-662-45373-5

Gessler, M. (2010). *Kompetenzbasiertes Projektmanagement (PM3): Handbuch für die Projektarbeit, Qualifizierung und Zertifizierung*. GPM Deutsche Gesellschaft für Projektmanagement, Okt 2010. ISBN 978-3-924-84140-9.

Häfele, W. (2007). *OE-Prozesse initiieren und gestalten, ein Handbuch für Führungskräfte, Berater/innen und Projektleiter/innen*. Bern: Haupt. ISBN 3-258-07552-2.

Heintel, P., & Krainz, E. (2000). *Projektmanagement: eine Antwort auf die Hierarchiekrise?* Wiesbaden: Gabler. ISBN 3-409-43201-9.

Hobbs, P. (2001). *Professionelles Projektmanagement*. Frankfurt: mvg. ISBN 3-478-86002-4.

Holland, D. (2002). *Projektmanagement für die Chefetage*. Weinheim: VCH Verlagsgesellschaft. ISBN 3-527-50041-3.

Jenny, B. (2006). *Projektmanagement*. Zürich: vdf Hochschulverlag. ISBN 3-7281-3043-5.

Jenny, B. (2009). *Projektmanagement – das Wissen für den Profi* (2. Auflage). Zürich: vdf Hochschulverlag. ISBN 978-3-7281-3290-1.

Kerzner, H. (2003). *Projekt Management*. Bonn: mitp-Verlag. ISBN 3-8266-0983-2.

Lombriser, R., & Abplanalp, P. (2005). *Strategisches Management*. Zürich: Versus. ISBN 978-3-03909-049-5.

Oestereich, B., & Weiss, C. (2007). *APM – Agiles Projektmanagement*. Heidelberg: dpunkt.verlag. ISBN 978-3-898-64386-3.

Patzak, G. (2004). *Projekt Management*. Wien: Linde. ISBN 3-7143-0003-1.

Petersen, D., Witschi, U., et al. (2011). *Den Wandel verändern: Change-Management anders gesehen*. Wiesbaden: Gabler. ISBN 978-3-834-92672-2.

Pfetzing, K. (2004). *Ganzheitliches Projektmanagement*. Giessen: Verlag Dr. Götz Schmidt. ISBN 3-921-313-678.

Rattay, G. (2003). *Führung von Projektorganisationen*. Wien: Linde. ISBN 3-7073-0433-7.

Rietiker, S. (2006). *Der neunte Schlüssel – vom Projektmanagement zum projektbewussten Management*. Bern: Haupt. ISBN 978-3-258-07044-5.

Scheuring, H. (2007). *Der www-Schlüssel zum Projektmanagement*. Zürich: Orell Füssli. ISBN 3-85743-727-8.

Zink, K. J., Kötter, W., Longmuss, J., & Thul, M. J. (Hrsg.). (2009). *Veränderungsprozesse erfolgreich gestalten*. Heidelberg: Springer.

3. Methodology

Ahlemann, F. (2008). *Project management software systems*. Wurzburg: BARC.

Berkel, K. (2002). *Konflikttraining, Konflikte verstehen und bewältigen*. Heidelberg: Sauer. ISBN 3-79387-280-7.

Böhm, R. (2005). *Methoden und Techniken der System-Entwicklung*. Zürich: vdf Hochschulverlag. ISBN 3-7281-2956-9.

DeMarco, T. (2003). *Bärentango, Mit Risikomanagement Projekte zum Erfolg führen*. Wien: Hanser. ISBN 3-446-22333-9.

Gomez, P. (1999). *Die Praxis des ganzheitlichen Problemlösens*. Bern: Haupt. ISBN 3-258-05575-0.

Heitger, B. (2002). *Harte Schnitte, neues Wachstum: Die Logik der Gefühle und die Macht der Zahlen im Changemanagement*. Wien: Redline, Wirtschaft bei Überreuter. ISBN 3-8323-0913-6.

Kerzner, H. (2001). *Project management maturity model*. New York, NY: Wiley. ISBN 0-471-40039-4.

Königswieser, R., & Keil, M. (2000). *Das Feuer grosser Gruppen: Konzepte, Designs, Praxisbeispiele für Grossverantstaltungen*. Stuttgart: Klett-Cotta. ISBN 3-608-91026-3.

Königswieser, R., Sonu, E., & Gebhard, J. (2006). *Komplementärberatung, das Zusammenspiel von Fach- und Prozess-Know-how*. Stuttgart: Klett-Kotta. ISBN 978-3-608-94142-5.

Königswieser, R., & Exner, A. (2002). *Systemische Intervention, Architektur und Designs für Berater und Veränderungsmanager*. Stuttgart: Klett-Cotta. ISBN 3-608-91938-4.

Lippmann, E. (2009). *Intervision, Kollegiales Coaching professionell gestalten*. Heidelberg: Springer. ISBN 978-3-540-78852-2.
Osterloh, M. (2006). *Frost Jens, Prozessmanagement als Kernkompetenz*. Wiesbaden: Gabler. ISBN 3-409-33788-1.
Schwarze, J. (2001). *Netzplantechnik*. Herne: Neue Wirtschafts-Briefe. ISBN 3-482-56068-4.
Seifert, J. W. (2003). *Moderation & Kommunikation*. Offenbach: Gabal. ISBN 3-89749-003-X.
Seiwert, L. J. (2001). *Das 1 × 1 des Zeitmanagement*. Frankfurt: mvg Taschenbuch Business Training. ISBN 3-478-81125-2.
Sell, R. (2002). *Probleme lösen, In komplexen Zusammenhängen denken*. Berlin: Springer. ISBN 3-540-43687-1.
Stroebe, R. W. (2004). *Motivation, Arbeitshefte Führungspsychologie*. Heidelberg: Recht und Wirtschaft. ISBN 3-80057-316-4.
Varga von Kibéd, M., & Sparrer, I. (2004). *Ganz im Gegenteil: für Querdenker und solche, die es werden wollen*. Heidelberg: Carl-Auer-Systeme. ISBN 3-89670-235-1.

4. Teams, Team Management, Conflicts

Antons, K. (2000). *Practising group dynamics*. Göttingen: Hofgrefe. ISBN 978-3801713706.
Backhausen, W., & Thommen, J.-P. (2003). *Coaching: Durch systemisches Denken zu innovativer Personalentwicklung*. Wiesbaden: Gabler. ISBN 3-409-12005-X.
Bohm, D. (2005). *Der Dialog – das offene Gespräch am Ende der Diskussion*. Stuttgart: Klett-Cotta. ISBN 3-608-91857-4.
Cohn, R. (2011). *Introduction to TZI*. Basel: Beltz. ISBN 978-3407229212.
Cronenbroeck, W. *Handbuch internationales Projektmanagement*. ISBN 3-589-23600-0.
Fisher, R., et al. (2009). *Das Harvard-Konzept: Der Klassiker der Verhandlungstechnik*. Frankfurt: Campus. ISBN 978-3-593-38982-0.
Frank, G. (2004). *Work Life Balance – Leistung und liebe Leben, Work Life Balance Expert Group*. Frankfurt: Redline Wirtschaft. ISBN 3-636-01122-7.
Fuchs, H. (2002). *Die Kunst (k)eine perfekte Führungskraft zu sein; 60 Denkanstösse für zukunftstaugliche Manager*. Wiesbaden: Falken & Gabler. ISBN 3-8068-2583-1.
Gellert, M., & Nowak, C. (2002). *Teamarbeit – Teamentwicklung – Teamberatung, ein Praxisbuch für die Arbeit in und um Teams*. Meezen: Limmer. ISBN 3-928922-13-0.
Glasl, F. (2004). *Konfliktmanagement: Ein Handbuch für Führungskräfte, Beraterinnen + Berater*. Stuttgart: Freies Geistesleben. ISBN 3-77250-954-1.
Greif, S., & Kurtz, H.-J. (1998). *Handbuch selbstorganisiertes Lernen*. Göttingen: für angewandte Psychologie. ISBN 3-8017-0837-3.
Gruhl, M. (2008). *die Strategie der Stehauf-Menschen, Resilienz – so nutzen Sie Ihre inneren Kräfte*. Freiburg im Breisgau: Herder. ISBN 3-7831-3444-7.
Hofstede, G. *Cultural dimensions*. ISBN 978-3-42350-807-0.
Janssen, C. (1996). *The four rooms of change (Förändringens fyra rum)*. Stockholm: Wahlström & Widstrand.
Kälin, K. (1999). *Sich selbst managen: die eigene Entwicklung im beruflichen und privaten Umfeld gestalten* (Ausgabe mit CD-ROM). Thun: Ott. ISBN 3-7225-6680-0.
Kreyenberg, J. (2008). *Handbuch Konfliktmanagement*. Berlin: Cornelsen. ISBN 3-589-23604-3.
Lay, R. (1999). *Dialektik für Manager; Methoden des erfolgreichen Angriffs und der Abwehr*. Berlin: Econ & List. ISBN 3-612-26659-4.
Marti, S. *Toolbox Führung, nützliche Werkzeuge und Modelle für Führungskräfte und Projektleiter* (8. Auflage). Winterthur: Eigenauflage.
Mole, J. (2003). *Managing business culture*. London: Nicholas Brealey Publishing. ISBN 1-85788-314-4.
Pietschmann, H. (2002). *Eris & Eirene, eine Anleitung zum Umgang mit Widersprüchen und Konflikten*. Wien: Ibera. ISBN 3-85052-131-1.

Pöhm, M. (2001). *Vergessen Sie alles über Rhetorik; mitreissend reden – ein sprachliches Feuerwerk in Bildern*. Frankfurt: mvg. ISBN 3-478-73231-X.

Ponschab, R., & Schweizer, A. *Kooperation statt Konfrontation*. Köln: Dr. Otto Schmidt. ISBN 3-504-18964-9.

Roth, G. (2001). *Fühlen, denken, handeln, wie das Gehirn unser Verhalten steuert*. Frankfurt: Suhrkamp. ISBN 3-518-58313-1.

Satir, V., et al. (1995). *Das Satir-Modell, Familientherapie und ihre Erweiterung*. Paderborn: Junfermann. ISBN 3-87387-167-X.

Schulz von Thun, F. (2000). *Miteinander reden – Störungen und Klärungen*. Hamburg: Rowohlt. ISBN 3-499-17489-8.

Schwarz, G. (2005). *Konfliktmanagement: Konflikte erkennen, analysieren, lösen*. Wiesbaden: Gabler. ISBN 3-40979-605-3.

Simon, F. B. (2004). *Gemeinsam sind wir blöd!?* Heidelberg: Karl Auer System. ISBN 3-89670-436-2.

Sprenger, R. (2010). *Mythos Motivation, Wege aus einer Sackgasse*. Frankfurt: Campus. ISBN 978-3-593-39200-4.

Stock, R. (2002). *Coaching von Teams, ein systematischer Ansatz zur Messung und Steigerung der Einflussgrößen des Teamerfolgs*. zfo, Zeitschrift Führung + Organisation, 71. Jg., Heft 2.

Tuckman, B. (1977). Stages of small group development. *Group and Organizational Studies*, *2*, pp. 419–427.

About the Authors

Jürg Kuster, dipl. Ingenieur ETH (M.Sc. Electrical Engineering) Studied electrical engineering at the ETH in Zurich (Swiss Federal Institute of Technology Zurich). Spent over 11 years working as a manager and project manager for various companies in the ICT sector. Expert in Swiss exams for ICT project managers and business information specialists. Lecturer at several higher education institutions. From 2008 onwards Director of the BWI Management Weiterbildung at the ETH Zurich. His main interests are management training, project coaching, the development and introduction of company-wide project management standards, and training.

Eugen Huber, dipl. Organisator (M.Sc. Business Planning) Dipl. phil. I, Université de Neuchâtel; Masters in individual psychological counselling; Swiss graduate organiser; systemic solution-focused coach FHNW, EFQM assessor. Coach and trainer specialising in stress management. Many years of experience in project management, organisational and team development, working with large groups, and coaching. Management experience in various projects, in a range of service-sector companies and training institutions. Since 2000, Eugen has worked as an independent consultant, coach and trainer in German, French and English. Main interests: Designing and moderating change processes, leading projects, coaching individuals and teams.

J. Kuster et al., *Project Management Handbook*, Management for Professionals,
DOI 10.1007/978-3-662-45373-5

Robert Lippmann, lic. oec. publ. (B.Sc. Economics) University of Zurich Business apprenticeship, studied business administration. Additional training in group dynamics (DAGG/SAAP) and industrial psychology, and a post-graduate qualification in business development (NDU SNU). Since 1990 working as a management trainer and consultant. Since 1996 as independent consultant, project manager, trainer and coach in the industrial, service and public sectors, working on project management, leadership, coaching, communication and moderation. In 2006 he took on a part-time role as coach and organisational development consultant for senior management in a large service-sector company.

Alphons Schmid, training manager Public service apprenticeship. Began working for Swissair in 1961. 1967–1991 held various management roles for cabin staff: Behavioural change manager, setting up and running a management training scheme and third-party training, divisional manager for cabin crew. 1991–1996 company-wide responsibility for: project management, moderation, team development, leadership and communication. He has worked as an independent consultant and trainer since 1996. Areas of expertise: Coaching individuals and teams, conflict resolution and team development processes.

Emil Schneider, dipl. Ingenieur ETH (M.Sc. Electrical Engineering) Masters in electrical engineering at the ETH in Zurich (Swiss Federal Institute of Technology Zurich). Additional training in project management, quality management, international industrial management. Post-graduate at BWI/ETHZ. Ten years experience as a developer and project manager. Fifteen years management experience running development departments in large companies. Since 1993 he has been an independent management consultant, coach and trainer focusing on project management and quality in business processes.

Urs Witschi, dipl. Architekt ETH (M.Sc. Architecture) Studied architecture. Post-graduate at BWI/ETHZ. Additional training in systems consulting. Worked for many years as a management consultant with BWI/ETH, specialising in leadership and organisation. In 1995 he was one of the founders and partners in the ONION network, and in 2002 he was one of the founders of DRIFT Consulting GmbH, which specialises in project management and change management. He is active in the GPM (German Association for Project Management), where his main area of interest is the further development of project management.

Roger Wüst, Ingenieur HTL, Betriebsing. STV (M.Sc. Engineering) Additional training in systemic organisational development, systemic consulting, coaching, organisational structures, management development. As a former BWI staff member, he worked in project management consultancy and training for the private and public sectors. Worked as an independent trainer, senior consultant and coach since 1996. Today, his preferred area of specialisation is organisational development, concept design and training of senior management, workgroups and organisations. He also provides assistance to various institutions and universities.

Appraisals

This book is a follow-up to the proven BWI guidelines, and it offers full updated information on project management. The structure of this book makes a refreshing change from many previous publications on this topic. The project management compass helps you navigate your way through a realistic project flow. At the same time, this book offers in-depth information on models, methods and checklists. I am particularly impressed by the underlying reference to the current IPMA Competence Baseline.

This is an excellent manual for practical project management in today's world.

Prof. Dr. Heinz Schelle
Honorary Chairman of the GPM (German Project Management Association)

The authors' many years of practical experience in setting up, implementing and managing projects shines through in this book. Project management is being used ever more frequently as a management approach in a wide range of companies and organisations. The book also reflects the current trend towards increased social competence.

I am therefore pleased to recommend this book as a basis for certification in project management.

Dr. Hans Knöpfel
Honorary President of the SPM (Swiss Project Management Association)

J. Kuster et al., *Project Management Handbook*, Management for Professionals,
DOI 10.1007/978-3-662-45373-5

Index

Note: Entries with page numbers shown in **bold** text are **main entries**, and indicate that the keyword is discussed in depth. Where page numbers are printed in a regular font, the entries are minor entries and refer to additional information or comments about the keyword.

J. Kuster et al., *Project Management Handbook*, Management for Professionals,
DOI 10.1007/978-3-662-45373-5

D

Kolibri 360 GmbH
Ferdinand-Jühlke-Straße 7
99095 Erfurt, Deutschland
produktsicherheit@kolibri360.de